TELECOMMUNICATION ENGINEERING

Line Communications

Volume I

N N DEB
Fellow IETE
Retired DET (Transmission)
Telecom Projects, Eastern Zone
Kolkata

NEW AGE REFERENCE
An Imprint of
NEW AGE INTERNATIONAL (P) LIMITED, PUBLISHERS
LONDON • NEW DELHI • NAIROBI
Bangalore • Chennai • Cochin • Guwahati • Hyderabad • Kolkata • Lucknow • Mumbai
Visit us at www.newagepublishers.com

Published by New Age International (P) Ltd., Publishers
First Edition : 2000
Reprint: 2018

GLOBAL OFFICES

- **New Delhi** **NEW AGE INTERNATIONAL (P) LIMITED, PUBLISHERS**
  7/30 A, Daryaganj, New Delhi-110002, (INDIA)
  Tel.: (011) 23253771, 23253472, **Telefax:** 23267437, 43551305
  E-mail: contactus@newagepublishers.com • Visit us at www.newagepublishers.com
- **London** **NEW AGE INTERNATIONAL (UK) LTD.**
  27 Old Gloucester Street, London, WC1N 3AX, UK
  E-mail: info@newacademicscience.co.uk • Visit us at www.newacademicscience.co.uk
- **Nairobi** **NEW AGE GOLDEN (EAST AFRICA) LTD.**
  Ground Floor, Westlands Arcade, Chiromo Road (Next to Naivas Supermarket), Westlands, Nairobi, KENYA
  Tel.: 00-254-713848772, 00-254-725700286
  E-mail: kenya@newagepublishers.com

BRANCHES

- **Bangalore** 37/10, 8th Cross (Near Hanuman Temple), Azad Nagar, Chamarajpet, Bangalore- 560 018
  Tel.: (080) 26756823, **Telefax:** 26756820, **E-mail: bangalore@newagepublishers.com**
- **Chennai** 26, Damodaran Street, T. Nagar, Chennai-600 017, **Tel.:** (044) 24353401, **Telefax:** 24351463
  E-mail: chennai@newagepublishers.com
- **Cochin** CC-39/1016, Carrier Station Road, Ernakulam South, Cochin-682 016, **Tel.:** (0484) 2377303, **Telefax:** 4051304
  E-mail: cochin@newagepublishers.com
- **Guwahati** Hemsen Complex, Mohd. Shah Road, Paltan Bazar, Near Starline Hotel, Guwahati-781 008,
  Tel.: (0361) 2513881, **Telefax:** 2543669, **E-mail: guwahati@newagepublishers.com**
- **Hyderabad** 105, 1st Floor, Madhiray Kaveri Tower, 3-2-19, Azam Jahi Road, Near Kumar Theater, Nimboliadda Kachiguda, Hyderabad-500 027, **Tel.:** (040) 24652456, **Telefax:** 24652457
  E-mail: hyderabad@newagepublishers.com
- **Kolkata** RDB Chambers (Formerly Lotus Cinema) 106A, 1st Floor, S N Banerjee Road, Kolkata-700 014
  Tel.: (033) 22273773, **Telefax:** 22275247, **E-mail: kolkata@newagepublishers.com**
- **Lucknow** 16-A, Jopling Road, Lucknow-226 001, **Tel.:** (0522) 2209578, 4045297, **Telefax:** 2204098
  E-mail: lucknow@newagepublishers.com
- **Mumbai** 142C, Victor House, Ground Floor, N.M. Joshi Marg, Lower Parel, Mumbai-400 013
  Tel.: (022) 24927869, **Telefax:** 24915415, **E-mail: mumbai@newagepublishers.com**
- **New Delhi** 22, Golden House, Daryaganj, New Delhi-110 002, **Tel.:** (011) 23262368, 23262370, **Telefax:** 43551305
  E-mail: sales@newagepublishers.com

ISBN: 978-81-224-1253-6
C-17-07-10619

Printed in India at Star Print-O-Bind, Delhi.

NEW AGE INTERNATIONAL (P) LIMITED, PUBLISHERS
7/30 A, Daryaganj, New Delhi-110002
Visit us at **www.newagepublishers.com**
(CIN: U74899DL1966PTC004618)

Dedicated to the sacred memory of late Indira Gandhi & late Rajiv Gandhi, the Ex-Prime Ministers of India, who took keen interest in the expansion of Radio and T.V. transmitting stations as also Telecommunications including Satellite Communications in India.

Foreword

In these days of information explosion and revolution in telecommunication, it has become necessary to train a large number of service engineers/technicians to instal and maintain the modern telecommunication systems and networks. These technical personnel are being trained in our polytechnic schools and training centres of DOT and Defence. Unfortunately there are very few suitable books for the students in the area of Line and Radio Communication with enough emphasis on circuits and practice. Sri N.N. Deb, Fellow IETE and a former Lecturer at the Telecom Training Centre, Calcutta, has attempted to fulfil this need in two volumes, viz., *Line Communications* (Vol. I) and *Digital and Radio Communications* (Vol. II). The topics in the two volumes generally cover the syllabi for the Junior Engineer trainees in the Telecom Training Centres of DOT and those of the Diploma Courses of IETE and the polytechnics. The introductory portion of the course for the Grad. IETE is also covered by the two volumes. The book not only discusses the classical topics of line transmission, telegraphy, telephony and radio communication, but also extends to modern areas, *e.g.*, digital switching, fibre-optics and satellite communication as well.

The book attempts to cover a wide variety of topics in Telecom Engineering in two comprehensive volumes, so that the students may study the basic principles in a connected way without referring to too many books. The chapters contain necessary illustrations and sets of model questions to facilitate self study. The author has a long experience of teaching the subject and this has been reflected in the well-considered treatment of the topics.

I am sure the two volumes will meet the needs of the Telecom students and their teachers as well and the book will become quite popular in our training establishments.

J. Das
D.Sc., FNA, DFIETE
Formerly Professor of Communication Engineering
and Head of the Electronics and Comm. Engineering
Department, IIT Kharagpur & Kanpur.

Preface

For optimising telecommunication applications, one requires specialised training in theory and practice in order to properly handle different types of machines and instruments in various systems of telegraph and telephone equipment, both in landline communications and radio communications. It has been found that even science and engineering graduates entering the telecommunication department as junior engineers require specialised training in theory and practice for an year or more before they are found suitable to join the maintenance duty for these apparatus and plants. Truly said late S.N. Bose, the national professor, "The subjects of telecommunication engineering should be taught in the polytechnique institutes and B.Sc. colleges furnished with practical laboratories for demonstrations of working principles of components of telegraph and telephone instruments, radio, T.V. and teleprinter closed circuits, etc."

While working as a lecturer in the telecommunication training centre at Calcutta from 1967 to 1971, I observed that there was not any single Indian book covering all the subjects to be taught to the junior engineer-trainees. This prompted me to write a book covering the subjects required to be taught to the telecom trainees.

This book has been written in two volumes, namely (1) Line Communications, and (2) Digital and Radio Communication. The two volumes have been written as per the syllabus of junior engineer-trainees in the telecom training centres in India. This syllabus not only covers the syllabus for the diploma course of IETE Examination but also the introductory portions of graduatship examination.

In writing this book I have consulted many standard books and journals on the subjects.I would specially like to acknowledge my debt to W.H. Freebady for Telegraphy, J. Atkinson for Telephony, W.L. Everitt for Communication Engineering, Departmental publications for construction and testing of lines and cables, F.E. Terman for Radio Communication and F. Cardiol for Microwave Theory and Technique and Dumas and Sands for Microwave Systems Planning. My thanks and gratitude are also due to Dr. J. Das for Digital Communications, Melvin and Leach for Digital Principles and Applications. Dr. M.C. Bhattacharya for his help in Electronic Exchanges, R.R. Gulati for Monochrome TV, W.H. Bushbaum for Coloured TV, G. Moral for Satellite Communication, and journals like *Telecommunications, Microwaves* and *Technical Review* by IETE *etc.* for cellular phones.

In Volume I., i.e., *Line Communications* the eight chapters have been arranged as follows:

Chapter I : Transmission requirements for telecommunication circuits (including network theorems and solid state devices).

Chapter II : Construction and testing of lines and underground cables (including open wire local telephone and trunk lines, coaxial and submarine cables).

Chapter III : D.C. telegraphy with simplex and duplex circuits.

Chapter IV : Manual telephone exchange with junction and trunk line working.

Chapter V : Automatic telephone systems strowger, cross-bar (Penta-conta) and auto-trunk exchanges.

Chapter VI : Carrier systems working on C-8 open wire trunk lines, underground symmetrical cable and coaxial cables.

Chapter VII : A.C. telegraphy with voice frequency telegraphy systems (AM and F.M.) teleprinter and telex systems.

Chapter VIII : Submarine cable telegraphy.

To avoid unwieldy size of volume I, the details of digital electronic exchange system and fibre optic communications have been described in Volume II, alongwith radio communications.

The ten chapters in Volume II i.e., *Digital and Radio Communications* have been arranged as follows:

Chapter I : Linear Integrated circuits

Chapter II : Pulse Switching circuits

Chapter III : Digital Integrated circuits

Chapter IV : Processing of signals for digital communications

Chapter V : Electronic Digital Exchange

Chapter VI : Fibre Optic communications

Chapter VII : Radio Waves and Antennas

Chapter VIII : Radio Telecommunication systems, Broadcasting and microwave systems.

Chapter IX : Radar, Telephoto and Television

Chapter X : Satellite Communications and launching of Satellites from (1) ground-to-ground and (2) space-to-ground via satellite repeater for space investigation by astronauts.

For explaining the principle of operations of a particular circuit, the necessary diagrams have been modified as far as possible for the sake of simplicity and easy understanding. As airconditioning of telecommunication equipment is a must for the efficient functioning of highly sensitive transisters, valves and other equipment, an outline for the same has been given in Chapter V of Vol. I. A number of worked-out examples along with set of model questions on each chapter has been given at the end of each volume for the guidance of the students.

Now, if the students find this book useful for them, I shall feel my purpose served and my labour rewarded. My thanks are due to my wife Anjali for her encouragement and also to my son Anjan for his help.

6th August, 2000 **N. N. DEB**

Contents

1

Transmission Requirements of a Telecommunication Circuit

1.1 Introduction

The word 'telegraphy' means transmission and reception of coded telegraph signals in direct current and alternating current from a distance through a medium of transmission of an open-wire line or underground or submarine cable or through a radio channel. Similarly, 'telephony' means speaking from a distance through similar mediums of transmission, and 'television' means seeing from a distance the movements of the speaker with the background scenery and simultaneously hearing his speech, song or any sounds made by him or others. When only the picture is transmitted without speech or sound, it is called 'telephoto'. Thus, all the four terms, namely, telegraphy, telephony, television and telephoto are implied in the general term 'tele-communication'.

The communication made through a medium of transmission made of open wires or underground cable or underwater or submarine cable is called 'line communication', whereas a communication made through a wireless medium is called 'radio communication'.

Telecommunication engineering involves conversion of acoustical energy into electrical energy by a telephone transmitter, by varying the resistances of the carbon granules through compressions and rarefactions produced by the vibrations of air on the diaphragm of the carbon-granule chamber. It also involves the conversion of mechanical energy into electrical energy by a telegraph transmitter, by sending coded signals in the form of dots and dashes by pressing and releasing the send key. It also encompasses the conversion of light energy of different intensities into electrical energy by emission of different frequencies proportional to these intensities of light by a TV camera. In each case, the conversion of original type of energy into electrical energy of alternating current consists of a band of frequencies with specific shapes of waveforms which reach the distant receiving station where these are reconverted back to their original shapes. Thus, the study of telecommunication engineering is based on alternating current of different frequencies. Even direct current can be considered as an alternating current of zero frequency.

Though human voice consists of frequencies ranging from 300 H_z to 20,000 H_z, for commercial purposes, the bandwidth is taken from 300 H_z to 3400 H_z. In case of A.C. telegraphy, the bandwidth for each channel in voice frequency telegraphy is taken as 160 H_z for amplitude modulated channel, and 120 H_z for frequency modulated channel. The bandwidth for a TV channel may, however, be as wide as 5 to 7 Mega H_z.

These bandwidths are to be maintained throughout the transmission and reception equipments as also the medium of transmission networks whether a pair of open wire lines or underground cable or a carrier system channel. This is done by the design of networks consisting of unidirectional control components like transmitters, relays, oscillators, amplifiers, modulators and demodulators, filters, etc. and by the design of bi-directional components like electromagnetic receivers and impedances, such as resistances, capacitances and inductances, etc. For land-line transmission, the medium may be of open wires of metals of low specific resistance, e.g. copper etc., with transposition at fixed distances to overcome the effects of inductive interference; while for transmission through underground and submarine cables, the cable pairs are properly loaded to overcome the effect of capacitance unbalance. For carrier systems, modulation of transmitted band at the send end as also demodulation of the received band at the distant end is required. The sidebands are separated by filters and equalisers and properly amplified. In fact, an open wire transmission line on an underground cable pair may be considered as a filter network of resistance, inductance, capacitance and leakance joined as T or Π section to provide a definite bandwidth and to produce different attenuation for different frequencies. Thus, the different sections of lines and cables must be properly balanced to overcome the effects of inductive and capacitive interferences, as also to decrease the unequal attenuation of various frequencies and chances of radiation and cross-talk due to limb unbalance and reflection.

The sound waves are sinusoidal in nature and are expressed in cycles per second or Hertz called frequency/sec. The wave velocity in different media is given by the relationship,

$$\text{Sine wave velocity} = \sqrt{\frac{\text{Elasticity of medium}}{\text{Density of medium}}}$$

If the distance between any two points on the sine curve of a wave motion is just one complete cycle, then the phase difference between these two points is called a wavelength (λ). Then, if f is the frequency in cycles per second or Hertz , the time required for one Hertz = $1/f$ second. And if V is the velocity of propagation of waves in cm. per sec., the distance travelled in 1 sec. = Vcm. Then the distance travelled in $1/f$ sec = $V/f = \lambda$, i.e. the wavelength $\lambda = V/f$.

If the transmission line and the measuring instruments are of the same impedances, then for direct current transmission, the voltage at any point can be measured by a voltmeter. But in the case of AC transmission, the measuring instruments are calibrated in the root-mean square value of the pressure, and for a pure sinusoidal wave, it is given by $\frac{1}{\sqrt{2}}$ of the maximum value. The energy transmitted from one point to another is measured by a logarithmic unit, called Bel or decibel. This is defined in common logarithm (i.e., to the base 10) of the ratio of two values of power, viz. P_i = input power and P_o = output power. Thus, in a transmission network the number of Bels = $N = \log_{10} \frac{P_i}{P_o}$, where the power ratio $P_i/P_o = 10^N$. As Bel represents a power ratio of 10:1, it is too big a unit for which a decibel (db) is used as a standard unit, i.e., n = No of db = $10N = 10 \log \frac{P_i}{P_o}$ or power ratio $\frac{P_i}{P_o} = 10^{n/10}$. Thus for 20 db, the ratio $\frac{P_i}{P_o} = 10^2 = 100$ and for 30 db, the ratio = 10^3 = 1000 and so on.

Now if I_i and I_o are input and output currents corresponding to voltages V_i and V_o, then $\frac{I_i}{I_0} = \frac{V_i}{V_0}$ holds good. Therefore

$$\frac{I_i}{I_o} = \frac{V_i}{V_o} = \sqrt{10^{db/10}} = 10^{db/20}.$$

Thus for 20 db, ratio $\frac{I_i}{I_o}$ = 10 and for 40 db, the ratio $\frac{I_i}{I_o}$ = 100 and so on.

This power ratio attenuation unit based on common logarithm is used in England and America. But in European continent, the attenuation unit used is Neper based on the current ratio in Naperian logarithm (i.e., to the base e = 2.7183).

$$\text{Thus } \frac{I_i}{I_o} = e^N \text{ or } N = \log_e \frac{I_i}{I_o}$$

where N is the attenuation of current in Neper.

∴ $\log_{10}{}^{e} = .4343$ or $10^{.4343} = e,$

∴ $10^{.4343} = 10^{db/20}$ and Attenuation factor

$$A = \frac{db}{8.686} = .1151 \text{ db.}$$

Or we may say that the attenuation factor A of a line with a transmission equivalent of one db = .1151 Neper and conversely 1 Neper = 8.686 db.

In India, we follow the common logarithmic procedure of measurement of attenuation of power in db in transmission as is prevalent in U.K. and U.S.A.

1.2 Direct Current Signal Transmission

For the transmission of direct current coded telegraph signals, we are to consider the impulses resulting from the sudden change in voltages at the transmitting station. If the time required for the formation of a steady value at the receiving station, i.e. the transient time, is less than the duration of the shortest signal element, no telegraph distortion occurs in the line. But if the formation time is longer than the shortest signal element, there is a distortion, depending on the ratio of the operating current to the steady state current.

The telegraph signals are composed of transients arising out of capacitance effects. All frequencies do not undergo equal reduction in amplitude nor do they undergo the same change in phase. So in D.C. telegraph transmission, it is not economically possible to design the line to eliminate or decrease the attenuation or phase distortion to a minimum. A direct current telegraph circuit is shown in Fig. 1.1(a) for basic telegraph transmission and is based on transient condition.

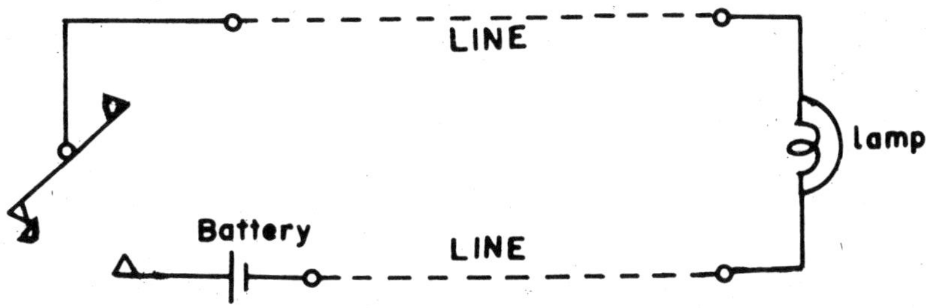

Fig. 1.1(a) A DC Telegraph Circuit

Here, space signals are sent for no glow and mark signals for glow. The Morse code signals consist of dots and dashes. Their difference is in time duration — a dash being three times longer in duration than that of a dot of uniform length. Each letter or figure is made up of a number of dots and dashes, each separated from the other by the time interval of one dot. The time interval of three dots is given between letters of a word or between digits of a number, while the time interval between two words or numbers is of seven dots duration, as internationally agreed in 1949. The Morse Code telegraph signals are shown below:

Letters

a= . — , b= — . . . , c= — . — . , d= — . . , e= . , f= . . — , g= — — . , h= . . . . , i= . . , j= . — — — , k= — . — , l= . — . . , m= — — , n= — . , o= — — —, p= . — — . q= — — . —, r= . — . , s= . . . , t= —, u= . . —, v= . . . —, w= . — —, x= — . . —, y= — . — —, z= — — ..

Digits or figures

1= . — — — — , 2= . . — — —, 3= . . . — — , 4= . . . . — , 5= . . . . . , 6= —. . . . , 7= — —. . . , 8= — — — . ., 9= — — — — . , 0 = — — — — —

Full stop= . — . — . — , Comma= — — . . — — , Colon= — — — . . . , ?= . . — — . . , Hyphen= — . . . . — , Fraction bar= — — . . — — , Bracket= — . — — . — , Inverted comma= . — . . — . , Double hyphen= — . . . — , Error= . . . . , End of message= . — . — . , Invitation to transmit= — . — Wait= . — . . . , End of work= . . . — . — , Start signal= — . — . —

N.B. A single hyphen must be sent between a whole number and a fraction.

An average English letter is taken as equivalent to nine dots and an average English word consisting of five letters. Thus the speed of *N* words per minute is equivalent to N {(5x9)+7}= 52N dots/min,and the duration of a dot $= \frac{60}{52N} \text{x}1000$ m sec.

$$= \frac{1154}{N} \text{ m sec.}$$

The speed of one elementary interval/sec is termed Baud, after the name of Emile Bund, a French telegraph engineer. The telegraph speed is expressed as inverse of the value of elementary intervals in seconds. In Morse code, the elementary interval is one dot while in start-stop code (7½ unit) one elementary interval = 20 m sec or .02 sec. Hence telegraph speed for TP = $\frac{1}{.02}$ = 50 Baud which is taken as equivalent to $66\frac{2}{3}$ words per minute for an average word of five letters and one space signal or 400 letters/min. when the speed of the transmitting shaft is 400 R.P.M.

A telegrapah transmitter is a device for transmitting these coded signals through the lines to the receiving instruments at the other end. The codes sent by the transmitter produce electrical changes in the circuit, which are repeated in sequence of corresponding electrical changes at the receiving end apparatus. By recognition of these code signals, the letters and figures are identified and recorded with proper punctuation.

For sending these signals, a Morse key is used at the send end, and for receiving these an electromagnetic relay called sounder is used at the receive end. Such electromagnetic relays are also used in telephone transmission and reception circuits as also in long-distance telephone carrier systems and Voice Frequency Telegraph (VFT) systems.

Telegraph modulation is a term used to denote the operation by the transmitting apparatus, and restitution is the term used to denote the operation performed by the receiving apparatus. The term restitution delay means the delay between a characteristic instant of modulation and the corresponding instant of restitution. By drawing a graph for the telegraph send and receive signals, showing the so-called built-up or modulation time at the send end and restitution delays of signals, we can make a comparative study of the delays of the received signals with respect to the corresponding positions of the sent signals, as shown in Figure 1.1(b).

The delays of the received signals are denoted by $r_1, r_2, \ldots, r_6$ and their displacements are denoted by $d_1, d_2, \ldots d_6$ respectively. It is observed that the restitution delays vary appreciably from one instant to another. Thus the stability of operation of a telegraph circuit will depend on the margin of duration upto which the received signals can be tolerably recognised to be recorded. This deviation or distortion of telegraph signals is expressed as a fraction or percentage of the duration of unit element of modulation indicated in Fig. 1.1(b) above by t. Thus, the distortion of the first two corresponding modulation instants is given by the difference of restitution delays namely, $r_1 - r_2 = d_1 + d_2$ or $\frac{r_1 - r_2}{t} \times 100\ \%$.

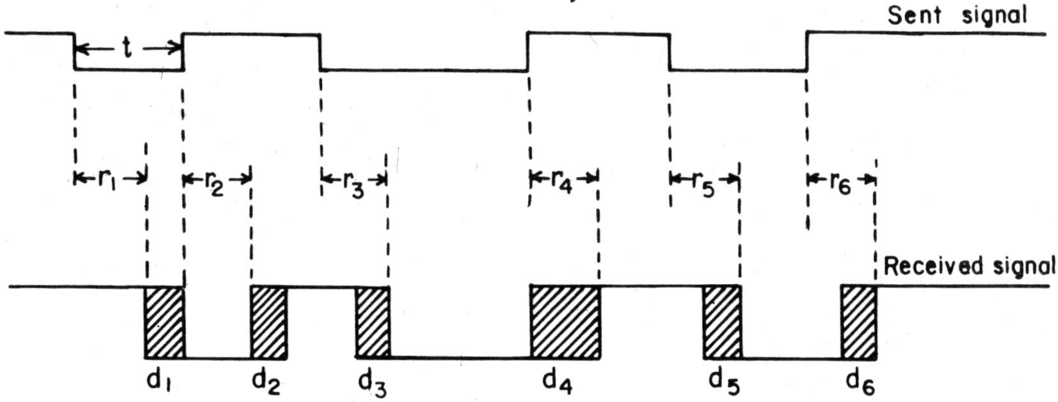

Fig. 1.1(b): Graphs of Sent and Received Signals.

As r_1 represents the greatest and r_5 the smallest restitution delays as shown, the factor $\frac{r_1 - r_5}{t}$ x100% is the maximum limit of distortion here. Thus, distortion of received telegraph signals is given by the maximum difference in the restitution delays of specific instant of modulation, and can be expressed as a ratio of this difference to the duration of unit element of modulation.

The transmission of DC telegraph signals depends on four elementary constants, viz., resistance, capacitance, inductance and leakance. The charge and discharge of a capacitor plays an important part. When a battery is connected to a non-inductive resistance, the current at once rises to the value E/R where E = voltage and R = resistance, i.e. there is no transient effect. But if a condenser is joined in series with the resistance, the current at first rises to the value E/R and gradually falls, as shown in Fig. 1.2.

This happens as there is no voltage on the condenser initially for which the whole voltage is impressed on the resistance. But when the condenser gets a gradual charge, the voltage on the resistance is decreased and the current is reduced to zero when all the voltage is impressed on the condenser. Thus the expression for the current at any instant is given by

$$I = E/R\, e^{-t/CR} \qquad (1)$$

where I = current in amperes, R = resistance in ohms, E = applied voltage, C = capacity in farads, t = time in secs. from switch-on, and e = base of Naperian logarithm = 2.7183 approximately.

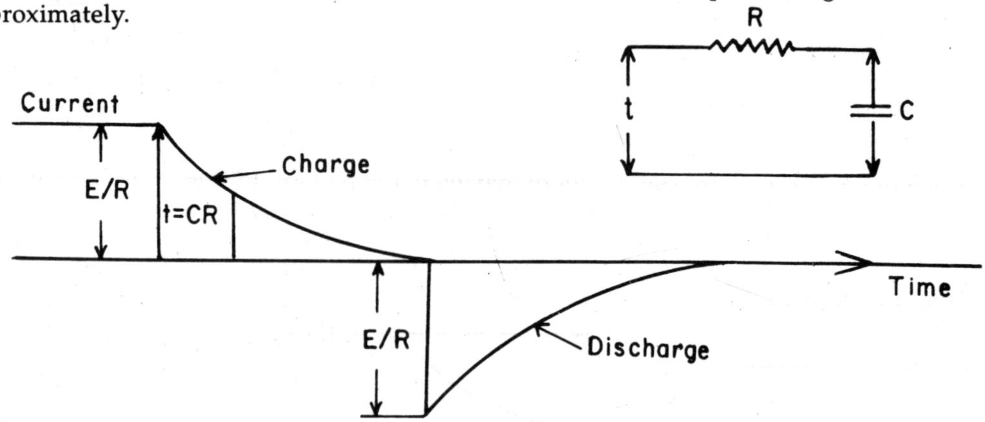

Fig. 1.2: Charge-discharge Current Curve of a Resistor-capacitor Circuit.

At the instant of switching $t = 0$ and $I = E/R$. When t is high I gets low. The product CR is called the time constant of the circuit and when $t = CR$, the value of $I = \frac{1}{2.7183}$ times the initial value. When the condenser is charged, the battery is removed and the free ends are short-circuited; the condenser discharges through the resistance. But this time, the discharge current flows in the opposite direction and is equal in magnitude to the charging current.

Case I — The equation for send and receive currents for a single section T-networks artificial line of resistor and capacitor with negligible inductance and leakance can be worked out as follows. The send and receive currents for a single section T-networks are shown in Fig. 1.3.

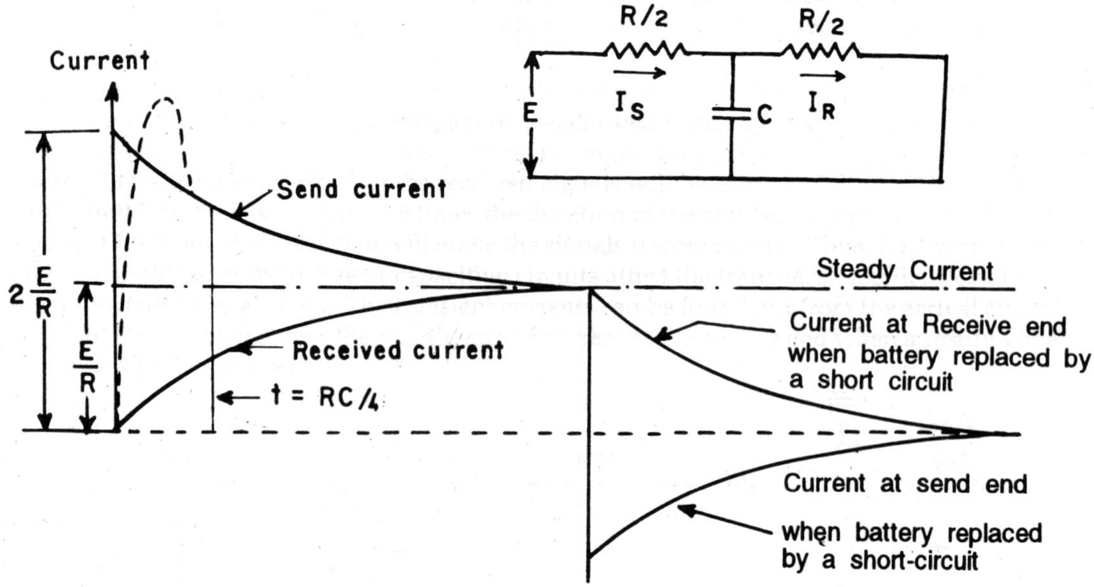

Fig. 1.3: Send and Receive Current for a Short T-section
(The dotted curve shows the effect of distributed capacitance near send end).

When a voltage E is applied at one end of such a line, with the other ends being short-circuited, if V is the voltage across the condenser at an instant, we have:

$\frac{E-V}{R/2} = I_s = c\frac{dV}{dt} + I_R$, where $c\frac{dV}{dt}$ is the current flowing into the capacitor at that instant and $I_R = V/\frac{R}{2} = \frac{2V}{R}$

$$\therefore \quad \frac{E-V}{\frac{R}{2}} = c\frac{dV}{dt} + \frac{2V}{R}$$

$$\text{or} \quad E - V = \frac{CR}{2}\frac{dV}{dt} + V$$

or $\quad \frac{E}{2} = V + \frac{CR}{4}\frac{dV}{dt}$, the solution of which is $\frac{E}{2} + A_e - \frac{4}{CR}t = V$

$V = 0$, when $t = 0$.

$$\therefore \quad A = -\frac{E}{2} \text{ and } V = \frac{E}{2}\left(1 - e^{-\frac{4}{CR}t}\right) \quad (1)$$

$$\therefore \quad I_s = \frac{E}{R}\left(1 + e^{-\frac{4}{CR}t}\right) \quad (2)$$

$$\therefore \quad I_R = \frac{E}{R}\left(1 - e^{-\frac{4}{CR}t}\right) \quad (3)$$

where $\frac{RC}{4}$ is the time constant, i.e. the time required by the receive current to build up the final value proportional to the product CR of the line. If the line is short-circuited at the send end, the condenser discharges with a current at each end given by $I = \frac{E}{R}e^{-\frac{4}{RC}t}$ and during the discharge period, the direction of the send end current is reversed, while at the receive end the same is not reversed.

Case II. Similarly, the equation for current can be deduced for a circuit containing resistance and inductance. If an emf of E volts is applied to a circuit of resistance R ohms, and inductance L henries, then

$E = R_i + L\frac{di}{dt}$ where i = current at any instant.

Then $i - \frac{L}{R}\frac{di}{dt} = \frac{E}{R}$, the solution of which is

$$i = \frac{E}{R} + Ae^{\frac{R}{L}t}$$

For $t = 0, i = 0$, so that $A = -\frac{E}{R}$ and

$i = \frac{E}{R}\left(1 - e^{-\frac{R}{L}t}\right)$, for the rise and fall of current in inductive circuit, as shown in Fig. 1.4(a).

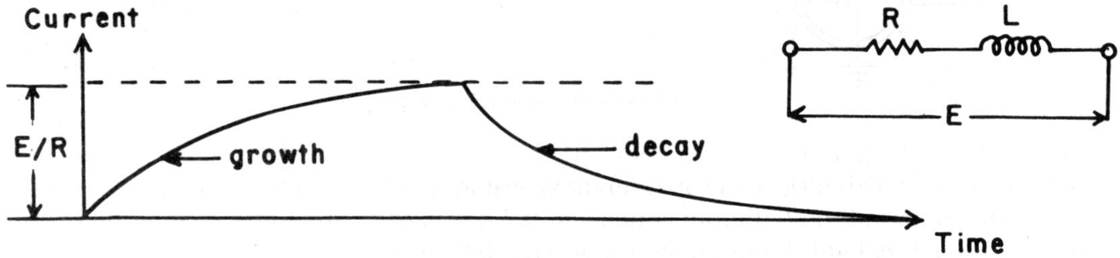

Fig. 1.4(a): Rise and Fall of Current in an Inductive Circuit.

If the battery is replaced by a short-circuit after a relatively long interval of time, an emf is induced in the inductor L, reducing the current gradually to zero. Like the case of a capacitor, the energy stored in the dielectric is dissipated in the circuit when the condenser discharges, so the energy stored in the magnetic field of the inductor is dissipated when the applied emf is removed and the field is collapsed. Then the current in given by $I = \frac{E}{R}e^{-\frac{R}{L}t}$ and the time constant for the inductive circuit is given by L/R, which can be reduced by increasing the value of R and thereby reducing the time of operation of receive relay, and thus the speed of transmission can be controlled.

If the transit time, i.e. the time required for building up the current to a steady state value at the receive end, is less than the duration of the shortest signal element, then there will be no telegraph distortion. But if the build-up time is longer than the shortest signal element, then there will be some distortion but the received signals will be commercially acceptable so long as the build-up time is less than 1.5 times the duration of the shortest signal element. But if it is equal to 1.5 times, the distortion will make the signals uncommercial. Thus, it is found that the transit conditions of inductive or capacitive circuits affect the transmission efficiency of a long direct current telegraph line. The transient currents can be found out from the arrival curve, as given in Fig. 1.4(b). It shows the building up of current at the receive end when a steady voltage is applied at the send end.

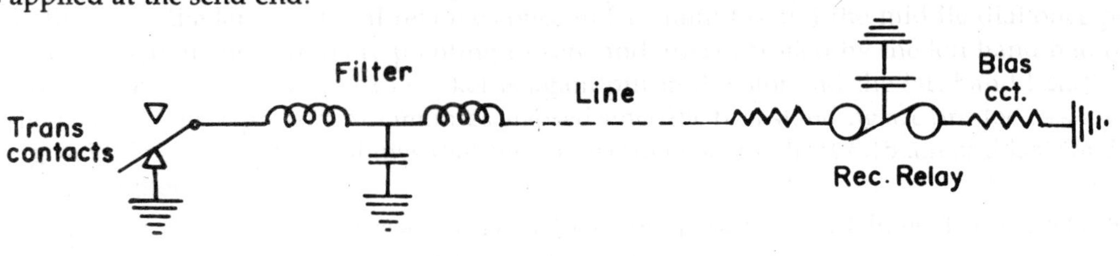

Fig. 1.4(b): Arrival Curve for Unloaded Line.

The response of a system to an applied voltage is given by the shape of an arrival curve. The propagation time is the time that elapses before any effect occurs at the receive end and the transient time is indicated in the above figure, showing the steady rise of current along with time.

In a single current D.C. circuit as shown below in Figure 1.4(c), the line is looped or earthed at the send end during no current periods through a resistance equal to that in the battery circuit.

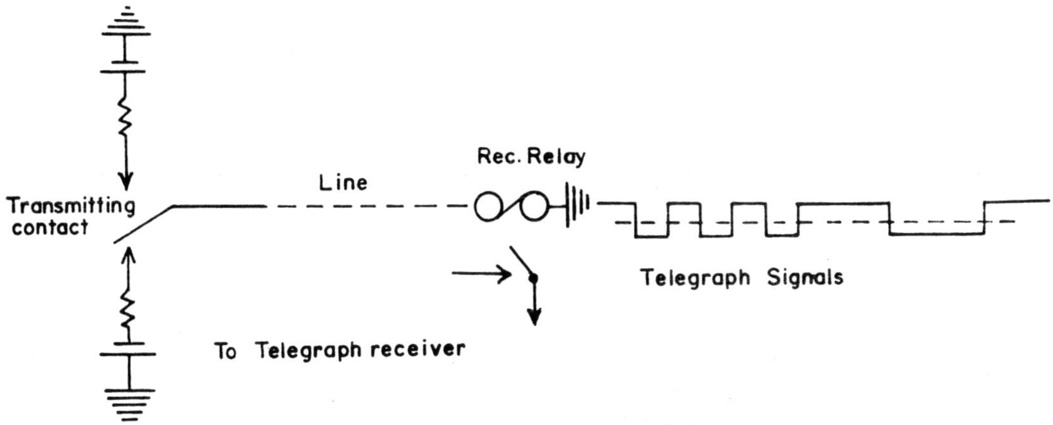

Fig. 1.4(c): Single Current Working in DC Circuits.

In a close-circuit working on a send leg, as shown in Fig. 1.4(d) below, the transmission is by opening and closing the circuit at the out office, where no battery is used. Thus when the circuit is closed $I = E/R$ and where it is open, the current at the battery end is given by

$$I_1 = \frac{E}{R} e^{-\frac{2}{RC}t}$$

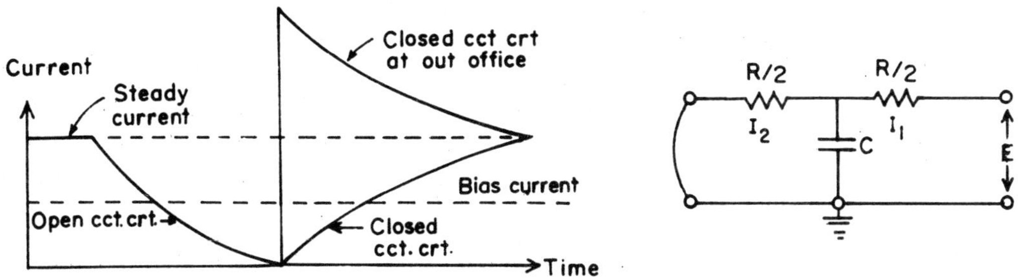

Fig. 1.4(d): Open CCT and Closed CCT Currents with Time.

When the short-circuit is made again, the current at the battery end is given by $I_1 = \frac{E}{R}\left(1 - e^{-\frac{4}{RC}t}\right)$ and that at the other end is given by $I_2 = \frac{E}{R}\left(1 + e^{-\frac{4}{RC}t}\right)$ which are the same as those given for artificial lines.

Alternating Current Signal Transmission: For the transmission of alternating current signals, let the voltage be applied to an open wire line consisting of uniformally distributed resistance, inductance, capacitance and leakance, i.e. the primary constants R, L, C and G per unit length, as shown in Fig. 1.5(a) for single wire circuit and in Fig. 1.5(b) for loop wire circuit where the resistance and inductance are equally divided.

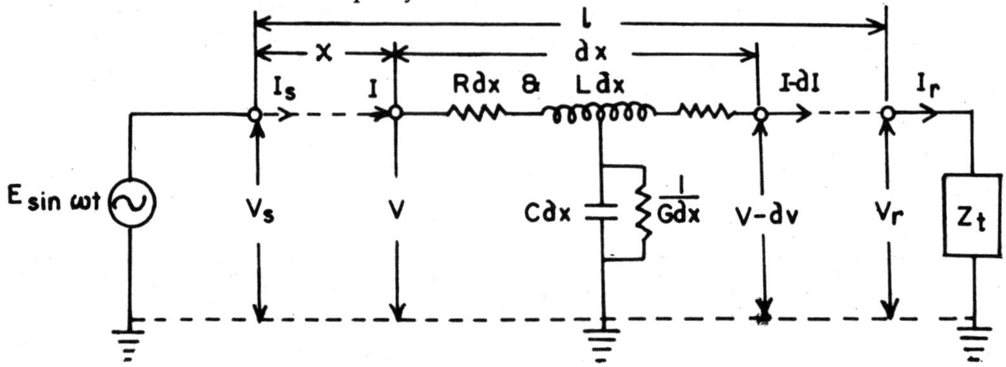

Fig. 1.5(a): Transmission of A.C. Signals in Single-wire Ciruit.

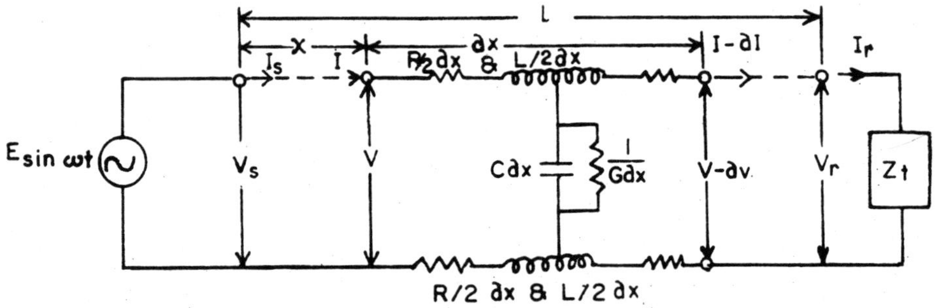

Fig. 1.5(b): Transmission of AC Signal in Loop wire Circuit.

Let a sinusoidal emf E sin ω be applied to one end of a circuit, while the other end is closed by an impedance Zt. Now to determine the voltage and current at any point in the circuit for steady state conditions, let us consider an elementary length of the line ∂x at a distance x from the sending end where the primary constants of this elementary length are $R\partial x$, $L\partial x$, $G\partial x$ and $C\partial x$.

Then the change in line current due to capacitance and leakance of this elementary section is given by

$$-\partial I = VG\partial x + \frac{d^c}{dt} C\partial x$$

as the current is proportional to the rate of change of voltage passing into the capacitor.

Similarly, change of voltage in the line due to voltage drop across the resistance and the opposing emf is proportional to the rate of change of current and is given by $-\partial V = IR\partial x + \frac{dI}{dt} L\partial x.$

In the limit when ∂x is infinitely small, we get

$$-\frac{dI}{dx} = VG + \frac{dV}{dt} C$$

and $$-\frac{dV}{dx} = IR + \frac{dI}{dt}L$$

As V and I are sinusoidal, $\frac{dV}{dt} = j\omega V$

and $$\frac{dI}{dt} = j\omega$$

Hence $$-\frac{dI}{dx} = V(G + j\omega C) \text{ and}$$

$$-\frac{dv}{dx} = I(P + j\omega L) \qquad (4)$$

Differentiating we get $$-\frac{d^2I}{dx^2} = \frac{dV}{dx}(G + j\omega C)$$

and $$-\frac{d^2V}{dx^2} = \frac{dI}{dx}(R + j\omega L).$$

Then substituting the values of $\frac{dV}{dx}$ and $\frac{dI}{dx}$ in the equation (4) we get,

$$-\frac{d^2V}{dx^2} = \frac{dI}{dx}(R + j\omega L).$$

and $$-\frac{d^2V}{dx^2} = V(R + j\omega L)(G + j\omega C).$$

or $$-\frac{d^2I}{dx^2} = p^2I \text{ and } \frac{d^2V}{dx^2} = p_2V \qquad (5)$$

where $p = \sqrt{(R + J\omega L)(G + j\omega C)}$ is called the propagation constant.

The solution of this type of differential equation is given by

$$I = A_1e^{-px} + A_2e^{px}$$

and $$V = A_3e^{-px} + a_4\, e^{-px} \qquad (6)$$

The constants A_1, A_2, A_3 and A_4 can be found as follows:

When $x = 0$, $I = I_s$ and $V = V_B$

$\therefore$ $I_s = A_1 + A_2$ and $V_s = V_1 + V_2$ (7)

Hence differentiating equation (6), we get

$$\frac{dI}{dx} = -pA_1\, e^{-px} + pA_2e^{px} \text{ and}$$

$$\frac{dV}{dx} = 2 \ - pA_3 \, e^{-px} + \ pA_4 e^{px}.$$

Putting $x = 0$ and using Eq. (4),
we get $V_s(G + j\omega C) = pA_1 - pA_2$

$$I_s \, (R + j\omega L) = pA_3 - pA_4 \tag{8}$$

where $Z_o = \Upsilon\sqrt{\frac{R + j\omega L}{G + j\omega C}}$ is called the characteristic impedance. From Eqs. (7) and (8), we get, $A_1 = \frac{1}{2}\left(I_s + \frac{V_s}{Z_o}\right)$, $A_2 = \frac{1}{2}\left(I_s + \frac{V_s}{Z_o}\right)$,

$A_3 = \frac{1}{2}(V_s + I_s Z_o)$ and $A_4 = \frac{1}{2}(V_s - I_s Z_o)$. Substituting these values of A_1, A_2, A_3 and A_4 in Eq. (6), we get

$I = \frac{1}{2}\, I_s(e^{-px} + e^{px}) + \frac{1}{2}\, \frac{V_s}{Z_o}\, (e^{-px} + e^{px})$ and $V = \frac{1}{2} V_s \, (e^{-px} + e^{px}) + \frac{1}{2} I_s Z_o (e^{-px} - e^{px})$ From these,

we get, $I = I_s \cos px - \frac{V_s}{Z_o} \sin h \; px$

and $\quad V = V_s \, px - I_s Z_o \sin h \; px \quad (9)$

Equation (9) represents the fundamental AC transmission equation giving steady state value of voltage and current at any point from the send end on the transmission path, where the current and voltage are sinusoidal. The voltage and current at the receive end can be obtained by putting $x = l$ in equation (9). Thus if V_r and I_r represent the voltage and current at the receive end, then

$$\text{and} \quad \left.\begin{aligned} V_r &= V_s \cosh pl - I_s Z_o \text{sink}\; h\; pl \\ I_r &= i_s \cosh pl - \frac{V_s}{V_o} \text{sink}\; kpl \end{aligned}\right\} \tag{10}$$

It has been observed that if the line is terminated by an impedance other than the characteristic impedance, there is a reflective effect leading to heavy power loss. So the line is always terminated by its characteristic impedance, which is taken as $Z_0 = \frac{V_r}{I_r} = \frac{V_s}{I_s} = \frac{V_x}{I_x}$. If l = length of line, then $\frac{I_r}{I_s} = \frac{V_r}{V_s} = e^{-l(a+jb)}$ where a = attenuation constant and b = phase constant. Now the propagation constant $p = \sqrt{(r + jwL)(g + jwC)}$ is a vector quantity and may be written as $p = a + jb = \sqrt{(R + jwL)(G + jwC)}$. Separating the real and imaginary parts, we get

$$a = \sqrt{\frac{1}{2}\{\sqrt{(R^2 + \omega^2 L^2)(G^2 + \omega^2 C^2)} + (GR - \omega^2 L)} \tag{11}$$

and $$b = \sqrt{\frac{1}{2}\{\sqrt{(R^2+\omega^2L^2)(G^2+\omega^2C^2)} - (GR-\omega^2LC)} \tag{12}$$

Eq. (11) indicates that the attenuation constant depends on W (=2 π f), i.e. frequency of the applied voltage and when this frequency increases, the attenuation also increases. Similarly, equation (12) shows that the phase constant also depends on frequency of the applied voltage but the higher frequency is more attenuated in amplitude than the lower frequency causing distortion of the transmitted wave form. So, for a distortionless transmission we require the use of equalisers so that the overall attenuation is the same for component frequencies at the receive end appearing with the same relative phases as they were in the transmitted band of wave form. It was Heaviside who first saw that by making ratio $\frac{R}{L} = \frac{G}{C}$, it gives the necessary condition for distortionless transmission.

Now rewriting $p = \sqrt{LC}\sqrt{\left(\frac{R}{L} + jw\right)\left(\frac{G}{C} + jw\right)}$ and

$$Z_o = \sqrt{\frac{L}{C}\sqrt{\frac{\frac{R}{L} + jw}{\frac{G}{C} + jw}}}, \text{ we get } p = \sqrt{LC}\left[\frac{R}{L} + jw\right] = \sqrt{\frac{C}{L}}[R + jwL]$$

and $Z_o = \sqrt{\frac{L}{C}}$, which is independent of frequency. Also, $p = a + jb = R\sqrt{\frac{C}{L}} + jw\sqrt{CL}$ gives the attenuation constant $a = R\sqrt{\frac{C}{L}} = \sqrt{RGe}$ which is independent of frequency. The phase constant is given by $b = \omega\sqrt{CL} = 2\pi f\sqrt{CL}$ which is proportional to frequency. Thus, the condition for distortionless transmission is satisfied.

In case of underground cables, the diameter of conductors is kept very small for accommodating maximum number of cable conductors in a given overall diameter of the underground cable. Thus the conductors are of fairly large resistance per mile and due to the small spacing between the conductors, a large capacitance C with a small inductance and very small leakance per mile is produced. It has been observed that in a continuously loaded cable where the conductors are loaded by adding magnetic material, the resistance losses and capacitance are increased but in a coil loaded cable, resistance losses are increased, and above a certain point, the addition of inductances increases losses. It has also been observed that in case of unloaded underground cable at audio frequency, the value of a and b are directly proportional to frequency.

1.3 Development of Telephone Transmitters and Receivers

On observing the operations of duplex telegraph circuits on balanced physical line for resistance and capacitance components of the line, the scientists began to experiment on such identical circuits for transmission and reception of speech. In this respect, the pioneer scientists were

Oersted, Davy, Sturgeon, Faraday, Page and others. Their historic researches began as early as 1854, when Charles Bourseul, a Frenchman, discovered the germ of electric telephony by showing that a vibration made by the speech waves on a movable disc sufficiently flexible can alternately make and break current from a battery and reproduce identical vibrations on a similar disc placed at a distance. Later, in 1861, Philip Reis, a German physicist made further progress by making a transmitter of membrane with a contact at its centre by which the circuit was made and broken due to vibrations by sound waves on it. The receiver was made of an iron rod inside a solenoid. But Reis telephone could only reproduce the pitch of the sound waves and not their intensity or any other qualities of speech. It was Sir Alexander Graham Bell of the United States of America who ultimately discovered electric telephone. The principle of Bell's original electromagnetic telephone circuit is shown below in Figure 1.6(a).

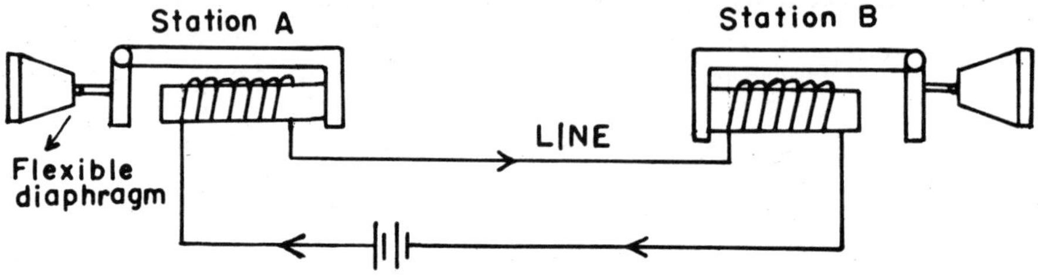

Fig. 1.6(a): Bell's Original Electro-magnetic Telephone Circuit.

The transmitter and receiver were the same electromagnetic instrument. The vibrations made by the speech wave on the metal diaphragm of the transmitter produced similar contact pressure between the diaphragm and the armature, giving rise to variable resistance and the transmitted current on the line reproduced similar vibrations on the diaphragm of the receiver from which the transmitted speech could be easily heard and understood. In 1877, Berliner and subsequently Hughes, Hummings and others found carbon granules as the most suitable material for use in a variable resistance transmitter. The resistance of the carbon granules decreases with increase in mechanical pressure and increase of temperature. It was found that when a pure sinusoidal sound wave was applied to a carbon microphone, the inward displacement of the diaphragm was less than its outward displacement due to greater stiffness of moving carbon granules when compressed. This nonlinearity of carbon microphone was found responsible for the production of harmonics. Thus, if a constant voltage E is applied in a transmitter of resistance R ohm, then the current in the transmitter is given by $I = \frac{E}{R}$. If the sound wave of a certain intensity changes the quiescent resistance by a peak value equal to x ohms, the alternating resistance changes can be expressed by $x \sin wt$, where $\omega = 2\pi f$. Then the effective resistance of the transmitter of any instant is equal to $R + x \sin wt$ and the instantaneous current

$$i = \frac{E}{R + x \sin wt}.$$

$$= \frac{E}{R\left(1 + \frac{x}{R} \sin wt\right)} = \frac{E}{R} \times \frac{1}{\left(1 + \frac{x}{R} \sin wt\right)}$$

The first factor E/R is the quiescent current and is constant while the second factor is alternating component, which can be algebraically expanded as $1-\left(\frac{x}{R}\right)\sin wt+\left(\frac{X}{R}\right)^2\sin^2 wt$, etc. As the expression $\sin^2\omega t$, $\sin^2\omega t$, etc. contain frequencies of 2ω, 3ω etc., harmonics, they are of little significance beyond the third harmonics.

In modern transmitters, the motion of the diaphragm by the pressure of the speech produces a variation of contact pressure between the granules and thus vary the electrical resistance between the front and back electrodes. The diagrams of a solid back transmitter and modern inset transmitter are shown in Figs. 1.6(b) and (c) respectively.

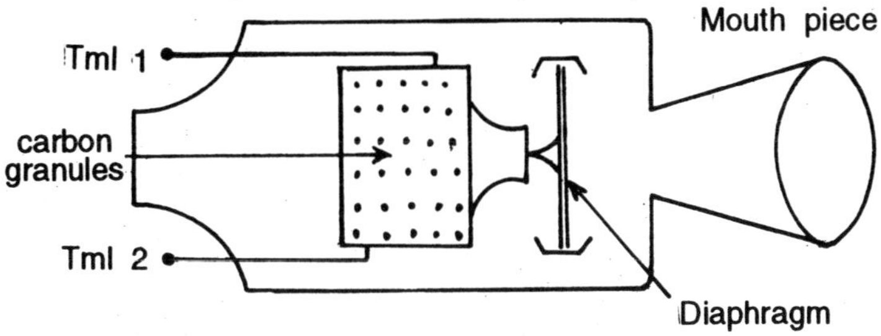

Fig. 1.6(b): Solid back Transmitter

In case of solid back transmitter as shown in Figure 1.6(b), the diaphragm is mounted on a soft rubber seating and its natural frequency is about 1200 c/s. The quiet resistance is of the microphone when no speech waves impinge upon the diaphragm and in case of solid back transmitter it varies with the angle at which the microphone is moved and with the current passing through it. Generally, for a quiet resistance of 40 ohms, a current of 80 mA is used and for the good performance of a telephone the DC should be more than 25 mA in the transmitter.

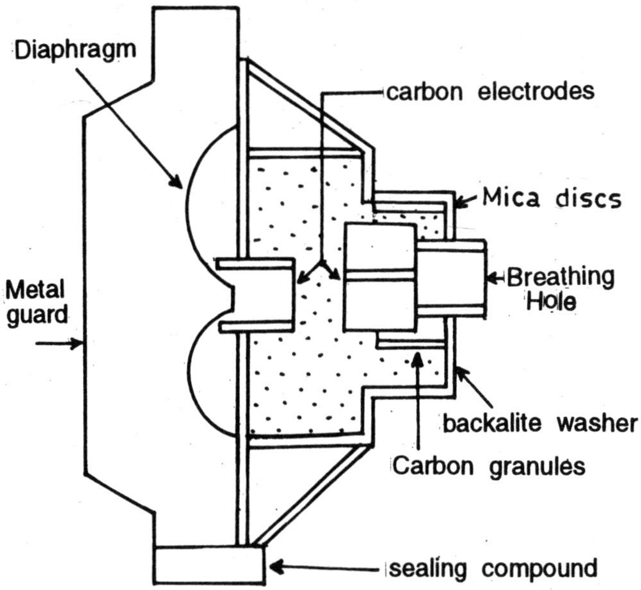

Fig. 1.6(c): Modern inset Transmitter

In case of the modern inset transmitter as shown in Figure 1.6(c), the electrodes are fully immersed in carbon granules so that in whatever position the transmitter is held, the pressure of the granules on the surface of the electrode remains within the required limit. The variation of pressure on the carbon granules remain more or less constant. The transmitter works with an energising current of 30 to 200 mA and its volume efficiency is 3 db, better than that of the solid back transmitter.

The telephone receivers are of electromagnetic type and consist of an electro-magnet in close proximity to the diaphragm so that when an alternating current of speech frequency passes through the winding, the variation of magnetic pull causes the diaphragm to vibrate in accordance with the received current. The arrangement of an electromagnetic receiver is shown in Fig. 1.6 (d).

In the modern inset receiver used in hand-micro sets, each of two coils is of about 100 turns with a nominal resistance of 40 ohms and the pole pieces are screwed to the case to clamp the permanent magnet to give a clearance of 12 to 14 mils of a diaphragm of stallory 9 mils thick. But the newer receivers (No 2P. type), as shown in Figure 1.6(e), are of much improved type, with increased sensitivities at higher frequencies and resonate at about 1700 c/s. But the resonance is damped by restricted air-space on each side of the diaphragm. The back air space is of larger volume and the resonant chamber in front of the diaphragm is slightly conical in shape to give uniform pressure distribution from the periphery to the centre and to the four ear-cap holes near the centre. This damping of resonance to equalise the output of the receiver tends to reduce the volume efficiency which can be compensated by putting pole pieces made of 36% Nickel-iron alloy and the air-gaps between the pole-pieces and the diaphragm should be reduced to minimum.

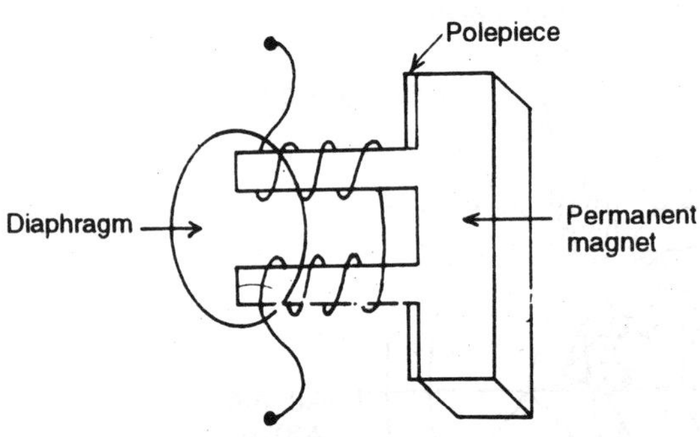

Fig. 1.6(d) Electromagnetic Receiver

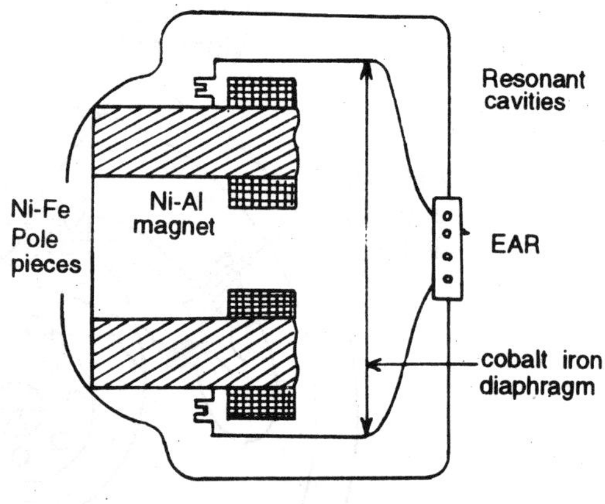

Fig. 1.6(e) Equalised Receiver (No 2P)

In a modern local battery telephone circuit, as shown in Figure 1.6(f), with three winding induction coils connected as an auto-transformer, an emf e in the primary coil of N_1 turns produces an emf E on the secondary coil of turns $(N_1 + N_2)$ of value $E = e\dfrac{(N_1 + N_2)}{N_1}$

Here, we find that the effective turn ratio of induction coil for coupling the line to the transmitter is 1500+400+1000 to 400, i.e. 7/1 while at the receiver, the turn ratio of the line to the receiver is 1500+400+1000 to 1000, i.e. 3/1.

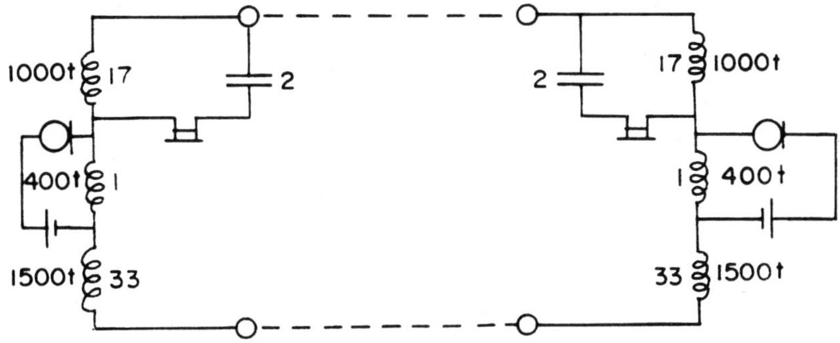

Fig. 1.6(f): Local Battery Telephone Circuit.

It can be seen from the above that in all telephone circuits of a C.B. system some parts of the transmitted energy is absorbed by the local receiver circuit, which is called the sidetone and which gives a disturbing effect to the ear of the user. The volume of this sidetone can be decreased by shunting away some of the transmitted energy appearing in the receiver by using a subsidiary transformer, as shown in Figure 1.7(a) below.

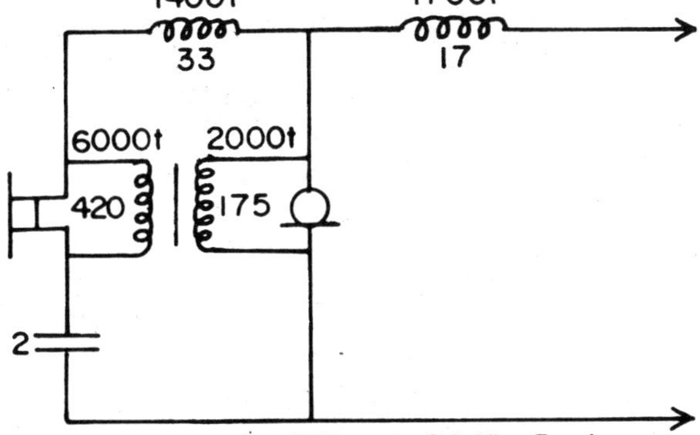

Fig. 1.7(a): Reduction of Sidetone by Subsidiary Transformer.

The subsidiary anti-sidetone transformer is connected with its two windings across the transmitter and the receiver, so that during transmission the emf applied to the receiver is equal and opposite to the normal P.D. across it. For this reason, the winding coil of 175 ohm of 2000 turns is connected across the transmitter, while the winding coil of 420 ohms of 6000 turns is connected across the receiver. In the practical anti-sidetone circuit, in addition to the three inductive windings, two more non-inductive resistances of 30 ohms are inserted, as is shown in Figure 1.7(b). The 30 ohm resistance adjacent to the condenser as a part of 800 turn winding gives signalling condition, while that in receiver circuit gives a P.D. across it, equal to emf in 75 ohm winding due to out of balance condition given by 2 MF condenser.

1.4 Network Theorems used in Communication System

So far, we have briefly discussed the transmission characteristics of physical lines in open wires and underground cables, serving as a medium of transmission and the developments of transmitters and receivers for a complete transmission network; a transmitter is required to transmit the codes or pulses at the send end of the medium and a receiver at the receive end to reproduce these codes or pulses as an exact replica of the transmitted signals. For this we require to understand the functions of various circuits.

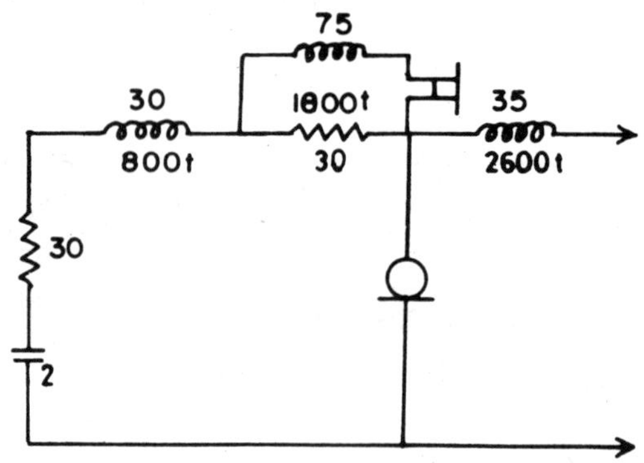

Fig. 1.7(b): Connection of Subsidiary Anti-sidetone Transformer.

In an electric communication system the energy is received initially in a mechanical form and after due transformation, the same is transmitted in an electrical form via the transmission medium. The received energy is retransformed in a mechanical form to reproduce the transmitted signals. The whole system may be considered as a complete network with electro-mechanical coupling between electrical and mechanical meshes. Thus carbon-grain microphone is unilateral as the mechanical movements of the diaphragms control the flow of electrical current, which does not react on mechanical system. But a telephone receiver is a bilateral device as it can be used either as a motor or generator. It can transfer energy either from electrical to mechanical system or vice versa. Due to this reason, in Bell's original telephone circuit the transmission and reception was done through the same instrument.

The main components of communication network are of resistance, inductance and capacitance; of valve and transistors used as oscillators, amplifiers, modulators etc., of which the mutual impedances of resistance, inductance and capacitance are bilateral. They transmit energy equally in both directions while the mutual impedance of valves, transistors, relays, rectifiers, etc. are unilateral, i.e. they transmit energy in one direction only. An electrical network can be defined as a number of resistances and batteries connected together to form an electrical circuit, which is governed by the application of Kirchhoff's laws which states that (i) the algebraic sum of circuits meeting at any point in a network is zero, (ii) the algebraic sum of the product of resistance and current in each conductor of a closed circuit is equal to the algebraic sum of EMFS in that circuit.

The four terminal networks have two input and two output terminals. If by interchanging the input and output terminals, the electrical properties of the network remain unaffected, it is said to be a symmetrical network, otherwise it is called an unsymmetrical network. A symmetrical network has got two characteristics, namely the characteristic impedance and the propagation constant. If in two networks Z_0 and ρ are the same, they are said to be equivalent. A few network equivalence theorems in respect of electrically equivalent T and Π sections are stated below:

Theorem 1: At any one frequency, a T-section can be interchanged in any network with a Π section and vice versa, provided that certain relations are maintained between the elements of the two sections. This in a T or π network shown below in Figure 1.8(a) and (b) connected between a generator of impedance Zs and generated voltage E and a load impedance Z_r, it can be shown that if proper relations are maintained between Z_1, Z_2, Z_3 and Z_A, Z_B and Z_C, then there will be no difference in current I_1, and I_2 for these two networks.

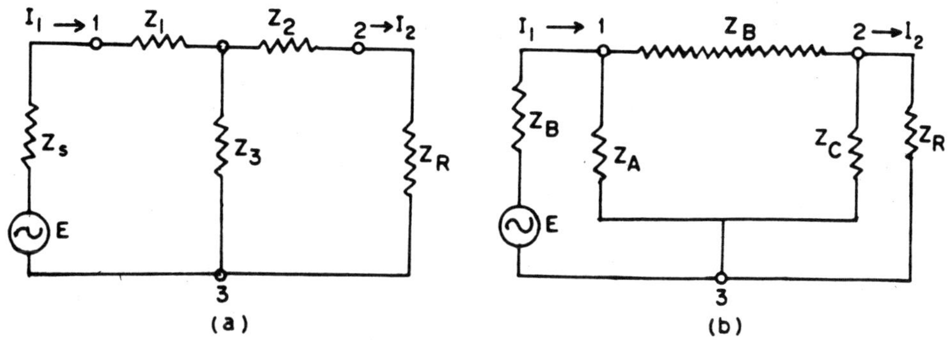

Fig. 1.8(a): T, and **(b)** Π Network Connecting a Generator and a Load.

Thus, if the impedance between two terminals 1 and 2 with 3 disconnected is equated for the two cases, then $Z_1 + Z_2 = \dfrac{Z_B(Z_A + Z_C)}{Z_A + Z_B + Z_C}$. Similarly, keeping the terminal 1 disconnected and measuring between 2 and 3, we get $Z_2 + Z_3 = \dfrac{Z_C(Z_A + Z_B)}{Z_A + Z_B + Z_C}$. And keeping 2 disconnected and measuring between 3 and 1, we get, $Z_3 + Z_1 = \dfrac{Z_A(Z_B + Z_C)}{Z_A + Z_B + Z_C}$

$\therefore$ Solving for Z_1, Z_2, Z_3, we get

$$Z_1 = \frac{Z_A + Z_B}{Z_A + Z_B + Z_C}, Z_2 = \frac{Z_B + Z_C}{Z_A + Z_B + Z_C} \text{ and } Z_3 = \frac{Z_C + Z_A}{Z_A + Z_B + Z_C}.$$

Hence the equivalent of T and Π networks. The equivalence of a complex network to a simple configuration can be stated as follows.

Theorem 2: Any four-terminal network made up of linear impedances, however complex, can be represented at a single frequency by a simple T or Π section.

By successive transformation from T to Π and vice versa, a complicated network can be reduced to a single section of T or Π network.

Development of lattice sections with series impedances: Lattice sections are symmetrical and balanced and have therefore characteristic impedances and propagation constants for which a lattice structure can be redrawn as a bridge structure. Thus the lattice section shown in Figure 1.8(c) can be redrawn as shown in Figure 1.8(d) and that shown in Figure 1.8(e) can be redrawn as shown in Figure 1.8(f).

If we consider the lattice section shown in Figure (c) and the resulting section (d) when modified in this way and if these two sections when terminated on one side with any impedence Z_R and a voltage E is applied to other side terminals, then by application of Kirchhoff's second law in two sections, we get from original lattice section Figure (c):

Mesh 1432, $E = (i - i_1)Z_B - i_2Z_R + (i_1 - i_2)Z_B$

$$= iZ_B - i_2(Z_B + Z_R) \qquad (1)$$

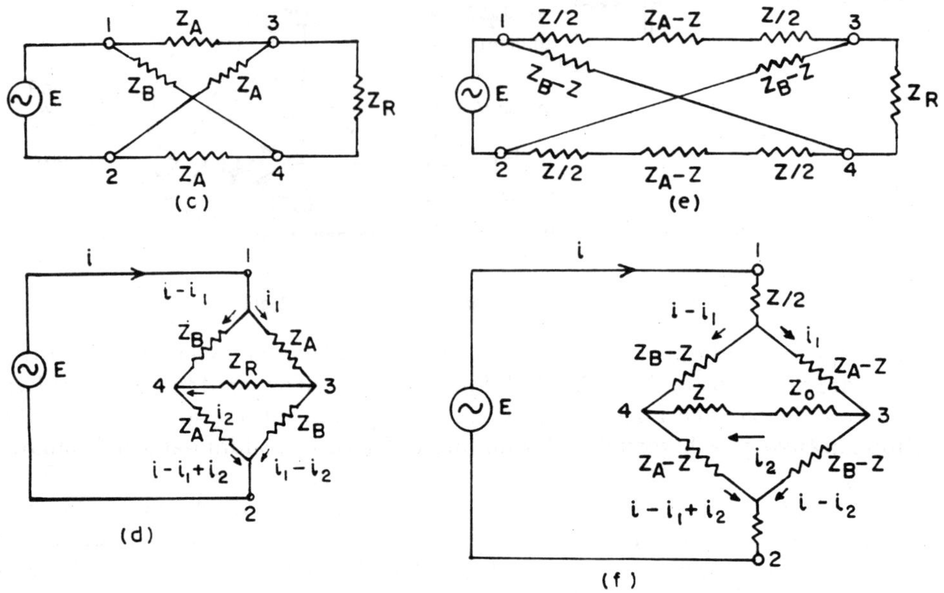

Fig. 1.8 (c), (d), (e) and (f): Development of Lattice Section with Series Impedance.

and Mesh 1342, $E = iZ_A - i_2Z_R + (i - i_1 + i_2)Z_A$

$$= iZ_A + i_2(Z_A + Z_R) \qquad (2)$$

Also, from modified lattice section Fig. (d), we get

Mesh 1432, $E = \frac{1}{2}Z_i + (i - i_1)(Z_B - Z) - i_2(Z + Z_R)$

$$+(i_1 - i_2)(Z_B - Z) + \frac{1}{2}Z_i$$

$$= iZ_B - i_2(Z_R + Z_B) \qquad (3)$$

Mesh 1342, $E = \frac{1}{2}Z_i + i_1(Z_A - Z) + i_2(Z + Z_R)$

$$+(i - i_1 + i_2)(Z_A - Z) + \frac{1}{2}Z_i$$

$$= iZ_A + i_2(Z_R + Z_A) \tag{4}$$

It shows that equations (1) and (2) are identical with6 Eqs. (3) and (4) respectively for which they will yield the same result for Z_o and p, i.e. the characteristic impedance and propagation constant. Hence the modified lattice section in Fig. (d) will be electrically equivalent to the original lattice in Fig. (c). Similarly the modified lattice section in Fig. (f) will be electrically equivalent to original lattice section in Fig. (e).

In this way, we can show the development of lattice with series impedances to prove the equivalence of a lattice to a T section. And from the development of lattice section with shunt impedances, we can similarly prove the equivalence of a lattice to a π section.

Theorem 3: Of the other network theorems, Thevenin's or Pollard's theorem states that the current in any impedance Z_R connected to two terminals of a network is the same as if Z_R were connected to a a simple generator whose generated voltage is open-circuited voltage at the terminals in question, and whose impedance is the impedance of the network looking back from the terminals with all generators replaced by the impedances equal to the internal impedances of these generators, as shown in Fig. 1.9(a).

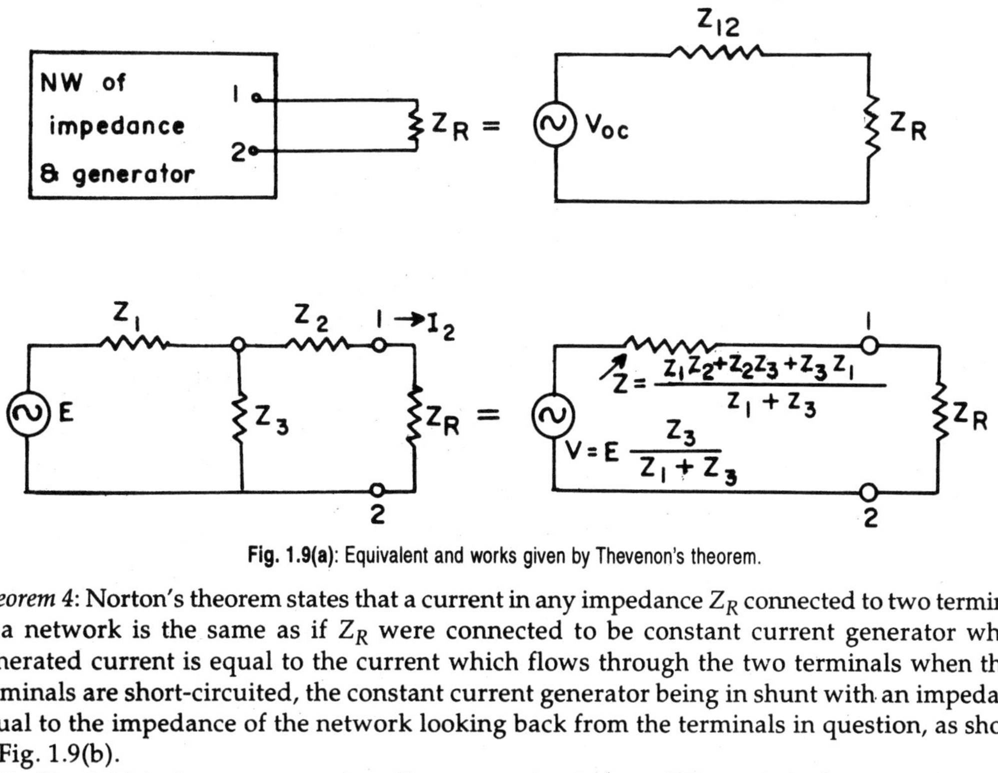

Fig. 1.9(a): Equivalent and works given by Thevenon's theorem.

Theorem 4: Norton's theorem states that a current in any impedance Z_R connected to two terminals of a network is the same as if Z_R were connected to be constant current generator whose generated current is equal to the current which flows through the two terminals when these terminals are short-circuited, the constant current generator being in shunt with an impedance equal to the impedance of the network looking back from the terminals in question, as shown in Fig. 1.9(b).

In Fig. 1.9(a), the upper part is a diagrammatic picture of Thevenin's theorem. V_{OC} is the voltage measured at terminals 1-2 with Z_R removed and Z_{12} is the impedance measured back from the terminals 1-2. In measuring this impedance it is assumed that the generators have been replaced by their internal impedance.

The network can be reduced to a simple T network as shown in the lower part of Fig. 1.9(a), where the solution of T network is given by

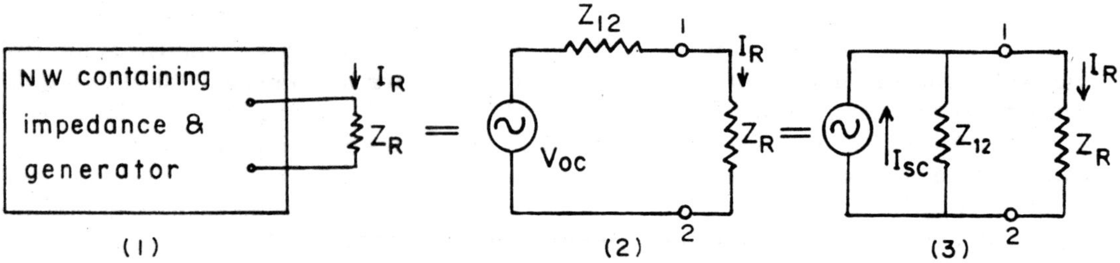

Fig. 1.9(b): Equivalent Network given by Norton's Theorem.

$$I_1 = \frac{E}{Z_1 + \dfrac{Z_3(Z_2 + Z_R)}{Z_2 + Z_3 + Z_R}}$$

$$I_2 = \frac{I_1 Z_3}{Z_2 + Z_3 + Z_R} = \frac{EZ_3}{Z_1Z_2 + Z_2Z_3 + Z_1Z_3 + Z_R(Z_1 + Z_3)}$$

Then by rearrangement, so that Z_R is in the denominator and is alone, we get

$$I_2 = \frac{E\dfrac{Z_3}{Z_1 + Z_3}}{\dfrac{Z_1Z_2 + Z_2Z_3 + Z_1Z_3}{Z_1 + Z_3} + Z_R} \tag{1}$$

Thus, I_2 will be the same for any value of Z_R connected to a generator whose generated voltage is $E\dfrac{Z_3}{Z_1 + Z_3}$ and internal impedance is $\dfrac{Z_1Z_2 + Z_2Z_3 + Z_1Z_3}{Z_1 + Z_3}$. If Z_R is removed, the voltage at the terminals 1-2 of the T network will be $V = E\dfrac{Z_3}{Z_1 + Z_3}$ (2)

If Z_R is removed and E is replaced by a short-circuit, the impedance at the terminals 1-2 will be

$$Z_{12} = Z_2 + \frac{Z_1Z_3}{Z_1 + Z_3} = \frac{Z_1Z_2 + Z_2Z_3 + Z_1Z_3}{Z_1 + Z_3} \tag{3}$$

It will be seen that Eq. (1) gives the current that will flow if a generator whose electromotive force is given by Eq. (2) and whose internal impedance is given by Eq. (3) is connected to a load Z_R. This proves Thevenin's theorem.

For Norton's theorem as illustrated in Fig. 1.9(b), it is evident from Thevenin's theorem that

$I_R = \dfrac{V_{oc}}{Z_{12} + Z_R}$, and on short circuit, the current $I_{sc} + \dfrac{V}{Z_{12}}$.

From Fig. 1.9(b)(3), the current through Z_R would be $I_R = \dfrac{I_{sc} \times Z_{12}}{Z_{12} + Z_R} = \dfrac{V_{oc}}{Z_{12} + Z_R}$

This shows the equivalence of Fig. (b)(2) and (b)(3) and hence the equivalence of (b)(3) and (b)(1), and hence the theorem.

In this way, we can prove and establish the following theorems generally used in a transmission network:

Theorem 5: According to the Maximum Power Transfer Theorem, the maximum power will be absorbed by one network from another joined to it at two terminals, when the impedance of the receiving network is varied, if the impedance looking into the two networks at the junctions are conjugates of each other.

Theorem 6: According to Reciprocity Theorem, in any system composed of linear bilateral impedances, if an electromotive force E is applied between any two terminals and current I is measured in any branch, then their ratio, called the 'transfer impedance' will be equal to the ratio obtained, if the positions of E and I are interchanged.

Theorem 7: According to Superposition Theorem, in any network consisting of generators and linear impedances, the current flowing at any point is the sum of the currents which would flow if each generator were considered separately, all other generators being replaced at the time by impedances equal to their internal impedances.

Theorem 8: According to Compensation Theorem, any impedance in a network may be replaced by a generator of zero internal impedance whose generated voltage at every instant is equal to the instantaneous potential difference produced across the replaced impedance by the current flowing through it.

Theorem 9: According to Impedance Transforming Theorem, if a group of four terminal networks containing only pure reactances are arranged in tandem to connect a generator to a load, then if at any junction there is a conjugate match of impandances, there will be a conjugate match of impedances at every other junction in the system.

1.5 Coupled Circuits, Interference, Resonance, Reflection, Transformers and Alternating Network

(a) *Coupled Circuit*: In a built-up communication circuit consisting of different meshes, each is connected to the other by one or more mutual impedance elements consisting of resistance, capacitance and self-inductance, mutual inductance or mutual capacitance. Such circuits are called coupled circuits.

(b) *Interference*: Inductive interferences occur between neighbouring telephone and other telegraph circuits and between power lines and communication lines. This is called cross-talk, which introduces undesirable noise in telephone circuits interfering with the transmission of speech. This can be minimised by (i) reduction of coupling by avoiding parallelism between the two alignments, (ii) keeping the distance between the two alignments to a safe margin, (iii) using underground cable for telephone alignment, (iv) transposing O.W. telephone trunk lines to equalise mutually-induced voltages, and (v) balancing telephone trunk lines to equalised induced currents.

(c) *Resonance*: We have already seen that the load will absorb the maximum power from the generator, if its impedance matches the impedance of the supply. If both are equally adjusted for magnitude and angle, then load impedance could be conjugate of generator impedance, i.e. the resistive components are equal and reactive components are equal and opposite. But as the load impedance does not usually match with the generator impedance, some methods of impedance transformation have got to be adopted. One such method is the use of resonant circuits to alter the magnitude of load resistance, where inductive reactance increases and capacitive reactance decreases as frequency increases. Thus the total reactance can be adjusted to zero only at one frequency by neutralising the capacitance by an inductance in parallel or by nullifying a series of inductive reactances by a series of capacitive reactances of the same value. This phenomenon is known as series and parallel resonance. Thus for series resonance, the total impedance $Z = R + j\left(\omega L - \frac{1}{wc}\right)$ gives Z as minimum when $\omega L = \frac{1}{wc}$ or $f_r = \frac{1}{2\pi\sqrt{LC}}$.

Hence for resonance $Z_r = R$ when the power factor of the circuit is unity. At lower frequencies $\frac{1}{wc} > \omega L$ and the total reactance is capacitive while at frequencies higher than the resonant frequency, $\omega L > \frac{1}{wc}$ and the circuit is inductive.

(d) *Reflection*: The *de* sign of a transforming network is made by the inductance-transforming theorem as already stated in the previous section. Now if a long transmission line is terminated by its characteristic impedance Z_0, a wave travelling down the line is completely absorbed at the receiving end. But if it is not so terminated, then all the energy that arrives at the far end will not be completely absorbed and the unabsorbed portion will set up a return wave. A similar return wave will occur at the point where there is an unequal cross-section of the line. Hence the unequal termination or cross-section is said to set up reflection of wave which will therefore interfere with the free transmission of a frequency or band of frequencies in that line. The idea of reflection at a junction may be carried over to any two impedances, as shown below in Fig. 1.10(a) and 1.10(b).

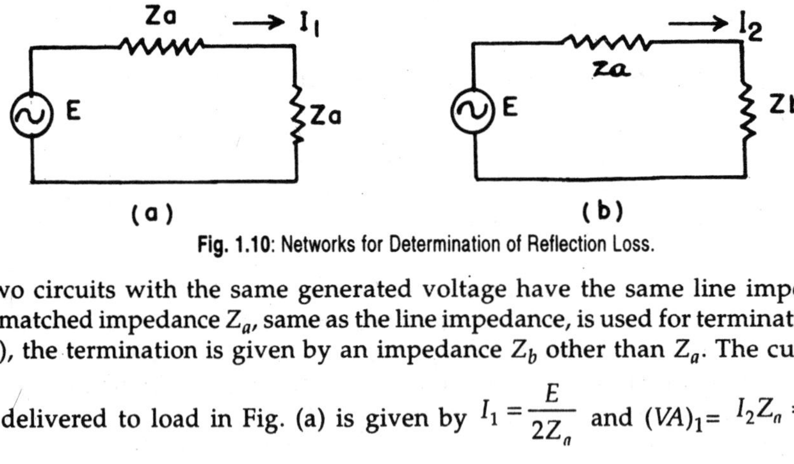

Fig. 1.10: Networks for Determination of Reflection Loss.

The two circuits with the same generated voltage have the same line impedance. In Fig. 1.10(a), a matched impedance Z_a, same as the line impedance, is used for terminating impedance; in Fig. (b), the termination is given by an impedance Z_b other than Z_a. The current and volt-amperes delivered to load in Fig. (a) is given by $I_1 = \frac{E}{2Z_a}$ and $(VA)_1 = I_2 Z_a = \frac{1}{4Z_a}$ and the current and volt amperes delivered to load in Fig. 1.10(b) is given by $I_2 = \frac{E}{Z_a + Z_b}$ and

$$(VA)_2 = I_2^2 Z_b = \frac{E^2 Z_b}{(Z_a + Z_b)^2}.$$

Now, the transfer constant of a properly terminated network is defined as $\frac{1}{2}$ the natural logarithm of the ratio of two input and output volt amperes. Thus the effect of mismatch expressed in terms of transfer constant θ of a network which is inserted between the generator and the load will produce the same drop in volt-amperes delivered as is produced by the mismatch. This gives

$$e^{-2\theta} = \frac{(VA)_2}{(VA)_1} = \frac{4Z_aZ_b}{(Z_a + Z_b)}$$

The term reflection factor is applied to the ratio $e^{-\theta}$ denoted by the symbol K. Then reflection factor $K = e^{-\theta} = \frac{2\sqrt{Z_aZ_b}}{Z_a + Z_b}$. The term $\log_e \left|\frac{1}{K}\right|$ is called the reflection loss.

The case where reflection loss $\log_e \left|\frac{1}{K}\right|$ is a negative number is said to have a reflection gain in nepers.

(e) *Transformer*: A transformer is required in a telecommunication circuit for voltage transformation, for isolating communication plant from electric mains and for impedance matching of two circuits of different impedances, so that a maximum transfer of power can take place from one circuit to the other. A transformer consists of two winding of coils wound round common laminated iron strips core so that a voltage E_p sent through one coil called the primary coil induces a voltage on the other coil called the secondary coil by mutual induction. The larger the secondary coil, the larger is the voltage induced in it. The secondary coil voltage $E_s = nE_p$, where n is the turns ratio of the secondary coil to the primary coil, as shown below in Figure 1.11(a) and 1.11(b) for symbolic and vector representation.

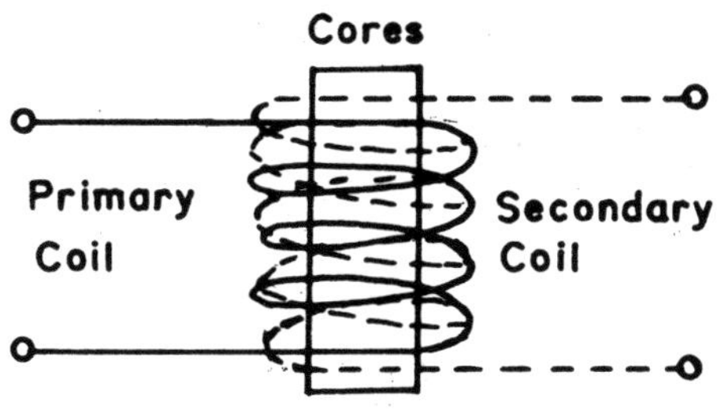

Fig. 1.11(a): Symbolic Diagram of a Transformer

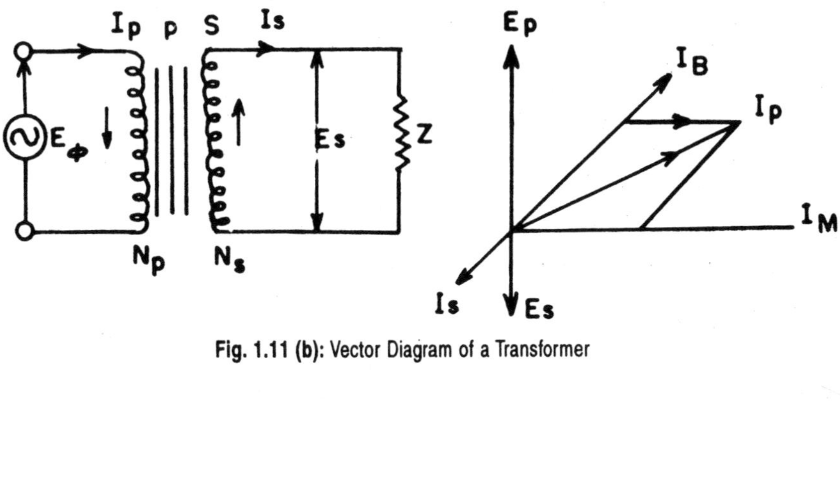

Fig. 1.11 (b): Vector Diagram of a Transformer

If the primary and secondary coils have N_p and N_s turns, then

$$\frac{E_p}{N_p} = \frac{E_s}{N_s} \text{ or } \frac{E_p}{E_s} = \frac{N_p}{N_s} \quad (1)$$

If the secondary winding is open circuit, then the primary winding is simply an inductor and a low magnetising current $I_M = E_p / (R - \omega L)$ flows in the secondary winding, 90^0 behind the primary voltage E_p. The magnetic flux θ on the iron core will be in phase with the current and hence is 90^0 behind E_p. When a load is applied to the secondary, a current I_s will flow in the secondary circuit and according to Lenz's law, this induced current produces in the core an opposing magnetic flux reducing the original flux. But if the primary voltage is applied at a constant value, this flux reduction cannot occur and an increased primary current I_B must flow to neutralise the counterflux of the secondary. If the low magnetising current be ignored, the ampere-terms NI in the primary and secondary windings are always equal in magnitude and opposite in direction

$$\therefore \quad N_p I_p = N_s I_s \text{ or } \frac{I_p}{I_s} = \frac{N_s}{N_p} \quad (2)$$

But from (1) $\frac{N_p}{N_s} = \frac{E_p}{E_s}$. Hence $\frac{I_p}{I_s} = \frac{N_s}{N_p} = \frac{E_s}{E_p}$ or $\frac{E_p}{E_s} = \frac{I_s}{I_p}$ (3)

Now, from the vector diagram as shown in Fig. 1.11(b), we find that the low magnetising current I_M indices a secondary EMF $E_s = E_p \times \frac{N_s}{N_p}$, which is in phase with the primary back EMF, i.e. 180^0 out of phase with the applied voltage E_p. The secondary load current $I_s = E_s/Z$ has the phase angle Φ of the load impedance. This secondary current causes an increased (balancing) current $I_B = I_s \times \frac{N_s}{N_p}$ to flow in the primary winding in opposite phase to I_s. Thus the vector sum of I_B and I_M gives the primary current I_P.

Impedance matching is discussed in Fig. 1.12(a) and (b) below.

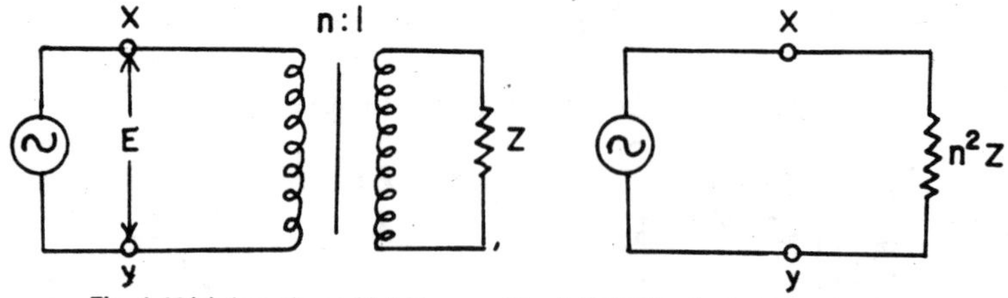

Fig. 1.12(a): Impedance Matching **Fig. 1.12(b):** Equivalent Impedance

In Fig. 1.12(a) is shown the generator with constant output voltage E applied across the primary winding of a transformer, when the secondary is closed with an impedance Z. The turns ratio $N_p/N_s = n$, the induced e.m.f. in the secondary= E/n and the secondary current $I_s = E/nz$. So if the magnetising current is neglected, the current in the primary coil is $I_p = \frac{1}{n}\left(\frac{E}{nZ}\right) = \frac{E}{n^2 z}$.

Here, n^2Z is equivalent to impedance of the transformer plus secondary load impedance Z when viewed from the input terminal (xy) as shown in Fig. 1.12 (b). Thus by using a transformer, the impedance of a given circuit can be altered to any desired value by selecting the value of the turns ratio n and hence by connecting a transformer between two impedances Z_1 and Z_2, the impedance ratio can be expressed as $n = \sqrt{\frac{Z_1}{Z_2}}$, where n is greater than 1 and $Z_1 > Z_2$.

For maximum transfer of power, the two circuits must have equal impedances. This is a very important problem in telecommunication systems, e.g. in telephone system where power from microphone is transferred to the distant receiver through switching equipments, amplifiers and various sections of filters, etc. having specific impedance for each link. So it is necessary to use transformers for impedance matching between each pair of links. For this, various types of audio frequency transformers are used. (i) Repeating coil or line transformers to match the line impedance to the exchange equipment impedance, (ii) amplifier input/output transformers to balance the impedance and to prevent leakage induction of power on both sides, (iii) inter-valve transformers and, (iv) auto-transformers, etc. for matching the impedances on both sides for the maximum transfer of power.

(f) *Attenuating networks*: At times it is desirable to attenuate the current and voltage in a transmission network at some stages to bring them within the required limit. For this, all frequencies are required to be attenuated to some extent, to prevent distortion or over-amplification. An attenuator should consist of purely resistive components so that for each such component the phase constant $b = 0$ and the propagation constant p becomes equal to the alternation constant a. Such fixed attenuators are known as pads. The unbalanced and balanced type of T, L and Π networks or pads are diagrammetically presented below in Fig. 1.13.

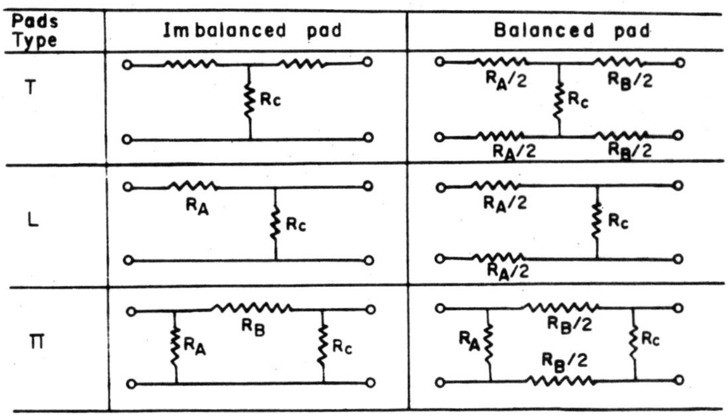

Fig 1.13: Unbalanced and Balanced T, L and Π pads.

1.6 Filters and Equalisers

(a) A filter is a network designed to attenuate a band of frequencies and pass other frequencies without loss. Such frequencies that separate a pass band from an attenuation band is called cut-off frequencies. These filters are constructed from purely reactive elements, as otherwise the attenuation could never become zero for some frequencies and change from zero to some other value for other frequencies. This is the difference of a filter from an attenuator, in which the attenuation is purely resistive and which absorbs power from any generator connected to it. It is constant in nature and independent of frequencies.

An electric wave filter may be defined as an arrangement of electric components, generally inductors and capacitors having two input and output terminals each, through which a particular band of frequencies can pass without any appreciable loss of energy, while all other frequencies beyond this particular band will experience heavy loss. These filters are used in different types of telecommunication networks, such as carrier and VFT systems, TV and satellite communications, etc. These filters can be divided into four classes, namely, low-pass, high-pass, band-pass and band-stop filters. A low-pass filter passes freely all the frequencies from zero up to the cut-off frequency f_c given by $f_c = \frac{1}{2\pi\sqrt{LC}}$. A high-pass filter passes all the frequencies freely above the cut-off frequency given by $f_c = \frac{1}{2\pi\sqrt{LC}}$. A band-pass filter passes frequencies in a band between the upper and lower cut-off frequencies and is, in effect, a combination of a low pass and high-pass filter in such a way that the cut-off frequency of the low-pass filter overlaps that of the high-pass filter. If f_1, and f_2 are the cut-off frequencies of high pass and low-pass filters, then $\sqrt{f_1 f_2} = \frac{1}{2\pi\sqrt{LC}}$ and $\frac{f_2 - f_1}{\sqrt{f_1 f_2}} = n$ where n is the number of sections of prototype or basic section of filters. A band-stop filter passes freely all the frequencies beyond a particular band outside the upper and lower cut-off frequencies of that band. The designs of a single T-section and Π-section and prototype sections of these filters are shown in Fig. 1.14 along with their frequency-attenuation curves.

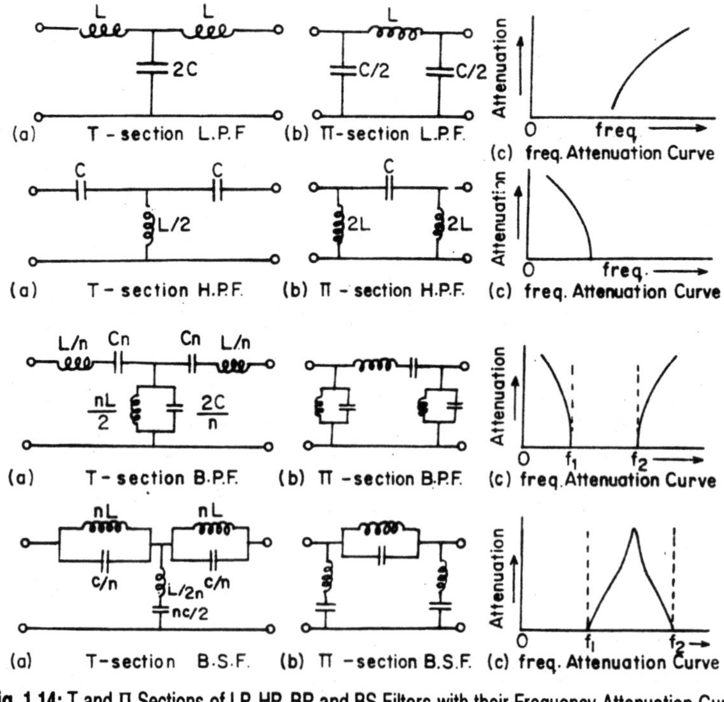

Fig. 1.14: T and Π Sections of LP, HP, BP and BS Filters with their Frequency Attenuation Curves

To make a complete filter for practical use, we require a lot of full sections and half sections within the desired limits of frequencies. If we take Z_1 and Z_2 as the series and parallel impedances of inductors and capacitors respectively, then we can fix their value in such a way that $Z_1Z_2 = Ro^2$, where *Ro* is a non-reactive constant resistance not depending on the frequency. *Ro* is known as the design impedance. The open-circuit and short-circuit impedances can be deduced as shown in Fig. 1.15(a) and (b) below:

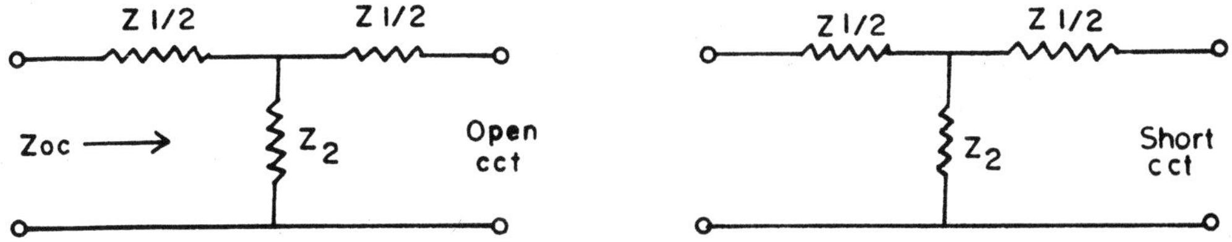

(a) (b)

Fig. 1.15: Symmetrical T-section Network on (a) Open-circuit and (b) Short-circuit.

Then $\quad Z_{oc} = Z_{1/2} + Z_2 Z$ and

$$Z_{sc} = Z_{1/2} + \frac{\frac{Z_1 Z_2}{2}}{Z_{1/2} + Z_2} = \frac{Z_1 Z_2 + \frac{Z_1^2}{4}}{\frac{Z_1}{2} + Z_2}$$

$$= \frac{Z_1 Z_2 + \frac{Z_1^2}{4}}{Z_{oc}}$$

$$\therefore \; Z_{ac} \times Z_{sc} = Z_1 Z_2 + \frac{Z_1^2}{4} = Z_o^2$$

$$\therefore \; Z_o = \sqrt{Z_{oc} \times Z_{sc}}$$

M-derived filter sections: To derive m-derived section of a filter from a prototype T-section, we can construct a new section form a prototype section with a series arm of the same type but of different value mZ_1, where m is a constant, as shown in Fig. 1.16 below.

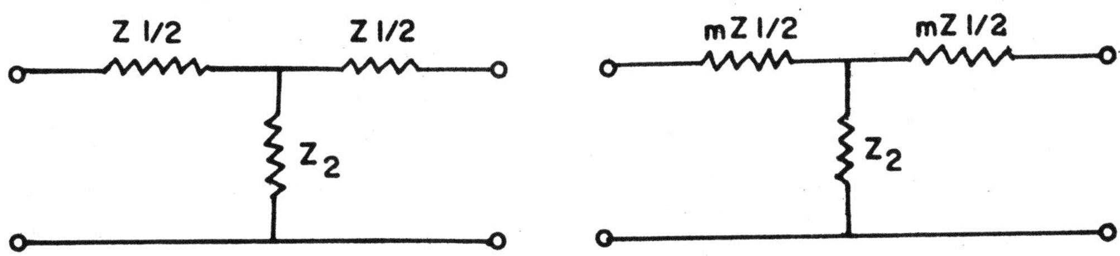

(a) Prototype section (b) m - derived section

Fig. 1.16: Derivation of *m*-derived T-section.

For prototype section, $Z_{OT} = \sqrt{\frac{Z_1^2}{4} + Z_1 Z_2}$ and for *m*-derived T section, the newly derived

$Z_{OT} = \sqrt{\frac{m^2 Z_1^2}{4} + m Z_1 Z_2^1}$. The two will be same if $\frac{Z_1^2}{4} + Z_1 Z_2 = \frac{m^2 Z_1^2}{4} + m Z_1 Z_2^1$

i.e. if $Z_2^1 = \frac{Z_2}{m} + \frac{1 - m^2}{4m} Z_1$.

This means that $Z_2{}^1$ must be an impedance $\frac{Z_2}{m}$ in series with an impedance $Z_1 \frac{(1 - m^2)}{4m}$ and both these impedances can be constructed if 0<m< 1 as shown from a comparison of prototype and m-derived T-section in Figs. 1.17(a) and (b).

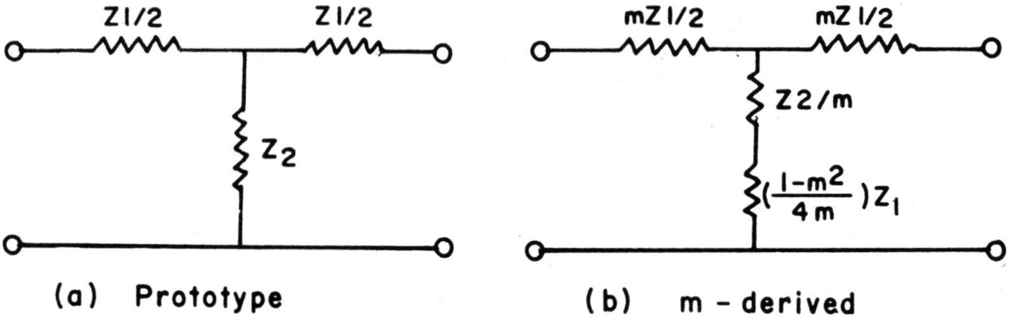

Fig. 1.17: Comparison of Prototype and m-derived T-section.

Similarly, we can derive a new section for Π section prototype having the same Z_o at all frequencies as shown below in Figure 1.18.

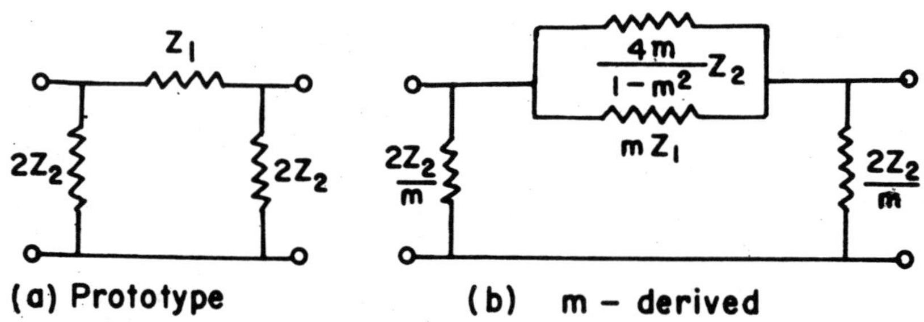

Fig. 1.18: Prototype and *m*-derived Π Sections.

(b) *Equalisers*: It is observed that when a desired frequency band passes through a filter, the end frequencies are distorted to a very great extent but not the mid frequencies. For this reason, after a band of frequencies pass through a filter, the output from the filter is passed through another filter, which distorts these frequencies in the opposite direction and thus equalises the unequal attenuation of the frequency band as a whole. Such an opposite filter is called an equaliser.

The distortion caused by various filters due to variation of attenuation with the frequencies of a band is rectified by the equalisers called attenuation equalisers. Similarly, the phase distortion caused by the variation of velocity with frequency is rectified by the phase equalisers. Without the use of such equalisers, it is difficult to understand the muffled speech of a particular person due to his hoarse voice as heard due to distortion of the end frequencies of the speech path. Such distortion, to affects the specific pitch, quality and exactness of the voice of a particular person as heard at the distant receiver. From the frequency-attenuation curve of filters and equalisers, as shown in Fig. 1.19 below, we can understand how the end frequencies are brought back to their original shape by the equalisers. Such equalisers are therefore called inverse filters.

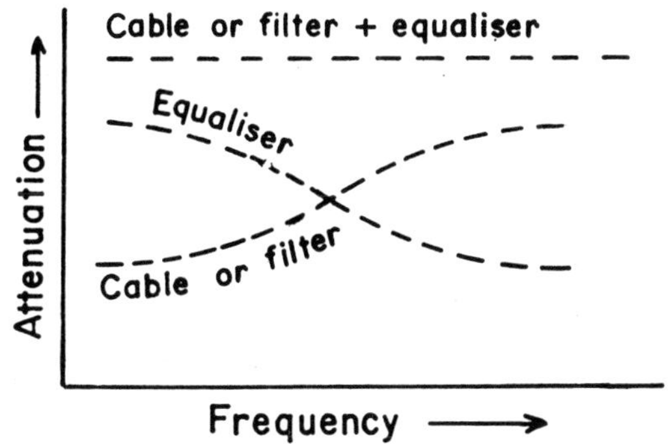

Fig. 1.19: Frequency-attenuation Curve of Filter and Equaliser.

To overcome such unequal distortion of end frequencies, there are mainly two types of equalisers used in practice, namely, (a) the constant resistance lattice type equaliser and (b) the two terminal shunt or series type equaliser.

In case of constant resistance lattice type equalisers with four terminals as shown in Figure 1.20(a) below, the impedance Z_1 and Z_2 must be constructed so that

$$Z_1 Z_2 = R^2 \quad (1)$$

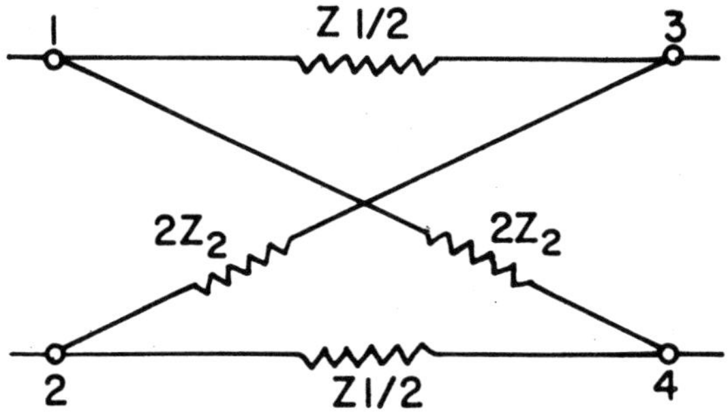

Fig. 1.20: Constant Resistance Lattice Type Network.

Where the resistance R is independent of frequency and the pairs of networks corresponding to Z_1 and Z_2 are called inverse networks. Thus, the two networks can be made inverse, if the impedance of one has the general form of admittance of the other and if definite ratios are kept between the corresponding elements of the two networks.

Thus, for each resistive element in Z_1 there should be a conductive element in Z_2, and for each inductive element in Z_1 there should be a capacitive element in Z_2 and vice versa. Also, if the two elements are in series in Z_1, the corresponding elements in Z_2 should be in parallel and vice versa.

The open-circuit and short-circuit are given by

$$Z_{oc} = \frac{Z_1}{4} + Z_2 \text{ and } Z_{sc} = \frac{2Z_1Z_2}{\frac{Z_1}{2} + 2Z_2} \tag{2}$$

$$Z_{sc} = \frac{Z_1Z_2}{\frac{Z_1}{4} + Z_2} \tag{3}$$

$$\therefore \qquad Z_O = \sqrt{Z_{OC}.Z_{SC}} = \sqrt{Z_1Z_2} \tag{4}$$

If Eq. (1) holds, then the characteristic impedance is a resistance independent of frequency and the input impedance of the network will be independent of frequency and equal to R, if the network is terminated by R. This is why the structure is called 'constant resistance' network.

The other type of equaliser is the two-terminal equaliser shown in Fig. 1.21(b), in which a network is connected in series or parallel with the sending or receiving end of the network to be corrected.

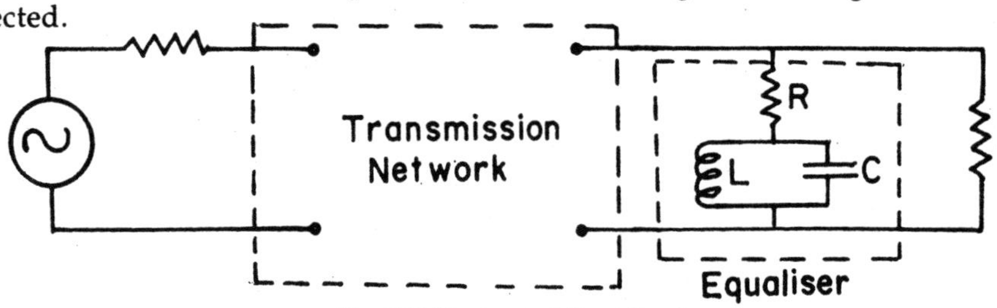

Fig. 1.21: Two-terminal Shunt Equaliser.

It is apparent that the loss introduced at the anti-resonant frequency of LC combination is negligible and the loss increases as the frequency is varied away from anti-resonance. Thus, for equalising a cable the anti-resonant frequency will have to be set at or above the highest frequency to be transmitted. If the system was originally terminated by an impedance Z_R, the same will be modified by the equaliser. Thus, from the above figure where the equaliser is connected in parallel at the receive end of the transmission network, we find that by Thevenin's theorem, the original network can be considered as a generator E_g in series with an impedance Z_s as shown in Fig. 1.21(c) in which the equivalent systems are shown with and without the equaliser Z_e in shunt.

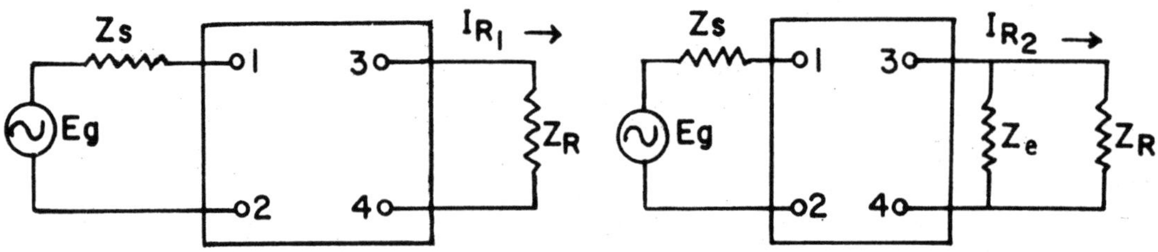

Fig. 1.22: Equivalent Networks for Determining Terminal Loss of Two-terminal Slant Equalisers.

In practice, it is found that after the high frequency line-up test in a transmission network, the end channel frequencies are sometimes beyond the permissible limit. In such cases, the two-terminal equalisers may be used in an easier way than the constant resistant type, where the four arms of the lattice structure are to be adjusted. But with two-terminal equaliser it is very common to make L and C fixed in a section called fixed equaliser and in a separate section called variable equaliser where R is made adjustable. In such cases, the insertion loss of the equaliser and the line is measured at a frequency near the anti-resonant frequency of LC and at a low frequency and an adjustment of variable resistance R in sections is made by trial and error method, till the losses at the two frequencies are more or less equal. Then, a test run of frequencies is taken and the resistance slightly adjusted to obtain the best test results.

1.7 Metal Rectifiers and Power Supplies

Metal rectifiers are used for normal supply of D.C. from A.C. mains and secondary batteries are kept as standby power plants in exchanges/telegraph offices.

(a) *Metal rectifiers*: In electric conductance, each conductor is found to have a constant resistance at a given temperature. A graph plotted with applied e.m.f. and resultant current is found to be a straight line and the relationship between the e.m.f. and current is called linear. But certain substances like red copper oxide do not conform to Ohm's law, as the applied e.m.f. and the resultant current when plotted in a graph is found to be a curve. Similar results are observed in case of thermo-ionic valyes when the output current is plotted against the input voltage. Such a type of relationship is known as non-linear.

The main characteristic of a non-linear material is that it passes current very easily in one direction while in the other direction it offers very high impedance or does not pass current at all. Such properties of a non-linear material is utilised in rectifying the alternating current into direct current. Silicon, germanium, copper oxide, etc. which show intermediate properties between conductors and non-conductors are called semiconductors. These semi-conductors along with copper oxide and selenium — are used in construction of metallic rectifiers for rectifying AC to DC and are used extensively in telecommunication circuits. The basic design of a copper oxide rectifier and a selenium rectifiers is shown in Figures 1.23(a) and (Fig. 1.24).

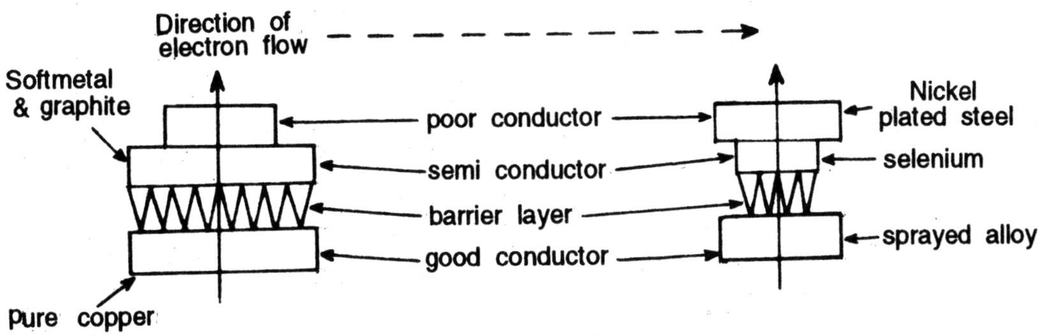

Fig. 1.23: Copper Oxide Rectifier

Fig. 1.24: Selenium Rectifier

Metal Rectifiers as Power Supply Units for Normal Work: These metal rectifiers are extensively used in telegraph and telephone networks in deriving direct current supplies from AC mains, in driving motors of telegraph instrument and for giving signalling supplies. Thus we can have power supply circuits providing 80+80 Volt and 12 Volt D.C. from AC mains or from 12 Volt DC as shown below in Fig. 1.25 for operation of telegraph equipments for normal work.

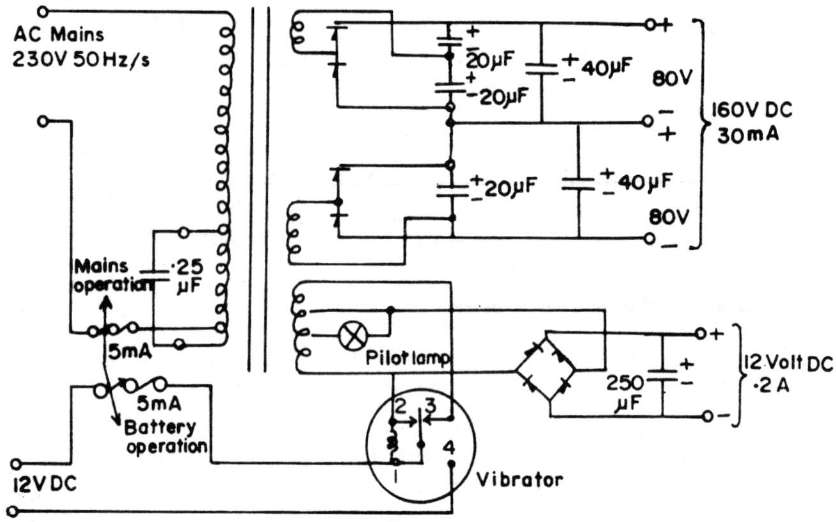

Fig. 1.25: 80+80V and 12V D.C. from A.C. or 12 V D.C.

Similarly we can have power supply unit providing 130V HT and 24V LT from A.C. mains for operation of carrier and VFT equipments, as shown in Fig. 1.26 below.

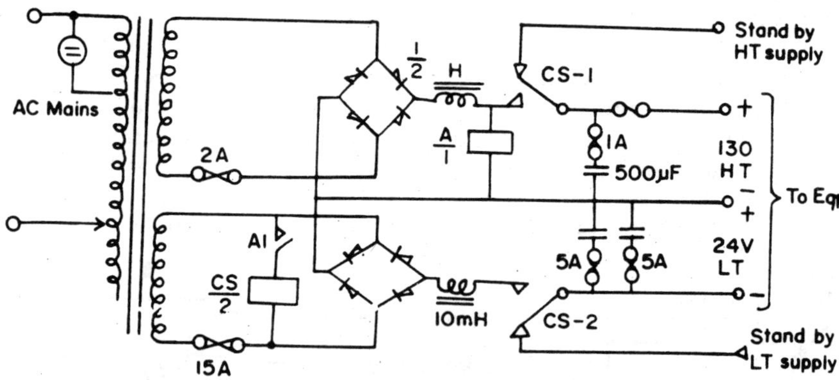

Fig. 1.26: Power Supply Unit Providing 130V HT and 24V LT from AC Mains for Normal Work.

Smoothing Circuits: The output from a rectifier is found to consist of DC pulses alongwith some AC components forming ripples, which are harmful to operation of equipment. So they are removed by smoothing circuits which are LP filters having their cut-off frequency lower than the lower ripple frequency, as shown in choke-input or condenser-input filters in Figs. 1.27 and 1.28 below.

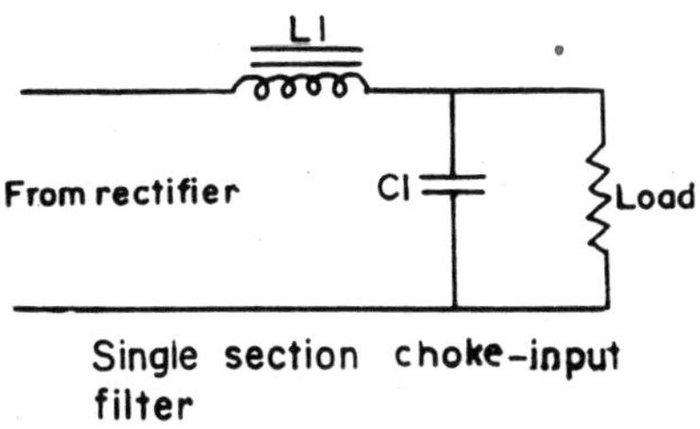

Fig. 1.27: Single Section Choke Input Filter

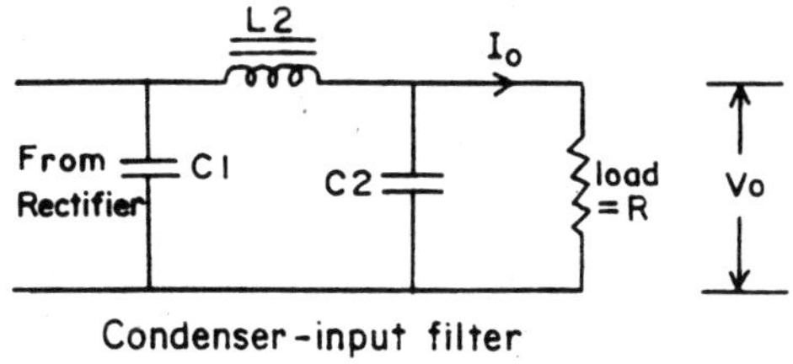

Fig. 1.28: Conduser Input Filter

A rectifier may also be used as a voltage limiter to prevent voltage from exceeding a peak value to check overloading as is done across a telephone receiver when voltage exceeds 0.25 volts.

Standby power supply: In case of mains supply failure or troubles etc., standby power supply by secondary battery is arranged for telegraph office and telephone exchange equipment. In these secondary cells, the chemical energy is converted into electrical energy during discharge of the cell while the electrical energy is converted with chemical energy during the charging period. The reactions during the charge is as follows:

Positive plates — $PbSO_4 + SO_4 + H_2O \rightarrow PbO_2 + 2H_2SO_4$;

Negative plates — $PbSO_4 + 2H \rightarrow Pb + H_2SO_4$.

The reactions during the discharge period is given below: $P60_2 + 2H= H_2S0_4= p6S0_4 + 2H_20$ for positive plates and $Pb + SO_4 \rightarrow PbSO_4$ for negative plates.

The number of negative plates exceeds that of the positive by one. The electrolyte in a solution of sulphuric acid in distilled water with a specific gravity of 1.140. Pure sulphuric acid is of specific gravity 1.84. In the fully charged battery, the specific gravity rises to 1.215 as measured by a hydrometer. Upto 300 ampere hour capacity battery, the container is of glass jar and beyond 300 (A.H.) the container is made of lead jars kept tightly filled in wooden boxes. The initial charge takes about 30 hours of continuous charge, first at half-normal rate till the specific gravity drops to the lowest and then at normal rate till gassing starts. The charge is stopped and the electrolyte is allowed to cool down. After a few hours if the sp.gr is below 1.215, the battery is given extended charge at half normal rate till gassing starts. This process of extended charge is continued till the sp.gr. is found steady at 1.215. Then the battery is completely discharged to sp. gr. 1.140. The battery is again charged at normal rate to a sp.gr. 1.215 and is kept ready for normal charge-discharge cycle. The charge is normally given by a motor-generator set or a full wave rectifier with centre-tapped transformer or a tungar rectifier arranged in a charge-discharge switch board, as shown in Fig. 1.29.

If T= time of charge in hours, t = time of discharge in hours, I and i are the average charge and discharge currents in amperes, then ampere-hour efficiency $= \frac{it}{IT} \times 100\%$ and watt-hour efficiency $= \frac{i \times v \times t}{I \times V \times T} \times 100\%$, where V and v are the average potential difference in volts during charge and discharge. Normally the ampere-hour efficiency is 90 to 95 per cent while the normal watt-hour efficiency is 70 to 80 per cent.

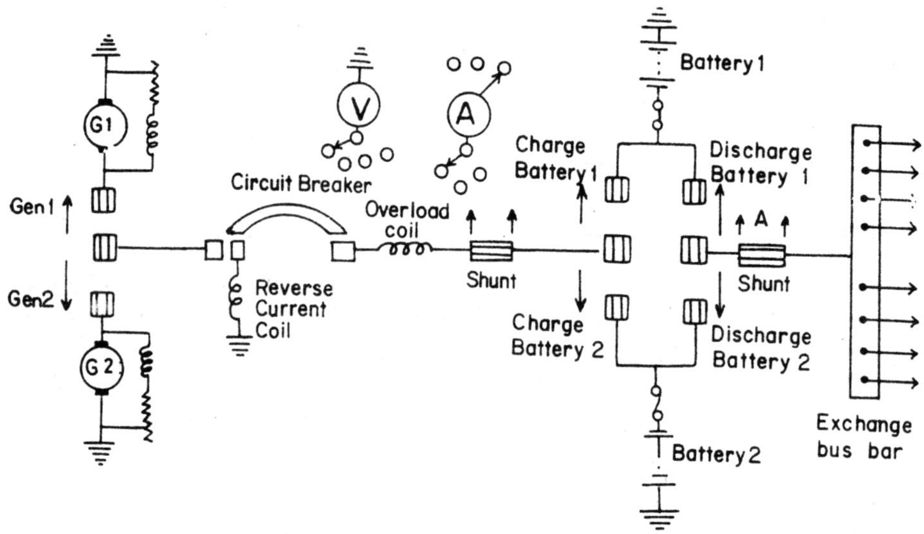

Fig.1.29: Diagram of Charge-discharge Power Switch Board.

Ringing Machines: The ringing machines are used to supply $16\frac{2}{3}$ Hz/s AC ringing current from a telephone exchange for ringing on subscribers telephone bells. For a small exchange, magneto-type hand generators are used but for exchanges with two positions or more power, ringing machines are used as shown below in Figure 1.30.

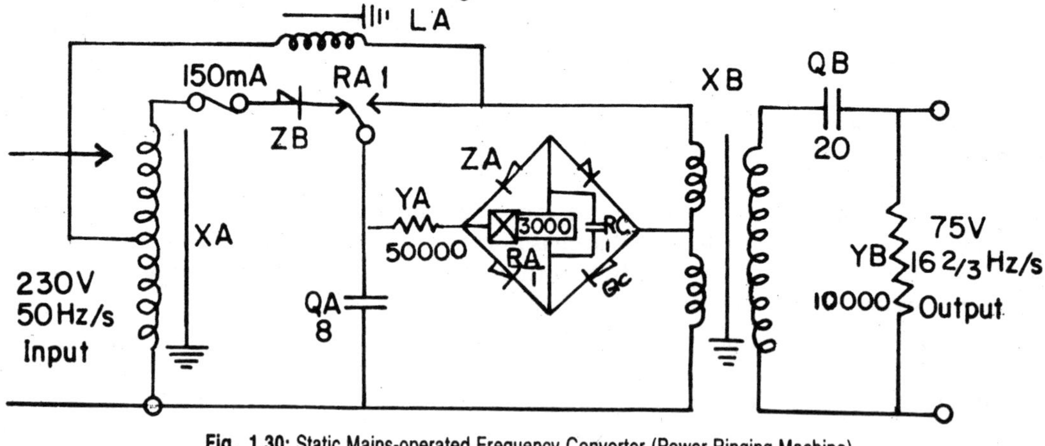

Fig. 1.30: Static Mains-operated Frequency Converter (Power Ringing Machine).

When the mains switch is on, the condenser *QA* is charged by the rectified current through the rectifier *ZB*. When the voltage across the condenser *QA* attains a high value, the relay *RA* operates. As relay *RA* is slow to operate, on operation *RA*1 contact holds and *RA* relay holds due to voltage across the oscillatory circuit which is rectified by the bridge after switching network *ZA*. $16\frac{2}{3}$ Hz/s oscillations of continuous ringing current in the timed circuit are maintained by periodical pulses of current through choke *LA*. By proper commutation, the periodicity of continuous ringing current is interrupted at 0.4 sec on, 0.2 sec off, 0.4 sec on and 2 sec off with the help of slow speed cam set fitted on the output of the ringing machines.

Valve Electronics: It has been observed that when an electrostatic field is super-imposed on a conductor, a drift of electrons from the outer orbits of the conductor flow from the low potential to high potential part of the electrostatic field constituting what is called an electric current in the reverse direction. Thus when a metal is heated, the normal motion of the conductor electrons is intensified so much so that some electrons leave the surface of the conductor altogether constituting what is called thermionic emission. The operation of a valve consisting of a suitably heated cathode or emitter of electrons together with one or more electrode for collecting or controlling the electron flow respectively called anode or grid, depend on whether the valve will function as a diode or triode. Thus the rate of emission of electrons in vacuum surrounding the cathode may be considered to be constant if the temperature of the cathode which varies with the heating current agrees with the formula $i = AT^2 e^{-b/t}$, where i= emission of electrons in amperes per sq.cm area, e = base of Naperian logarithm (= 2.71 828), A and b are characteristics of the cathode material. Thus for the pure tungsten A = 60, when b = 52400 and for thoriated tungsten A = 3, when b = 30500. Fig. 1.31(a) depicts a diode circuit, and Fig. 1.31(b) the tungsten emission curve.

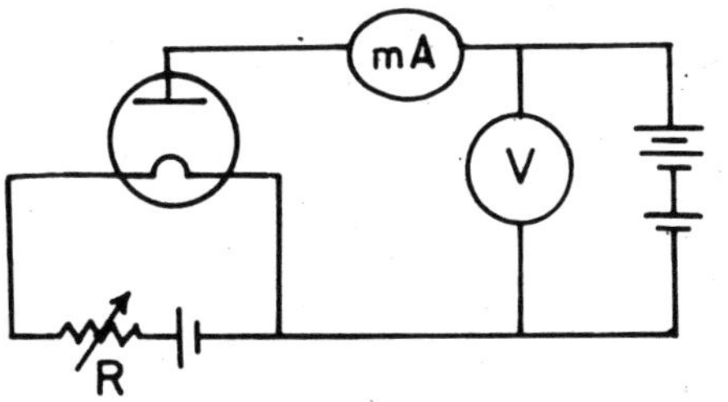

Fig. 1.31(a): Diode Circuit.

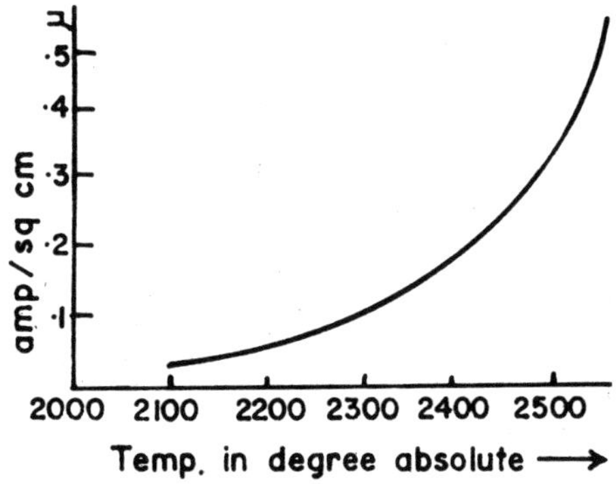

Fig. 1.31(b): Tungsten Emission Curve.

The anode current increases as the anode voltage rises alongwith the rise of temperature. If the voltage is raised to a potential where the anode attracts all the emitted electrons a saturation in anode current is reached. Beyond this point no increase of anode voltage is found to raise the anode current which is then called the saturation current, as shown in Fig. 1.31(e) below. This rise of anode current with fixed filament temperature and increasing anode voltage is found to follow what is called the 'Three halves power law' represented by the formula $I_a = \frac{K}{r} . V_a^{3/2}$, where K is a numerical factor depending on dimensions of constituents of valve material, r = radius of cylindrical anode surrounding the filament and I_a= anode current. When the anode current attains the saturation value, the emitted electrons in excess of demands remain in the neighbourhood of filament, forming a negatively charged cloud called the 'space-charge' which neutralises the attraction of anode or electrons just about to leave the cathode, and those electrons emitted in excess of the number are at once repelled back to the cathode.

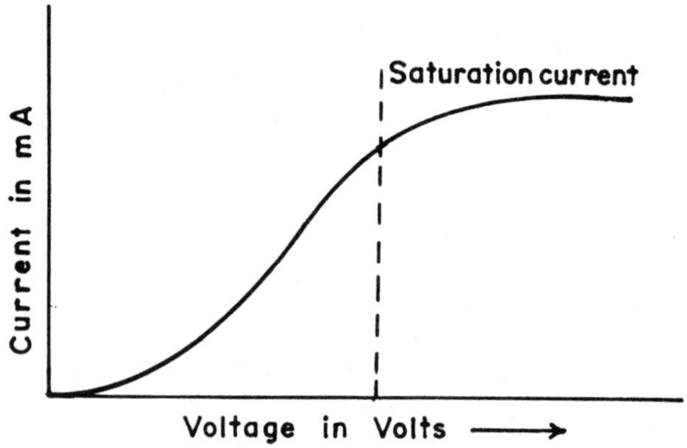

Fig. 1.31(c): Diode Circuit.

In a triode the flow of electrons from cathode to anode is controlled by means of an additional electrode called 'grid' in the form of a mesh inserted between the cathode and the anode. The grid is normally operated at a negative potential relative to the cathode and so it attracts no electron to itself and so no grid current flows. But it tends to repel those electrons which are being attracted to the anode, thus increasing the speed of electron flow from cathode to anode and decreasing the electron density due to space charge concentrated near the cathode.

The amplification'μ 'of a valve is defined as the ratio of the relative effectiveness of grid voltage to that of anode voltage in controlling the anode current. Thus

$$\mu = \frac{\partial I_a}{\partial E_g} / \frac{\partial I_a}{\partial E_a} \qquad (1)$$

The mutual conductance 'g' is defined as the rate of charge of plate current with respect to change of grid potential, i.e.

$$g = \frac{\partial I_a}{\partial E_g} \qquad (2)$$

Th.. has the dimension of conductance and is expressed in micro-mhos (μ mho).

AC resistance R_a (also called anode characteristic resistance) is defined as the rate of change of anode potential with respect to the change of anode current and is expressed

$$R_a = \frac{\partial E_a}{\partial I_a} \qquad (3)$$

when grid potential is kept constant.

Thus from Eqs. (1), (2) and (3), we find that the three value constants μ, g and R_a are interdependent and are connected by the relation

$$\mu = R_a \times g \qquad (4)$$

Uses of valves: Valves may be used as amplifiers, oscillators, rectifiers, etc. Thus the use of a diode as a rectifier is shown below in Fig. 1.32(a), (b) and (c).

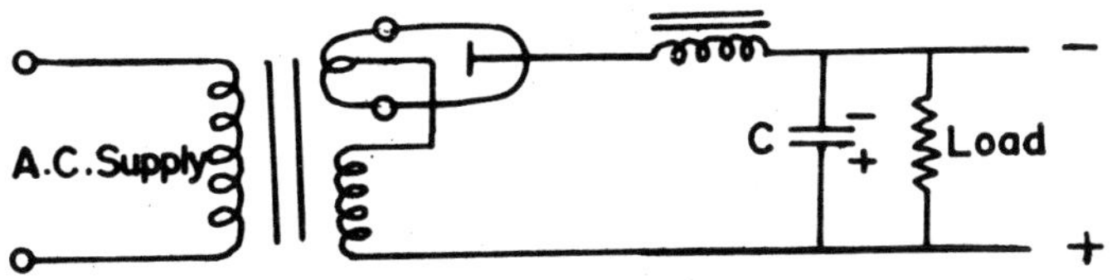

Fig. 1.32(a): Diode as a Halfwave Rectifier

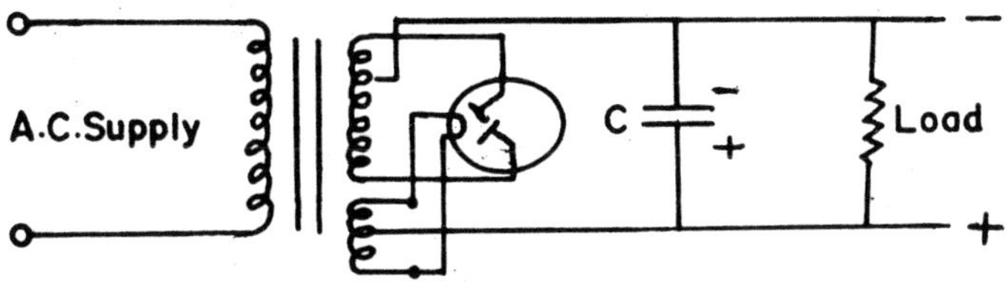

Fig 1.32(b): Diode as a Fullwave Rectifier

In Fig. 1.32(a) the diode will pass current from anode to cathode but not from cathode to anode. A directly heated cathode is used. It gives excessive mains hum which are smoothened by smoothing condenser C. Fig. 1.32(b) shows a double diode sharing a common cathode directly heated for fullwave rectification. Fig. 1.32(c) shows the voltage doubling by two diodes heated indirectly for fullwave rectification.

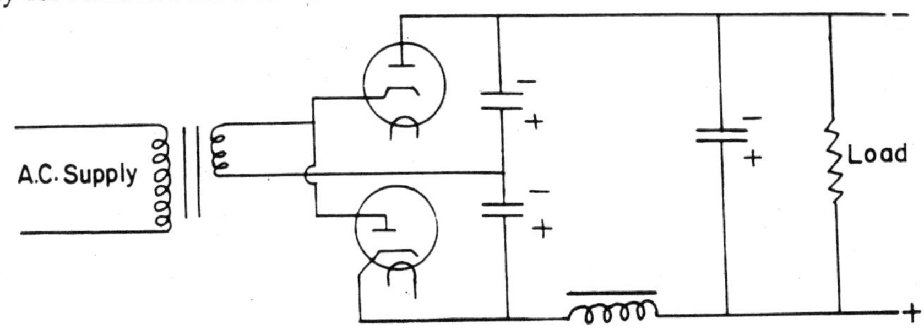

Fig. 1.32(c): Voltage Doubling by Two Diodes with Fullwave Rectification.

The diagram of a triode valve used as an amplifier is shown below in Fig. 1.33(a).

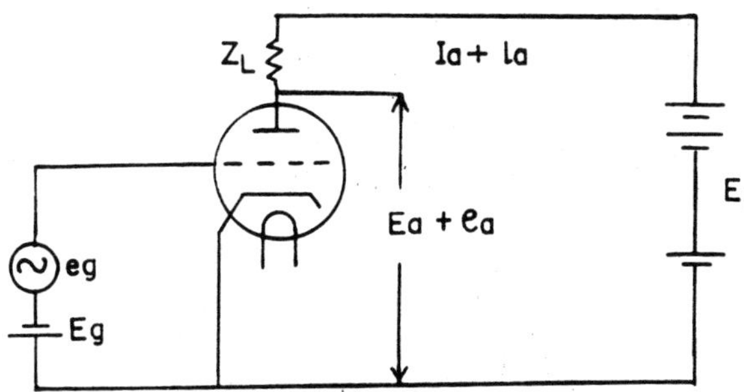

$E_a + e_a = E_b - (I_a + i_a) Z_L$, $E_a = E_b - I_a Z_L$, $e_a = -i_a Z_L$

Fig. 1.33(a): Triode Valve Used as an Amplifier.

The signal to be amplified is applied to the grid in the form of an alternating voltage. E_g is a steady voltage applied in series with the signal of such a magnitude as to ensure that the grid is always negative with respect to the cathode. The anode is maintained at a high positive potential relative to the cathode by a plate battery of voltage E_b in series with the load impedance Z_L. There will be a potential drop across the load impedance Z_L when anode current will flow and the potential of anode relative to cathode will therefore be less than E_b. When no signal is being applied to the grid ($e_g = 0$) the DC bias E_g will determine the value of the steady anode voltage E_a. If the small alternating voltage e_g is now applied to the grid circuit, an alternating component i_a is now applied to the anode circuit which produces an alternating voltage drop in Z_L resulting in an alternating component e_a in the anode voltage. This alternating component across the load impedance Z_L will be utilised for determining the gain of the amplifier.

The amplifiers are divided into two classes according to the use made for the voltage developed across the load impedance. In the simplest case, it is required merely to develop the maximum possible voltage across the load and apply the same to the grid of the second valve for further amplification. Since the grid circuit of a valve is voltage operated, without consumption of power, it is immaterial what power is developed in Z_L and such a stage of amplification is called "voltage amplifier".

In certain cases, however, (e.g., audio amplifier) the loudness of an audible sound depends on the power dissipated in the loud speaker or telephone receiver which will be acting as the anode load of the final stage of amplification. In this last stage power is the primary concern for which it is called "the power amplifier". In voltage amplifier the amplitude of the signal only is amplified. The equivalent circuit of a triode amplifier is shown in Fig. 1.33(b).

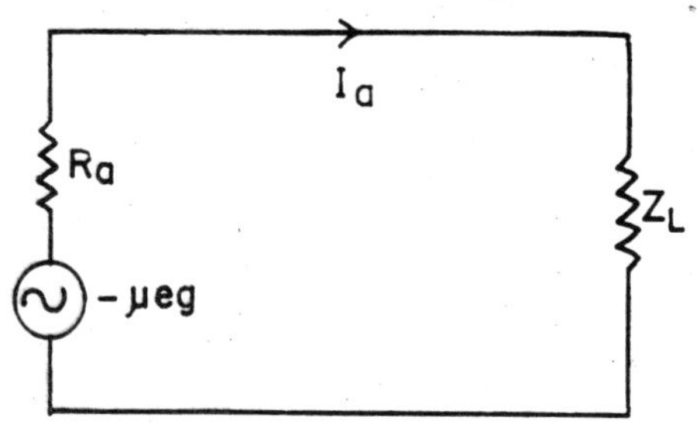

Fig. 1.33(b): Equivalent Circuit of a Triode Amplifier (constant voltage generator form).

Here it is usual to take I_a in a clockwise sense and regard the generator as having negative emf. Thus,

$$i_a = \frac{-\mu e_g}{R_a + Z_L} \tag{5}$$

Then the change in voltage produced across Z_L is given by

$$e_L = \frac{-\mu e_g Z_L}{R_a + Z_L} \tag{6}$$

Then the stage-gain M will be ratio of the magnitudes of e_l and e_g,

$$\therefore \quad M = \frac{\mu |Z_L|}{|R_a + Z_L|} \tag{7}$$

which shows $M < \mu$. But M can be made as great as μ by taking sufficiently large value of Z_L provided that there is no limit to the value of E_b.

From Eq. (5), by dividing both numerator and denominator by R_a, we have

$$I_a = \frac{\frac{-\mu}{R_a} e_g}{I + \frac{Z_L}{R_a}} \quad \text{i.e.} \quad I_a = -g e_g \frac{R_a}{R_a + Z_L}$$

$\therefore$ The voltage developed across anode load is

$$E_1 = -ge \frac{R_a Z_L}{R_a + Z_L} \tag{8}$$

This voltage is developed by a constant current $-ge$ across a current consisting of R_a and Z_L in parallel. Hence,

$$\mu = g \frac{R_a |Z_L|}{|R_a + Z_L|} \tag{9}$$

The diagrams of tetrode as an amplifier are shown in Fig. 1.34(a) and (b) and that of pentode as an amplifier in Fig, 1.34(c) below.

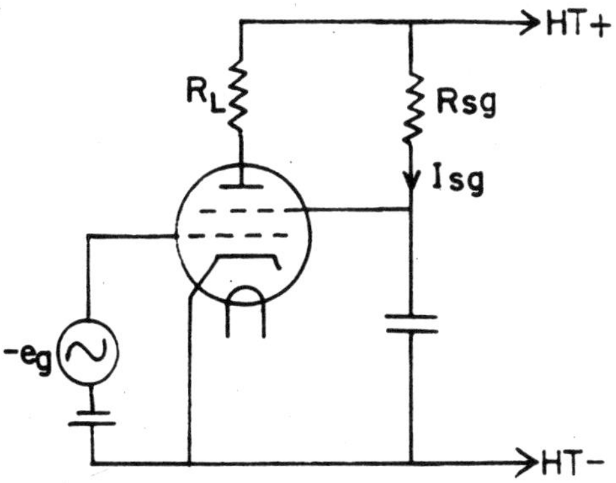

Fig. 1.34(a): Tetrode as an Amplifier (Screen grid Voltage Tapped on HT).

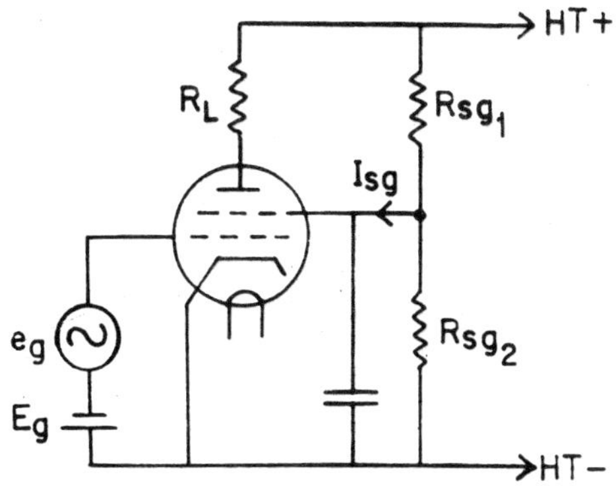

Fig. 1.34(b): Tetrode as an Amplifier (Screen grid Voltage by a Dropping Resistor).

Due to secondary emission, the screen grid tetrode is neither used as a power amplifier nor used as a voltage amplifier.

The diagram of a pentode used as a voltage amplifier is shown in Fig. 1.34(c) below with its equivalent circuit in Fig. 1.34(d).

$$\text{The gain } M = \frac{e_l}{e_o} = g \frac{R_a R_L}{R_a + R_L} = g \frac{R_a}{1 + \frac{R_a}{R_L}}. \text{ if } R_a >> R_L, M = g \cdot R_L$$

In power amplifiers the load resistance power is developed. The power amplifiers are divided into three classes: (i) In class A amplifier, the anode current flows for the complete input signals

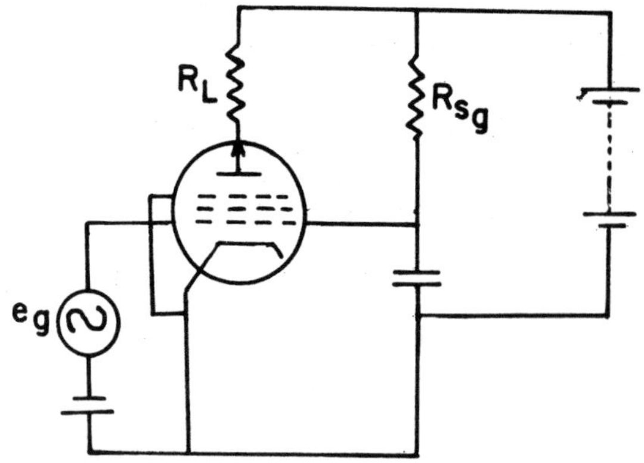

Fig. 1.34(c): Pentode used as a Voltage Amplifier with Constant Current Generator.

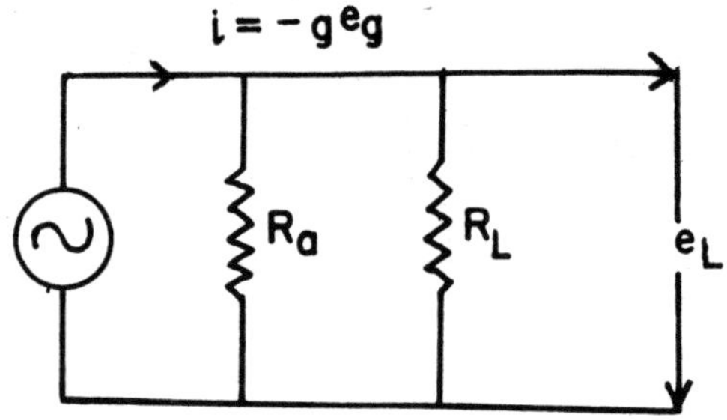

Fig. 1.34(d): Equivalent Circuit for Pentode as Voltage Amplifier.

(360[0]). In class *B* amplifier, the anode current flows for one-half of input cycle (180[0]) and, in class *C* amplifier, anode current flows for less than one-half of input cycle, for which this type is used only for R.F. amplifiers. The load impedance R_L is usually coupled to the valve of a class A power amplifier through a transformer as shown in Fig. 1.34(e), for a triode valve along with equivalent circuits.

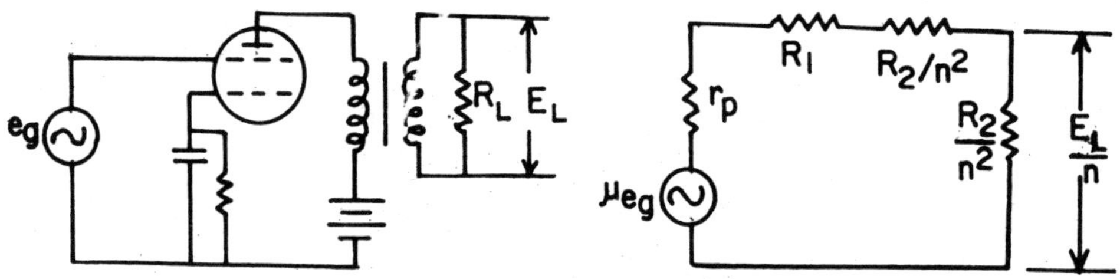

Fig. 1.34(e): Triode Power Amplifier class *A* type (Equivalent circuit for mid-frequency range).

Here for mid-frequency range $A_{mf} = \dfrac{\mu R_L}{n(\tau_p + R_L)}$ where $\tau_p^1 = \tau_p + R_1$ (plate resistance of valve

$\tau_p +$ primary winding of D.C. resistance R_1), $R_L = \dfrac{1}{n^2}(R_2 + R_L), n = \dfrac{N}{N_p^s}$ (DC resistance of secondary winding R_2 + load resistance R_L.

$$\frac{E_L}{n} = \frac{\mu e_g}{\tau_p' + R_L^1} \times \frac{R_L}{n^2} \therefore E_L = \frac{\mu e_g}{\tau_p + R_L'} \times \frac{R_L}{N} = \frac{\mu e_g R_L}{n(\tau_p + R_L')}.$$

Hence amplification or gain $A_{mf} = \dfrac{E}{e_g^L} = \dfrac{\mu R_L}{n(\tau_p' + R_L')}$.

Class *AB* amplifier is a push-pull system, where the operating point is between class A and class B and the plate current flows for more than half-cycle in each valve as shown in Figure 1.34(f).

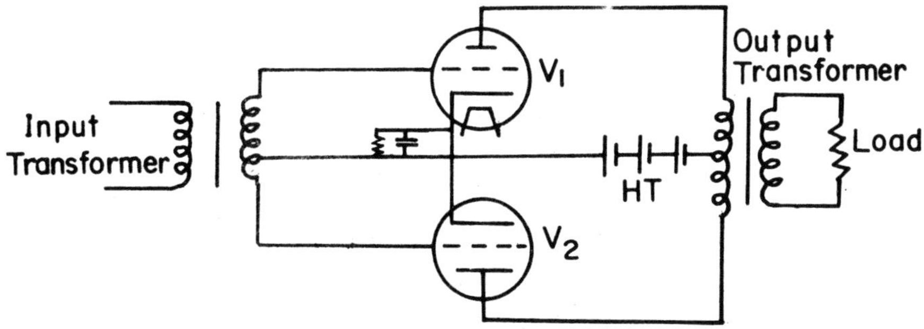

Fig. 1.34(f): Push-pull Arrangement of AB type Power Amplifier.

By using valves in push-pull, even harmonics are cancelled out, reducing distortion. Phase inversion is used as a substitute for interstage transformer in push-pull system.

The cathode-follower amplifier is widely used as a video frequency power amplifier. Here the load impedance is placed between the cathode and the ground for which it is called cathode-follower or grounded anode power amplifier, as shown below in Fig. 1.35(a).

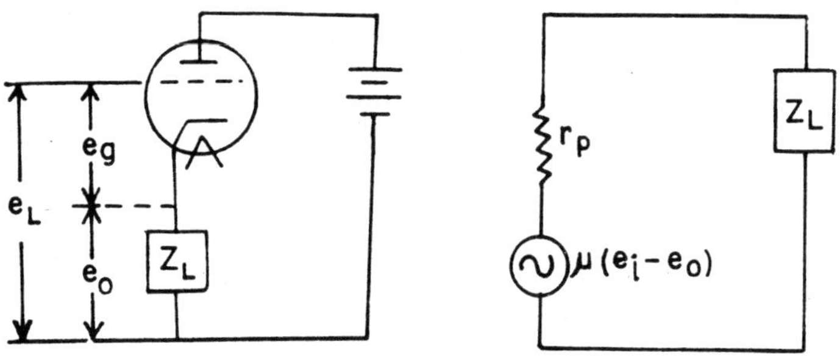

Fig. 1.35(a): Cathode-follower Power Amplifier. Equivalent Circuit of Cathode-follower Power Amplifier.

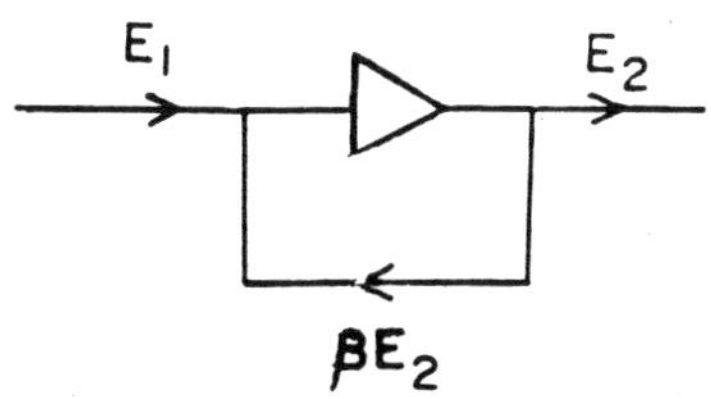

Fig. 1.35(b): Positive Feed-back Amplifier.

In a feed-back amplifier, a part of the output voltage if fed back in phase with the input voltage, is called a positive feed-back amplifier and, if it is fed back in opposition to the input voltage, then it is called a negative feed-back amplifier, as shown in Fig. 1.35(b) for a positive feed back amplifier. Thus if E_1, E_2 are the initial input and output voltages and a fraction B of E_2 is fed back in phase with E_1, then voltage fed back= βE_2 and the net total input= $E_1 + BE_2$ and the output voltage $E_2 = M(E_1+\beta E_2)$ or $E_2(1-\beta M)= ME_1$ where $M =$ gain of the amplifier.

Hence overall gain M_0 of the amplifier $= \dfrac{E_2}{E_1} = \dfrac{M}{1-\beta M}$ which shows an increase in gain but there will be a very high gain if $\beta M = 1$ and even without an input voltage, there will be output voltage, i.e. the amplifier will act as an oscillator.

In case of negative feed back amplifier, $M_o = \dfrac{E_2}{E_1} = \dfrac{M}{1+\beta M}$ which shows a decrease in gain.

If βM is large compared to 1, then $M_o = \dfrac{M}{\beta M} = \dfrac{1}{\beta M}$, which shows gain is independent of M as shown in Fig. 1.35(c), but stabilised for a wide range of frequency with reduced distortion.

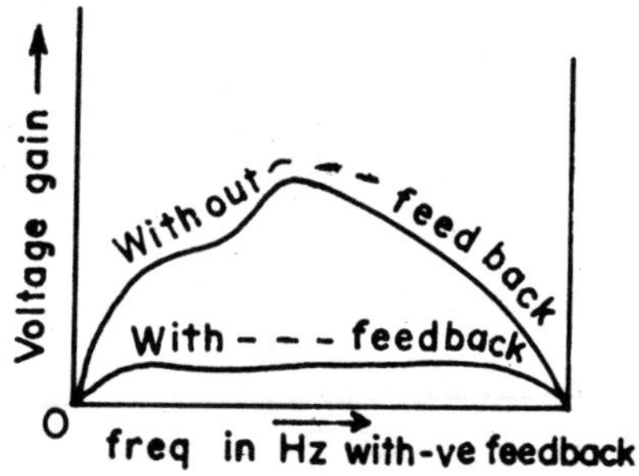

Fig. 1.35(c): Frequency vs. Voltage Gain Curve for -ve FB Amplifier.

Valves as oscillators: In tuned-anode oscillators, a transformer is used to give 180^0 phase-shift in the feed-back. It is connected between anode and grid circuits and to fix the frequency of oscillations, the anode winding is tuned with a parallel condenser. The series and parallel fed circuits are equivalent as shown in Fig. 1.36(a) and 1.36(b) below.

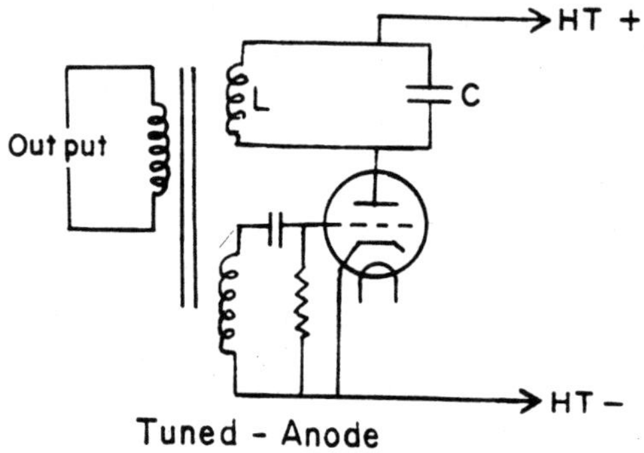

Fig. 1.36(a): Tuned Anode series fed oscillator circuit.

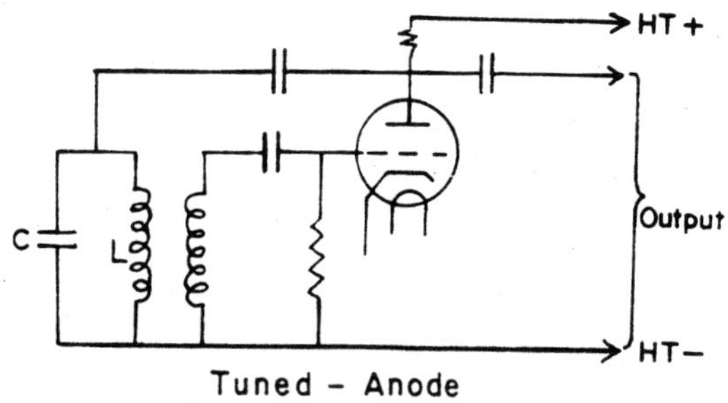

Fig. 1.36(b): Tuned Anode parallel fed oscillator circuit.

The advantage of the parallel-fed circuit is that the anode winding of the transformer does not carry the standing anode current. The frequency of oscillation is approximately the resonant frequency of L and C,

$$\text{i.e. } f = \frac{1}{2\pi\sqrt{LC}}$$

The condenser and resistance of the grid circuit constitute the grid leak bias circuit. The output may be taken from a third winding on transformer via a condenser from grid or anode as shown in Fig. 1.36 (a) or from across a resistance between cathode and HT–.

1.8 Modulation and Demodulation

Modulation is the process in which audio frequency is superimposed on a carrier frequency and the resultant wave is known as a modulated wave, which consists of two side bands and the carrier. There are three types of modulations, viz. the amplitude, frequency and phase modulations.

(a) In amplitude modulation, the modulated wave is obtained by varying the amplitude of the carrier waves to be modulated.

The modulation index is defined as the ratio of the modulating wave amplitude to the carrier wave amplitude. In a modulated wave, two-thirds of the total power is contained in the carrier and one third is on the side bands. Thus by suppressing the carrier and transmitting intelligence or a single side-band, a lot of power is saved.

(b) In frequency modulation, the intelligence is transmitted by varying the frequency with the modulating signal. The variation of frequency from the average value is called the frequency deviation. Modulation index of the frequency modulated wave is a ratio of the frequency deviation to the modulating frequency. This type of modulated wave contains a number of frequency components. Bandwidths is the difference of the two side frequencies on either side of the carrier frequency. In general, it is twice the modulating frequency f_m.

(c) In phase modulation, the intelligence is transmitted by varying the phase of the carrier wave. The modulation index m_p is independent of the modulating frequency, but depends on the amplitude of modulation.

Detection or demodulation is the process by which the modulating wave (audio frequency wave) is obtained from the modulated wave by again modulating the transmitted sideband with the same carrier frequency and recovering the intelligence from the transmitted modulated wave at the point of reception.

Amplitude Modulation: As the amplitude modulation is produced by varying the amplitude of the carrier wave to be modulated, this change in amplitude from the unmodulated wave depends upon the instantaneous value of the modulating wave. For simplicity, the modulating wave may be taken as a sine wave, say $e = B \sin pt$, where p = angular frequency which is a function of time t, and b is the depth of modulation. Unmodulated carrier wave is of constant amplitude but of varying frequency. Let it be represented by $A \sin wt$, where A = maximum amplitude of carrier wave and ω = angular frequency which is a function of time t. The modulating wave, carrier wave and the resultant modulated wave are shown in Fig. 1.37(a) to (c), respectively.

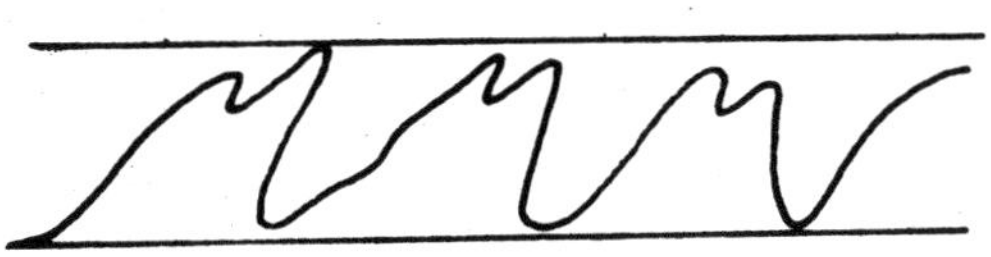

Fig. 1.37(a): Modulating wave.

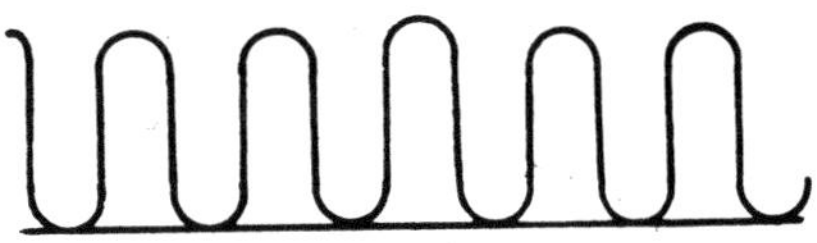

Fig. 1.37(b): Carrier wave

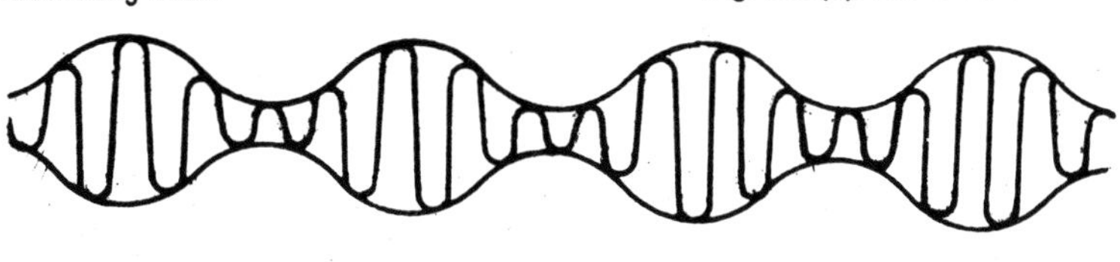

Fig. 1.37(c): Modulated wave.

Variation of the amplitude of the carrier wave is sinusoidal and takes place between the limits $A + B$ and $A - B$, where $A =$ the maximum amplitude of carrier, and $B =$ depth of modulation. B/A is called the degree of modulation and is represented by the factor m.

$\therefore B/A \times 100 = m \times 100$ is the percentage modulation.

Side Bands: The carrier wave is represented by $A \sin wt$ and the modulating wave by $B \sin pt$, hence the amplitude of the modulated wave $= A + B \sin pt$.

$\therefore$ the instantaneous value of the modulated wave $= A \sin wt + B \sin pt \sin wt$.

$$= A \sin wt + B \sin pt \sin wt$$

$$= A \sin wt + \frac{B}{2}\{\cos(w-p)t - \cos(w+p)t\}$$

$$= A \sin \omega t + \frac{B}{2}\cos(\omega - p)t - \frac{B}{2}\cos(w+p)t$$

$$= A \sin \omega t + \frac{B}{2}\sin\left\{(\omega - p)t + \frac{\pi}{2}\right\} + \frac{B}{2}\sin\left\{(w+p)t - \frac{\pi}{2}\right\}$$

= carrier + L.S.B. + U.S.B.

The power dissipation by each component is proportional to the square of the amplitude, i.e. power in carrier $\alpha A^2 = KA^2$ where K is a constant.

Also power in L.S.B. $\alpha \left(\frac{B}{2}\right)^2 = K\frac{B^2}{4}$

and power in U.S.B. $\alpha \left(\frac{B}{2}\right)^2 = K\frac{B^2}{4}$

$\therefore$ Total power in carrier and side bands

$$= KA^2 + K\left(\frac{B^2}{4} + \frac{B^2}{4}\right) = K\left(A^2 + \frac{B^2}{2}\right)$$

If the modulation is 100%, then A = B

$\therefore$ The total power $= K\left(A^2 + \frac{A^2}{2}\right) = \frac{3}{2}kA^2$

i.e., $\frac{3}{2}$ of carrier power. Hence the carrier power $= \frac{2}{3}$rd of total power and two side band power= $\frac{1}{3}$ rd of total power.

Thus, by suppressing the carrier and transmitting intelligence on a single side band, a lot of energy is saved. Balanced modulator provides secrecy also. S.S.B. transmission is free from adjacent channel interference.

Frequency Modulation: As in frequency modulation, the intelligence is transmitted by varying the carrier frequency with the modulating signal frequency, the amplitude of the wave is kept constant, while the rate of variation of frequency depends upon the modulating frequency, as shown in Fig. 1.38 below.

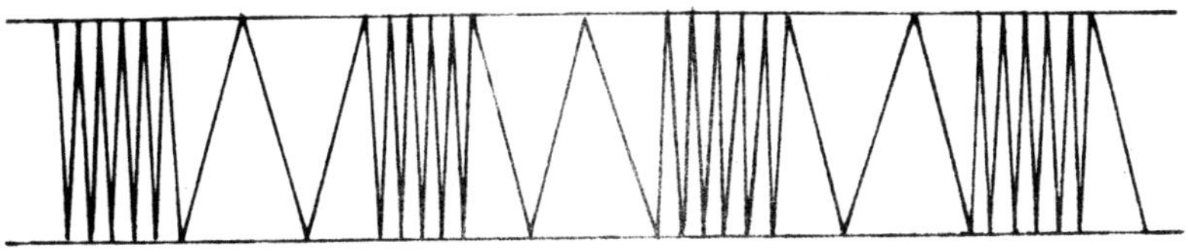

Fig. 1.38: Frequency Modulated Wave.

This variation of frequency from the average value is called the frequency deviation, and is proportional to the modulating signal amplitude.

Let the equation of unmodulated signal be $e = A \sin \phi$ (t), where e = instantaneous amplitude, ϕ (t) = angular displacement = wt. Then instantaneous angular velocity $\omega_i = \frac{d\phi(t)}{dt}$

or $d\phi(t) \;=\; \omega_i\, dt$

$$\therefore \quad \phi(t) = \int_o^t \omega_i dt$$

In a frequency modulated wave, with sinusoidal modulation,

$$\omega_i = \omega + 2\pi\Delta f \; \cos pt \tag{2}$$

where ω = angular velocity of carrier wave and Δf = maximum deviation of carrier frequency. Substituting equation (2) in equation (1), we get ,

$$\phi(t) = \int_o^t (w = 2\pi\Delta f \cos pt)dt$$

$$= \left[wt + \frac{2\pi\Delta f \sin pt}{p}\right]_o^t$$

$$= wt + \frac{2\pi\Delta f}{p} \sin pt$$

$$= wt + m'_f \;\sin\; pt$$

where m_f is the modulation index= $\frac{2\pi\Delta f}{p}$

$$= \frac{\text{frequency deviation}}{\text{modulating frequency}}$$

Then, the general equation for unmodulated carrier $e = A \sin \phi(t)$ becomes the general equation of frequency modulated wave and is given by A sin (wt + m_f *sin pt*).

Thus we have, e = A sin(wt + m_f sin pt). (3)

From Equation (3), we observe that the side bands (from trigonometrical expansion by Taylor's Theorem) are ω, $(\omega \pm p)$, $(\omega \pm 2p)$, $(\omega \pm 3p)$ ...where ω is the angular velocity and p is the modulating frequency. We further observe that the frequency modulated wave contains a large number of side frequencies for each audio component in the modulating signal. When the modulation index m_f is less than 0.5, i.e. the frequency deviation is less than half the modulating frequency, then the second and higher order side band components are relatively small for which the frequency band width is practically the same as that for amplitude modulation. But when m_f exceeds unity, i.e. frequency deviation is greater than modulating frequency, then necessary higher order side band components are contained in the wave.

Phase Modulation: As in phase modulation, the intelligence is transmitted by varying the phase of the carrier wave; the phase of carrier wave is varied by a factor $(l + m \sin pt)$ while its amplitude remains constant.

The general equation for the instantaneous value of unmodulated wave is given by

$$e = A \sin (wt + \phi) \tag{1}$$

Let the modulating wave be represented by B sin pt. Then the amplitude of the modulated wave = $A + B \sin pt$

$$= A\left(1 + \frac{B}{A}\sin pt\right) = A(1 + m \sin \text{pt}).$$

This equation justifies the statement that the amplitude is varied by the factor $(l + m \sin pt)$.

For the phase modulated wave, this gives

$$\text{e} = A \sin[wt + \phi(1 + m \sin pt)] \tag{2}$$

$$= A \sin[wt + \phi + \phi m \sin pt]$$

when $t = 0$, $e = 0$, then from Eq. (1)

$$\phi = 0$$

$$\therefore e = \text{A} \sin [wt + m\phi \sin pt]$$

$$= A \sin[wt + mp \sin pt] \tag{2a}$$

$$= A [\sin wt \cdot \cos (mp \sin pt) + \cos wt \sin (mp \sin pt)]$$

Now from Eq. (2) of frequency modulation and Eq. (2a) of phase modulation, we find that the shifting of frequency of one, i.e. m_f corresponds to the shifting of phase of the other, i.e. mp. Hence the two types of modulation can be grouped together as shown below.

In Eq. (2a) A = amplitude of wave, $\omega = 2\pi$ times the carrier frequency, $p = 2\pi$ times the modulating frequency, and m_p = angle in radius through which the phase is displaced at the peak of the modulation cycle away from the phase that exists before modulation. Hence in phase modulation, the modulation index *mp* is independent of the modulating frequency but depends on the amplitude of modulation and the frequency deviation Δf is proportional to modulating frequency, i.e. $\Delta f = m_p \times$ modulating frequency. But in frequency modulation, frequency deviation is independent of modulating frequency. Hence

$$\frac{\Delta f}{\text{modulating frquency } (f_m)} = m_f \quad (\therefore \; \Delta f = m_f)$$

will be the output available in a frequency modulated wave if the modulating voltage applied to the phase modulator is inversely proportional to frequency.

Hence in phase modulation,

$$\Delta f \ \alpha \ f_m \alpha \frac{1}{m_p} \therefore m_f \ \alpha \ \frac{1}{m_p}.$$

Therefore to get a frequency modulated wave from a phase modulator, the modulating voltage applied must be inversely proportional to the frequency.

1.9 Transistor Electronics

While discussing the principles of working of metal rectifiers, we have seen the properties of germanium, silicon, copper oxide, etc. which pass current very easily in one direction while offer high impedance in the reverse direction. These are called semi-conductors as they show a middle-path property between the conductors and non-conductors. We find that in the structure of atoms, the electrons are rotating in orbits or rings round the nucleus just like the planets are moving round a star, as in the solar system. The atomic number of an element is given by the total number of electrons in the orbits around the nucleus and the outermost ring called the valence ring contains the free electrons which can move away very easily to form covalent bond with the free electrons in the valence ring of an identical atom when an emf is applied. Thus if there is a deficiency of electrons from the stable state, then the semi-conductor is known as the positive or P-type and if there is an excess of electrons, it is known as the negative or N-Type (Fig. 1.39(a) and (b)).

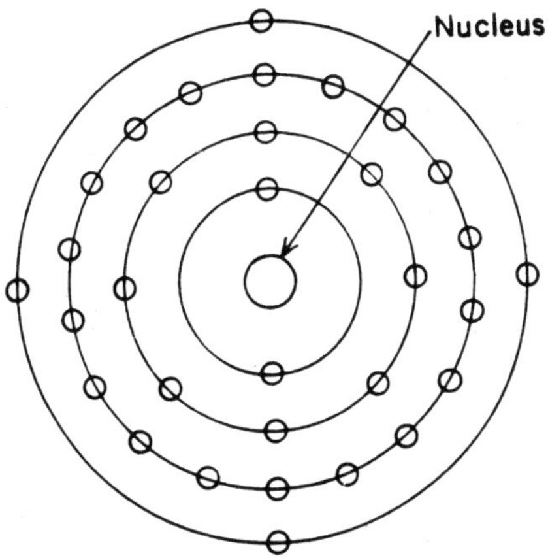

Fig. 1.39 (a): Atomic Structure of Germanium with 2+8+18+4= 32 Electrons.

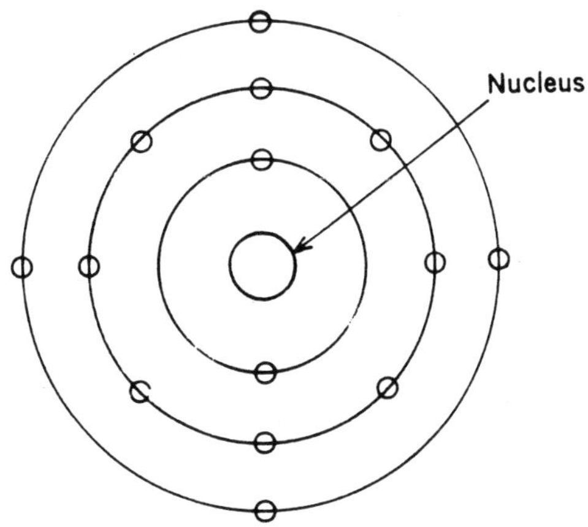

Fig. 1.39(b): Atomic Structure of Silicon with 2+8+4=14 Electrons

Thus, in case of germanium the ring wise number of electrons are 2, 8, 18 and 4 in the valence or outermost ring while in the case of silicon, they are in order of 2, 8 and 4. Similarly, in case of arsenic and Indium, they are respectively in the order of 2, 8, 18 and 5, and in the order of 2, 8, 18, 18 and 3 from the innermost to the outermost ring. The name 'transistor' has been given to such semi-conducting materials as can form such covalent bonds and is derived from the words 'Transfer resistor' as its value can be controlled by application of an emf or temperature.

To make pure germanium or silicon more suitable to be used as a transistor, an impurity like arsenic is added to a crystal of silicon. Of the five electrons in the valence ring of arsenic, four electrons from covalent bonds with the four electrons in the valence ring of silicon and the fifth electron of the valence ring of arsenic becomes free to move when an emf is applied. Thus arsenic is said to have donated its free electrons to silicon crystal which increases its conductivity and it is referred to as donor impurity. Silicon treated with arsenic is called N-type transistor and if a cell is connected to an n-type silicon crystal, free electrons are attracted to the positive and repelled from the negative terminal, thus forming a drift of electrons from the negative to the positive terminal, constituting what is called an electric flow of current through the crystal. When the terminal of the cell are reversed, the direction of the flow of current through the crystal is also found to be reversed without change of amplitude, thereby showing that the flow of current was due to the linear conductivity of the crystal.

Similarly, if an impurity of Indium is added to a germanium crystal, then the three electrons in the valence ring of Indium form covalent bonds with the three electrons in the valence ring of germanium. Thus to make a covalent bond of four, Indium is deficit of one electron and this bonded group behaves like a positively charged particle with a charge equal to that of an electron, called 'hole' carrying current through the crystal. So germanium treated with Indium is a P-type semiconductor and if a cell is connected across such a P-type germanium crystal, the holes having positive charge are attracted to the negative and repelled from the positive terminal and a drift of holes passes from the positive to the negative terminal through the crystal. If the battery connections are reversed, the drift of holes reverses in the crystal without change in amplitude. This shows that the flow of current is in the opposite direction to that of electrons and p-type crystal is a linear conductor also.

Types of transistors: The transistors are mainly classified into two basic types, namely (a) point-contact type and (b) the junction type.

Point-contact Type Transistors: If a contact with germanium is made from a thin metal wire, forming a germanium diode, the resistance of the combinations is either very low or high according to the directions of the applied p.d. Moreover, if a second point contact is made from another thin metal wire of a separate metal within .01 inch (.25 mm) of the first point contact, then it is found that there is a relation between the currents flowing in the two point contacts, provided that appropriate potential conditions are applied to the two point contacts. If the first contact point called the emitter is biased by a p.d. applied in one direction to produce a low resistance and the second contact point called the collector is biased in the reverse direction by a p.d. arranged to produce a high resistance, then collector current I_c is found to vary with the emitter current I_E and the ratio of the small changes in these currents is known as the current gain α. Hence $\alpha = dI_c / dI_e$ (1)

Since the input resistance at the emitter is low (a few hundred ohms) while the output resistance at the collector is high (some tens of thousands of ohms), an approximately equal current changes in both the contacts resulting in a power gain of up to 100.

The diagrams for the structure of a point-contact transistor and that of its bias potential for common-base mode are shown respectively in Figures 1.40(a) and (b) below.

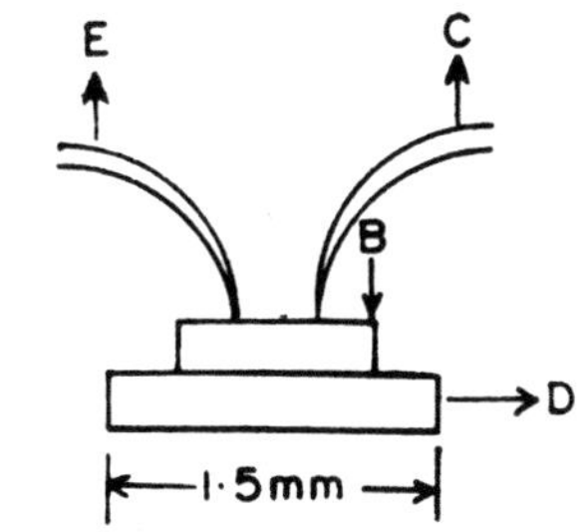

Fig. 1.40(a): Structure of Point-contact Transistor.

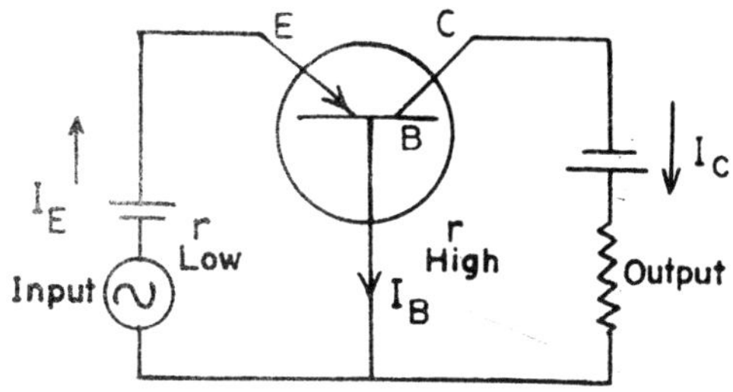

Fig. 1.40 (b): Bias Potential for (common base mode) Point-contact Transistor.

In Fig. 1.40(a), the emitter contact wire may be of beryllium-copper and the collector contact wire of phosphor-bronze, both of diameter of order .002 to .005 inch and they are suitably pointed by grinding and shearing process and the contact pressure is controlled to about one gram, by deflexion or compression of the wire into a L or S-shaped spring. The wires may be bonded by an adhesive resin. The germanium base with a top surface of 1mm x 1mm and a thickness of .5 mm is soldered to a metal support D which forms the base electrode. The entire assembly is moulded in resin and hermetically sealed to exclude moisture at the point contacts. In Fig. 1.40(b) is indicated the biasing potentials showing the symbols for the emitter E, the collector C and the third electrode for the germanium itself, termed base (B). Here the emitter has a positively applied potential, the collector has a negatively applied potential, while the n-type germanium base is common to both emitter and collector circuits.

Junction Type Transistors: The junction transistors can be made either p-n-p or n-p-n type. In both, the emitter is a rectifying junction biased in the direction of easy current flow (positive bias for a p-n junction) and the collector is a second rectifying junction biased in the direction of restricted current flow (negative bias for a n-p junction). The bias is for the purpose of temporarily injecting an excess of minority carriers of the opposite type to the majority carriers, which permanently characterise the region.

The action of the junction transistor depends on the current carriers originated at the emitter. These pass through the base in numbers determined by the emitter-base potential and they are absorbed by the collector which passes a proportional current into the external circuit. In fact, the transistor consists of two diodes connected back to back with potential barriers at each junction as shown in Figures 1.43(a), (b) and (c) below for p-n-p and n-p-n types for common base mode.

Fig. 1.41(a): Formal Representation of p-n-p and n-p-n type Transistors.

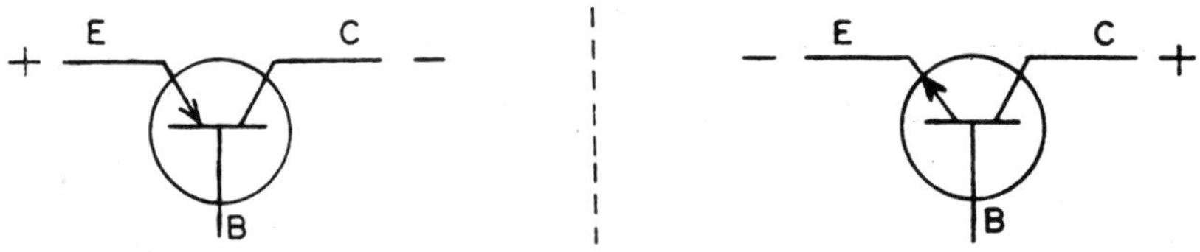

Fig. 1.41(b): Symbolic Representation.

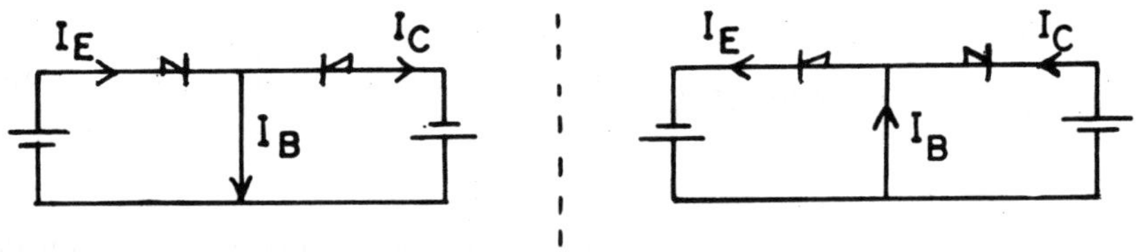

Fig. 1.41(c): Bias Potential Representation.

The transistors are classified as (a) common-base, (b) common-emitter and (c) common-collector amplifiers as shown below in Figures 1.42(a), (b) and (c) along with their equivalent circuits from which their performance can be calculated by applying Kirchhoff's laws.

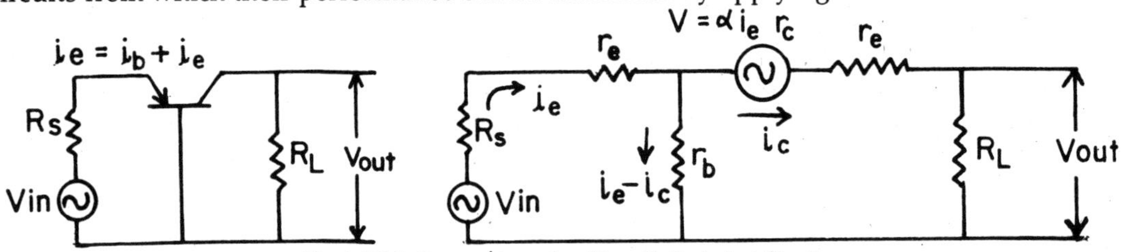

Fig. 1.42(a): Common-base Amplifier with Equivalent Circuit.

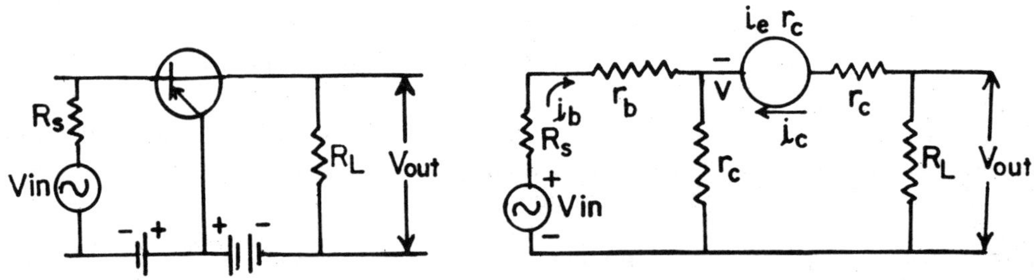

Fig. 1.42(b): Common-emitter Amplifier with Equivalent Circuit.

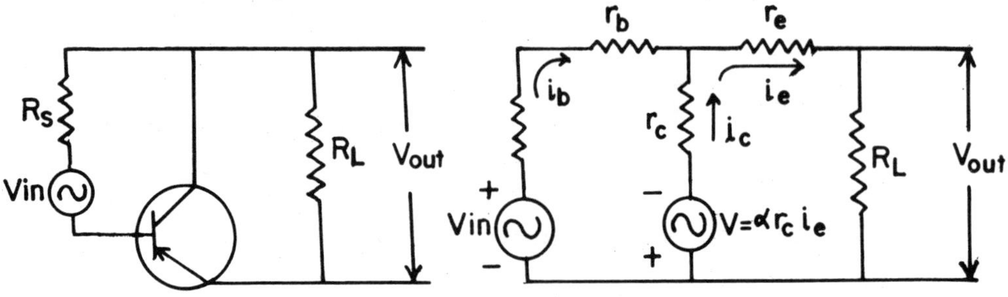

Fig. 1.42(c): Common-collector Amplifier with Equivalent Circuit.

The collector current is controlled by the base current and so by the base voltage. Three items are required for the calculations. They are the input resistance r_e, output resistance r_o and the voltage gain= $\frac{Vout}{Vin}$. If R_s= signal source resistance, R_L= collector load, r_i = emitter resistance, r_b= base resistance, r_c= collector resistance and α = current gain, then for:

(a) common-base amplifier $r_i = r_e + r_b(1-\alpha)$, when R_L is small. When R_s= 0, r_o= $r_c \left\{\frac{r_e + r_b(1-\alpha)}{r_e + r_b}\right\}$

and Voltage gain= $\frac{\alpha R_L}{R_s + r_e + r_b(1-\alpha)}$ if R_L when is very small voltage and Voltage gain $= \frac{\alpha l_c}{R_s + r_e + r_b}$ if R_L is very large.

(b) In case of common-emitter amplifier, if R_L is very small, $r_i = r_b + \frac{r_e}{1-\alpha}$ and $r_o = r_c \left\{\frac{r_b(1-\alpha)+r_c}{r_e + r_b}\right\}$ when R_s is very small. Voltage gain= $\frac{\alpha R_L}{R_s(1-\alpha)+r_b(1-\alpha)+r_e}$ if R_L is very small and Voltage gain= $\frac{\alpha r_c}{R_s + r_b + r_e}$, if R_L is very large. Current gain = $\frac{\alpha}{1-\alpha}$.

(c) For common-collector amplifier $r_i = r_c + \frac{r_e}{1-\alpha}, r_o = r_e + r_b(1-\alpha)$ in similar cases. Voltage gain= $\frac{R_L}{R_L + R_S(1-\alpha) + \frac{R_L R_S}{r_c}}$ and current gain= $\frac{1}{1-\alpha}$

Thus transistors work like valves as amplifiers, oscillators and modulators. The main factors are the simplicity of its power circuits, reduced power consumption and heat dissipation and instantaneous operation without any warming up delay and longer life. The main advantages are its small size, greater efficiency and low power consumption.

Its disadvantages are the sensitivity to temperature rise, limitations to applied voltage, and upper frequency.

The analogy in circuit operation between the valve and the transistor is in electrodes: emitter corresponds to cathode, base to grid and collector to anode. While valve is voltage operated, a transistor is current-operated, the emitter-base voltage being small. While valve shows no backward transmission (i.e., from anode to grid), all transistors have appreciable transmission in the backward direction even at low frequencies due to limited receptivity of base material. While the valves depend upon negatively charged carriers (electrons) only, the transistors use both electrons and positively charged carriers (holes).

Identification of tags on transistors: A transistor is provided with 3 or 4 tags. The collector end is marked with a dot of some colour, the base is in the middle tag and the other extreme tag is the emitter tag. In some types, the tag at a greater distance from base in the centre is the collector and the remaining is the emitter. In case of 4-tagged transistor, keeping the collector at the extreme left, the second lead is the body of the transistor, the third is the base and the fourth is the emitter.

Power transistors used in output stages has a large collector surface for heat dissipation. The collector in such cases may be the body itself and the emitter and base leads are marked for easy identification.

Precautions to be observed are that: (a) no repair work should start before switching off the power, (b) no heat should reach the transistor, (c) no ohm meter should be used for continuity test etc., and (4) no alteration of D.C. polarity should be made.

Field-effect transistor: It is a voltage-controlled semi-conductor which has very high input impedance at low frequencies and as it uses only one type of charge carriers, it has only one p-n junction as shown in the schematic Fig. 1.43(a) below.

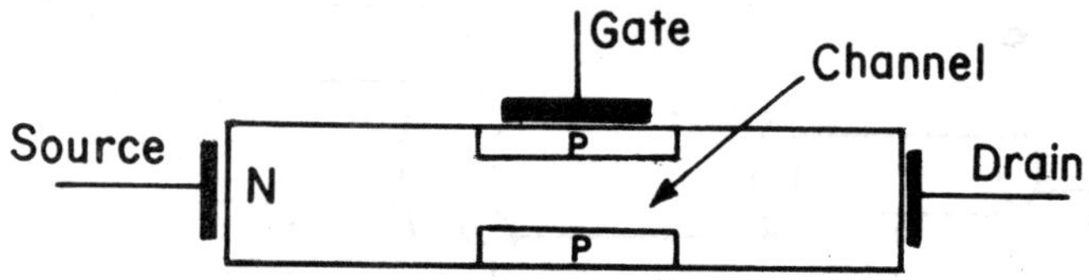

Fig. 1.43(a): Field Effect Transistor (FET) with N type Channel.

A narrow semiconductor channel provides a conducting path between the source and the drain and the channel may be either n or p type.

If extremely high input resistance is required in a solid state circuit, then metal oxide semi-conductor field effect transistor (MOSFET) may be used, whose diagram is shown in Figure 1.43(b) below.

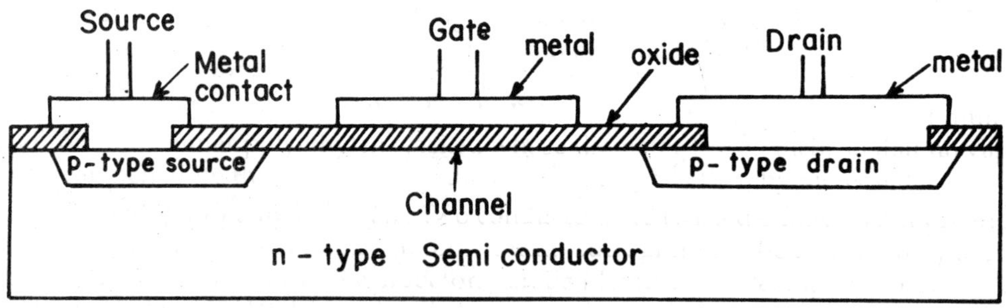

Fig. 1.43(b): P-type Channel MOSFET.

An electric field from gate produces a conduction channel whose thickness is enhanced by a higher field and the MOSFET operating in this manner is said to be in enhancement mode. If a channel is present when V_{as} is zero, the transistor is operating in the depletion mode. The channel of a depletion mode MOSFET is represented by a solid line while that of an enhancement mode is represented by a dashed line as shown in Fig. 1.43(c) below.

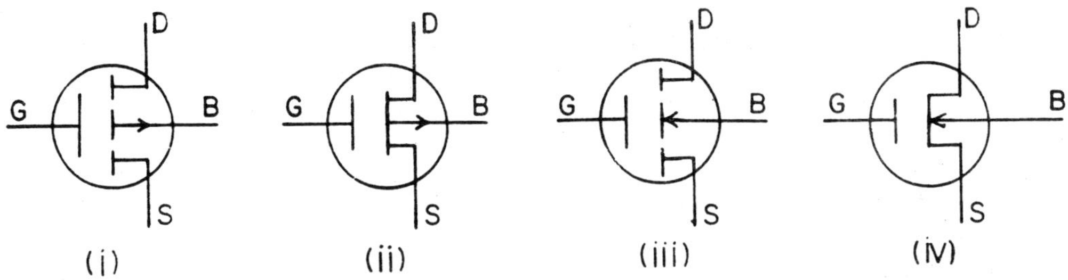

Fig. 1.43(c): (i) p-Channel Enhancement Mode, (ii) p-Channel Depletion Mode (iii) n-Channel Enhancement Mode, and (iv) n-Channel Depletion Mode.

An n-channel MOSFET diagram is shown in Fig. 1.43(d) below.

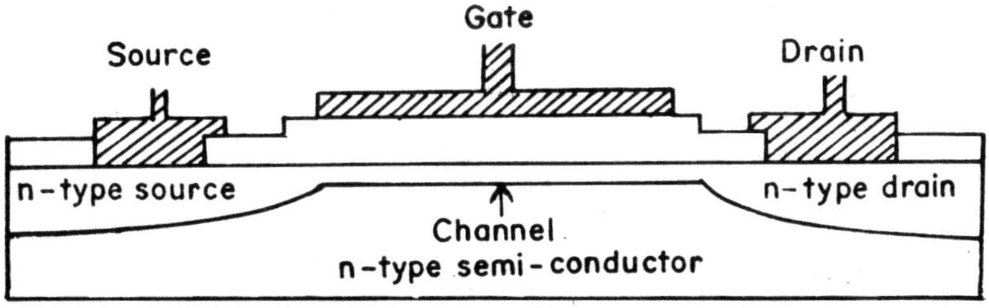

Fig. 1.43(d): An n-Channel MOSFET.

In common base amplifier, no protective circuits are required as the effect of leakage current due to thermal leakage of covalent bond increasing with temperature is negligible. But in common emitter amplifier, this is very prominent for which stabilising circuits are required so that the values of the stabilising factor K is decreased where

$$K = \frac{\text{change in total collector current in stabilised circuit}}{\text{change in total collector current in instabilised circuit}}$$

$$= \frac{\Delta I_c}{\Delta I_c'} = 1 \text{ if } \Delta I_c = \Delta I_c'$$

The voltage stabilisation for common emitter amplifier can be done by (a) collector-base resistance (b) by potential divider method, and (c) by two battery supply method. Functions of linear digital interpolated circuits and further details will be described in Vol. II before the description of electronic exchanges, where the transistor will be used in binary digital circuits.

1.10 Photo-electric Cells and Lasers

Photo cells: The photo cells work on electromagnetic radiation from the visible infra-red or ultra-violet regions of the light spectrum and may be classified into three types, namely, the photo-emissive cell, photo-conductive cell and the photo-voltaic cell.

In photo-emissive cell, when light falls on the photo-emissive surface, electrons are emitted. When an externally connected battery produces an electric field in the cell, the movement of electrons towards the anode is accelerated causing a flow of current.

The photo-conductive cell is formed by condensing selenium vapors in the form of a very thin film on a double grid of gold mounted on a glass plate. When light is allowed to fall on it, the electrical resistance decreases with the increase of illumination and photo-electrons move causing ionisation. But due to low velocity of heavy positive ions, the response of these cells to interrupted light is very poor and is different for lights of different wavelengths. Moreover, as the current capacity of photo-conductive cell is of a few milliamperes and its dark current value is appreciable, it is not commercially popular.

In photo-voltaic cell, the emf is generated internally by the action of the radiated energy, thus allowing the current to flow in the external circuit without the use of an external battery. The main characteristics of photo voltaic cells are given below: (a) no external voltage is required as in the case of photo-emissive and photo-conductive cells, (b) direct amplification by vacuum tube is not possible as these cells generate D.C. voltage, (c) these cells are very sensitive to high temperature, (d) they respond quickly to changes in light intensity falling on the coated surface, and (e) their output is proportional to light intensity.

The photo-cells are used in television and space research expeditions.

Lasers: These are basically a means of generating a special kind of light contained in a very narrow almost parallel beam. Now to understand the operation of laser, we are first to consider the various ways in which radiation interacts with matter which is considered to consist of a large number of atoms, light and other forms of electromagnetic radiations. Thus matter may be considered as a large number of particles called photons containing energy. The energy of a single photon is taken as equal to one quantum of light energy, i.e. frequency of light multiplied by Plank's constant. Thus a photon of blue light contains more energy than a photon of red light.

Process of absorption and emission of energy: An atom with the least energy is said to be on the ground state. When a photon of light strike a metallic surface, it transfers its energy to the atoms on the surface of the metal and excites them, and instantaneously looses its existence. In this process when the atom absorbs so much energy that the electrons move further from the nucleus, then it is said to be in an excited state. The photons disappear but after 10^{-8} seconds the atom emits another photon which excites another atom and so on. In gaseous conduction photons play an important part and in a mixture of gases, the photons emitted from one kind of atoms can ionise other atoms in the mixture if the exciting potential of the atoms of the first kind is greater than the ionisation potential of the other atoms. But some atoms loose their energy about a million time more slowly and are known as metastable atom.

For absorption it has been found that one photon+one atom on ground state → an excited atom. For emission, it observed that one excited atom → atom in a state of lower energy+one photon. If a photon strikes an excited atom, it will cause the atom to emit a photon. Thus an excited atom+one photon → an atom in a state of lower energy+two identical photons. This phenomenon is called stimulated emission as the photon produced is an exact replica of the original photon. The fact that the two photons have the same frequency, same plane of polarisation, same phase and same direction of travel renders the operation of a laser possible. Thus stimulated emission may be considered to be the amplification of light. If the photons meet further excited atoms, further amplification by stimulated emission will take place. Thus a laser signifies 'light amplification by stimulated emission of radiation'.

As in the case of radio transmitters, when required to convert a laser amplifier into an oscillator, feed-back method is used to generate electromagnetic oscillators by a device as shown in Fig. 1.44(a) below.

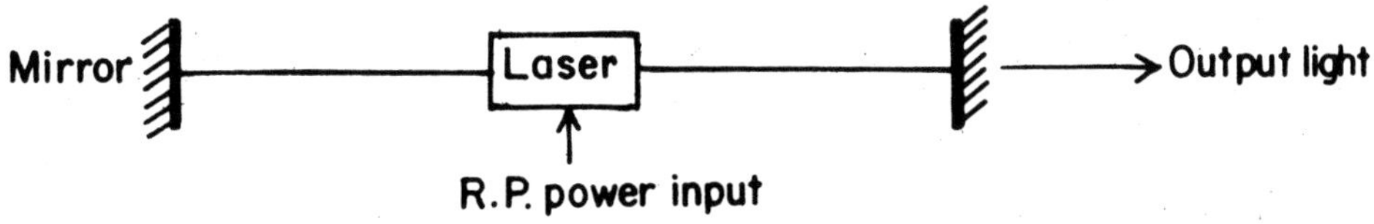

Fig. 1.44(a): Simplified Feed-back circuit of a Laser.

The two mirrors facing each other should be accurately parallel to each other. Any light passing along the axis of the system is reflected. The laser is placed at the middle between the mirrors as shown, so that the mirrors can be aligned easily. When the light is reflected back into the system, further amplification occurs. Light which deviates from the axis even by a very small angle soon passes out of the system and is lost. Now one of the mirrors of the system is only partially reflecting so that the output beam from the laser can pass through it. The divergence of this beam is extremely small. In other words, the output from the laser is obtained as a beam of light which is almost exactly parallel as in the case of a Mullard gas laser or a Ferranti gas laser.

For the laser action to occur, there should be a continuous energy supply known as 'pumping' which raises the atoms up from the ground state level E_0 to an excited state level E_2 when the atoms loose a part of their excess energy instantaneously and fall to energy level E_1, as shown in Fig. 1.26(b) below.

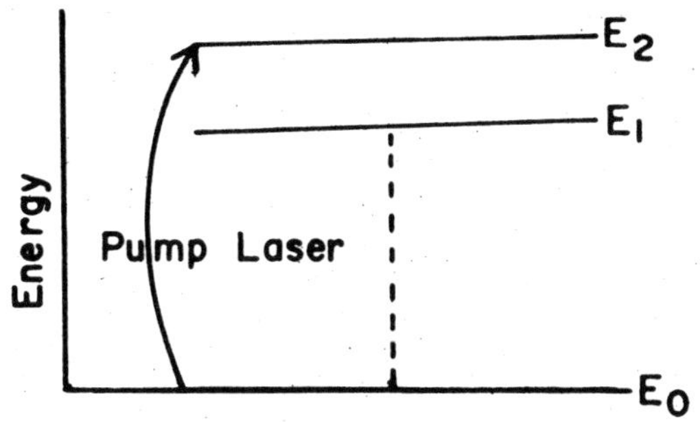

Fig. 1.44(b): Energy Level in a Simple Laser Material.

If the energy level E_1 is a metastable one, the number of atoms with this energy may exceed the number of atoms in the ground state E_0. Then laser action becomes possible. But actual laser materials normally have a much more complicated energy level diagram than that shown in Fig. 1.44(b). In particular the laser action is chosen so that atoms are left in a state of slightly greater energy than the ground, as shown in Fig. 1.44(c) below.

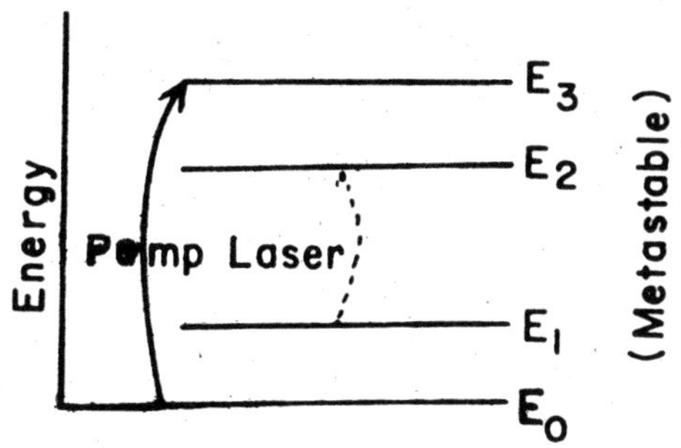

Fig. 1.44(c): Higher Energy Level for Better Result.

By pumping of energy, the level is left to state E_3 first. Laser material with the above type of energy level properties tend to be more satisfactory than that shown in Figure 1.26(b). It is easier to arrange for the atoms here in E_2 level to exceed those in E_1 level than it is for atoms in any of the excited state to exceed the number in the ground state E_0.

Gas lasers employ a gas discharge tube filled with a mixture of helium and neon at low pressure, as shown in Fig. 1.44(d) below.

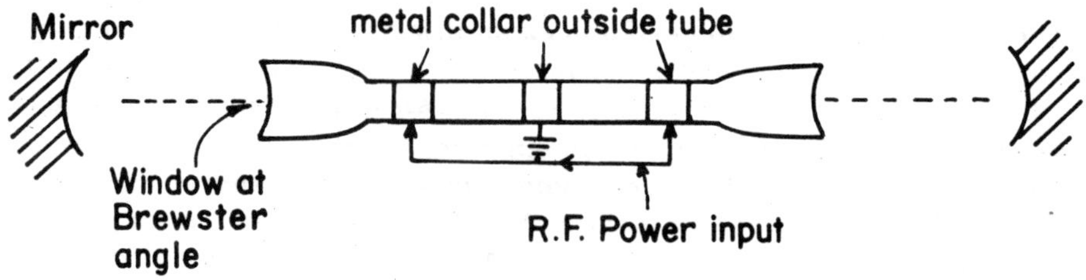

Fig. 1.44(d): A Gas Laser with External Curved .

Specially designed curved mirrors with suitable dielectric coatings make them reflect about 99 per cent of the incident light. The ends of the tube are fitted with windows cut to Brewster angle whose tangent is equal to the refractive index of the material of the window (in this case glass) to minimise reflection losses. Only light polarised in one particular plane can pass through Brewster window, for which light is amplified on this plane for laser action to occur.

The output from the helium helium-neon gas laser, operating in the visible region of the spectrum, appears as a bright red circle if it falls on a white paper. The output is some tons of milli watts of radio frequency power at about 27 Mc/s. Though it will not burn the skin, care should be taken to see that light does not enter any person's eye. Lasers can be used to carry information only between places which are in a direct line of sight to each other. As fog or mist present interruption to its propagation, so for communication links lasers are to be used in pipes or ducts.

1.11 Human Speech Transmission

The modification of transmitter and receiver design began by the development of variable distance transmitters and receivers with increased sensibility with carbon granules used in the microphone. Along with the development of transmitters and receivers, a corresponding development of the design of telephone exchanges were being carried out. For impedance matching, transformers were being used and both-way telephone circuit with a local battery and a tertiary winding transformer were used at each end, as shown in Fig. 1.45 below.

Fig. 1.45: Both Way Telephone Circuit.

In good old days, the signalling was done by magneto generator current by hand ringing. Later on such magneto systems were replaced by CBS1, CBS2 and lastly by central battery systems, in which the signaling was done by battery current. These systems will be described in the chapter on telephony. Similarly the various types of relays, etc. used in the transmission circuits will be described in the chapter on telegraphy and telephony.

Before we take up telegraphy and telephony it is worth while to describe the medium of transmission, i.e. the construction of land lines and cables with their testing.

2

Construction of Lines and Cables and their Testing

2.1 Construction of Open Wire Telecommunication Lines

The telegraph posts are made strong enough to bear the load of the wires carried on it so that they can stand against the severity of the storms and wind pressures. The alignment should be high enough to allow the pedestrians and vehicular traffic to pass below it with sufficient headway below the lowest sag of wires put on the lowest bracket of the posts. There should be a safety margin for the lowest sag of wires kept above the high flood water level in the areas usually inundated with flood water.

The wind pressure on an open – wire line or aerial cable can be calculated to keep the mean position of the load on the post within the carrying capacity. Thus for a line if $p =$ wind pressure in pound per sq. foot, l = length of the span in feet, $w =$ weight of iron wire in pounds per mile, the total wind pressure, $P = \dfrac{pl\sqrt{2w}}{3000}$ pounds.

The maximum wind pressure in India is taken as 25 pounds per sq. ft. for which $P = \dfrac{l\sqrt{2w}}{120}$

For copper wire the formula becomes $P = \dfrac{l}{90}\sqrt{w}$.

Thus for a single wire, the maximum wind pressure on a 600 lb wire of iron of span 220 ft (with 24 posts per mile) is given by $P = \dfrac{220\sqrt{2 \times 600}}{120} = 63.5\ lbs.$

To determine how many wires may be carried by a rail post under the maximum wind pressure, the bending moment is taken round the probable point of fracture, so that if p, p_1, p_2, etc. are the wind pressures for the wires and h, h_1, h_2, etc. are their heights above the assumed point of fracture, then the sum of their products, say $x = ph + p_1h_1 + p_2h_2 +$ etc. To this is added the bending moment of y due to the wind pressure on the post, so that $(x + y)$ must not exceed the safe working load of the post, which can be found from the standard chart.

To determine the same for a tubular post, let the wire load below the top of the post be p, p_1, p_2, etc. and that above top of post be q, q_1, q_2, etc. Now if their distances from the top of the post be d, d_1, d_2, d_3, etc. and d', d'_1, d'_2 etc., then the mean distance of the load from the top of the post

$$= \frac{(pd + p_1d_1 + p_2p_2 + ...) - (qd + q_1d_1 + q_2d_2 + ...)}{p + p_1 + p_2 + ... + q + q_1 + q_2 + ...}$$

Now if F is the sum of the various p's, being the wind pressures of all the wires, P be the wind pressure on the post itself, which may be taken as roughly acting half – way up the post as shown in Fig. 2.1 below

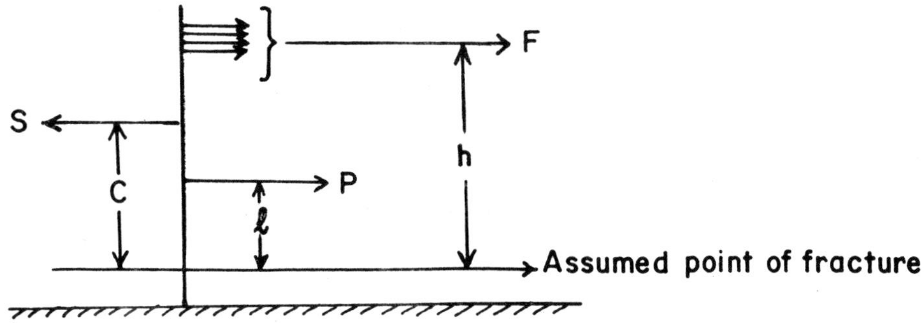

Fig. 2.1: Safe Working Moment if Resistance of Tube ; and S is the Horizontal Component of the Strength of the Stay Acting in a Direction Opposite to that of P and F,

then by taking moment round the assured point of fracture at 2 feet above the ground level, we find, at equilibrium, the algebraic sum of ($Fh + Pl - Sc$) must not exceed the safe working moment of resistance of the tube where,

F = sum of wind pressures on all wires

H = height of post – mean distance of load from top – 2′

$$l = \frac{h}{2}, p = \frac{2}{3}\left(\frac{\text{(mean diameter of post in inch)} \times \text{(height of post 1')}}{12}\right)p$$

S = 1200 lbs.

C = height of post – distance of x – stay from top – 2′.

In open country route, the maximum head way should be about 11 ft from the lowest sag of the span to the surface of the earth, but in town area it should be at least 15 feet and even higher in congested areas. The tubular posts are made of steel sheets of 8 ft length with different cross-sections and thicknesses and are categorised as *T, A, B, C, D, E, F*, etc. with increased thickness of sheets. As per suitability, even rail posts and wooden posts are used. These tubular posts are jointed together to give combined posts like *TA, AB, BC*, A_4BC, B_4CD, etc. if required with 4′ long pieces like A_4, B_4, etc. to give extended lengths.

Brackets are fitted to posts with U – backs and saddles made of channel iron of different sizes producing 2 wire, 4 wire and 8 wire telegraph brackets for single posts and for coupled posts. To increase the load-carrying capacity of posts, they are stayed and cross-stayed to raise the assumed section of fracture to the point of application of cross-stays. For this purpose, fresh calculation is required to see if the bending moment at the new assumed point of fracture is less than the safe working moment of resistance at that point and if the stay can carry the load.

It can be seen that the construction of open wire lines with iron, and copper and other metallic wires mainly depends on the principles of Applied Mechanics and theory of structures. Thus the strength of a rail post is found from the formula $M = F\frac{I}{Y}$ where M is the moment of resistance to rupture, F is the maximum stress the material will stand without breaking, I is the moment of inertia of the section, and Y is the distance of the furthest fibre from the neutral section. For rail posts, the assumed point of fracture is taken at the ground level and not 2′ above ground level as in the case of tubular posts. The maximum breaking stress is 20 tons per sq inch for iron and 25 tons per sq inch for steel. The safe working load on wire = ¼ breaking stress but in districts north of latitude 26^0 this is kept = 1/3 breaking stress, due to 70^0 fall in temperature, increasing load on wire. So the safe working load is always obtained from the standard chart. The maximum stress falls on the terminal post insulators, as it bears the full working load.

At angle points of the alignment, the stress depends on the size of the angle and it increases with fall in temperature. If ϕ is the supplement of the angle between the two wires at the angle point as shown in Fig. 2.2 below, then the resultant stress on the angle insulator $= 2s\cos\frac{(180-\phi)}{2}$, where S = maximum working load of wire.

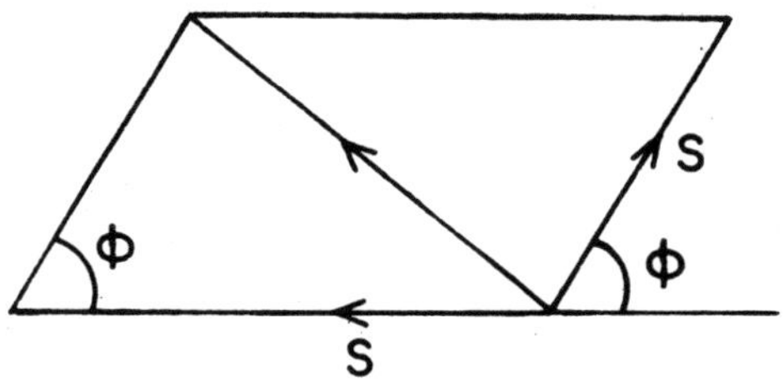

Fig. 2.2: Stress at an Angle Point

Hence the resultant load = $2s \sin\frac{\phi}{2}$. If P is the maximum wind pressure on the wires on the span if calculated from each gauge and span of wire, then the effect of P for the same angle $= p\cos\frac{\phi}{2}$.

Hence $2s \sin\frac{\phi}{2} + P\cos\frac{\phi}{2}$ must not exceed 325 lbs for telegraph stalks or 220 lbs for telephone stalks.

So at angle points, the stays are fitted exterior to the angle and strength of stays are taken as the strength of the straining screws, which have a working load of 2000 lbs. Then $\frac{5}{6}W$ is taken as the working load, where W is the weight per mile of the wire forming the stay. With this information, the gauge and the number of wires required for each stay and the number of stays required for each angle post or any terminal can be calculated.

The maximum working load is based on the tensile stress of 33,300 lbs per sq. inch, which is two-thirds of maximum strength of sheets making the tubes. The strength of iron tubes is about four-fifths of the steel-tubes. For trunk telephone lines as main routes, 242 lbs per mile of copper wires are erected and for local telephone lines 49 lbs per mile bronze wire or 365 lbs per mile steel wires are used where long spans are required. For telegraph lines on main route and short trunk telephone and canal lines, 150 lbs per mile iron wire and 75 lbs per mile iron wire is used for binding and making joints like Britania joints, etc. From terminal posts to the side of the wall of a telegraph office or telephone exchange, lead – in cables are joined to the open wires terminated on insulators and led to the commutator through some protective devices like fuse – strips, heat coils, and arrestors, etc. inside the office. For telephone exchange, the lines from the verticals containing the protective devices are first led to Intermediate Distribution Frame, whence they are jumpered to the Group Distribution Frame for joining to cables through racks leading to different positions of switch boards. In a telegraph office the lines are taken from the telegraph commutator to the different positions fitted with telegraph instruments. The method of binding iron wire with a 2 ft 6 inch binder is shown in Fig. 2.3(a); the termination of a single wire on Pot-head insulators is shown in Fig. 2.3(b); the single post BC with 4-wire brackets and attachment of stays and cross-stays is shown in Fig. 2.3(c), and coupled BCD posts, with 8-wire brackets for terminals is shown in Fig. 2.3(d).

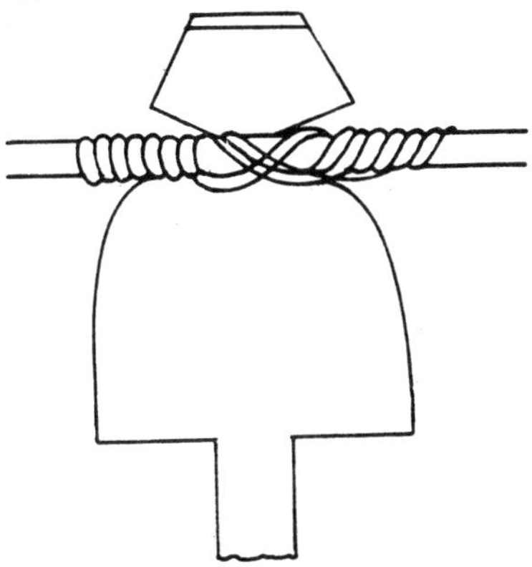

Fig. 2.3(a): Binding of Iron Wire on Insulators with 2½' Binder

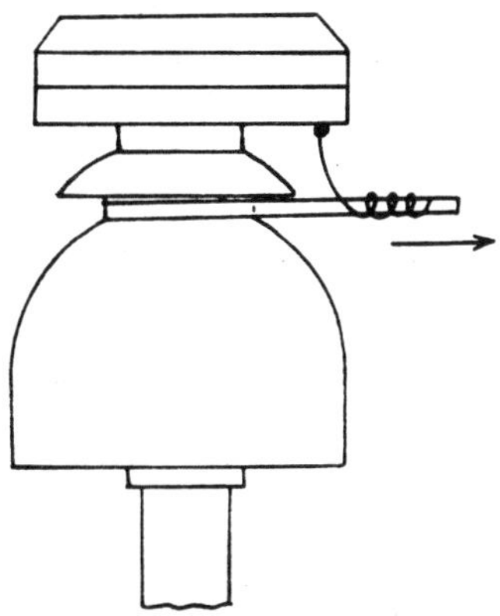

Fig. 2.3(b): Terminating a Single Wire on P.h. Insulator

Due to contraction of wires for varied temperature in districts south of latitude 26^0, wires are erected at ¼th of the breaking stress irrespective of temperature, but for places north of latitude 26^0, the tension of wires should not exceed 1/6th of the breaking stress. For this purpose, wires are pulled up to a tension as measured by a dynamometer greatly in excess of that at which it will ultimately rest and after a few minutes of pulling it is slacked out to the tension at which it will rest on the span. This process is called 'killing of wires'. For iron and bronze wires, the killing tension K_s is half the breaking stress and is given by $K_s = 5/3\ W$, where W = wt. per mile of wire. For copper, K_s = 1.3W. Thus, killing tension required for 300 lbs, 242 lbs and 200 lbs of copper wire are 390 lbs, 342 lbs and 280 lbs, respectively. At the time of erection, the tension of the wire is to be altered for every 5^0 variation above the standard temperature, but for variation of temperature below 5^0, no alteration of tension is required. The first wire is put and bound to the stalk fitted on the bracket and the subsequent wires are pulled out and put parallel to the first wire keeping the same dip as checked by a dip-stick as in the following chart.

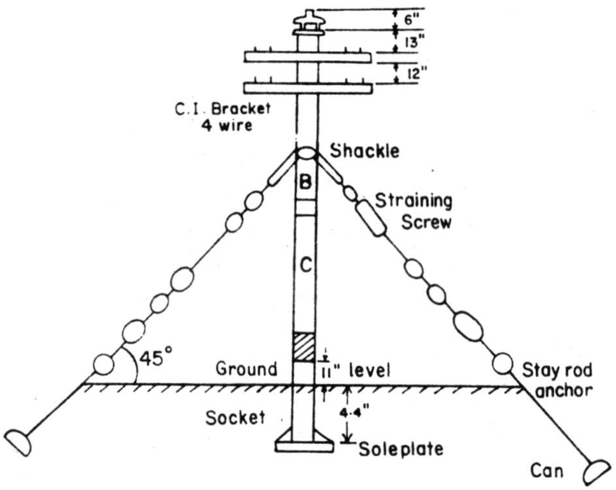

Fig. 2.3(c): Single BC Post with Attachment of Stays and Cross-stays

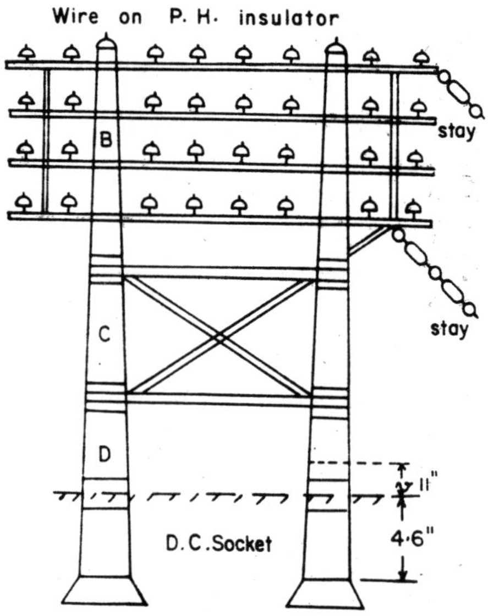

Fig. 2.3(d): Terminal Posts BCD Coupled Posts with 8-wire Brackets and Stays

Posts/mile	Separations in feet	Dip in feet and inches
30	176	10½"
24	220	1′ 4"
22	240	1′ 7"
20	264	1′ 11"
17	300	2′ 6"
15	352	3′ 4"
12	440	5′ 4"

In case of low tension power line crossing, the telecommunication alignment is erected below the power line alignment at right angle and guard circuits are provided with guard wires, ensuring that the space between them is not more then 32 inches and the outside guard wires are 8 inches beyond the outside telegraph wires. The guard wires should be erected at a height of not less than 24 inches above the telegraph wires and the minimum separation between the low tension lines and telegraph lines at the crossing should be 4 feet or more. An angle iron bracket supported by two angle iron arms is fitted to the top of the post and the guard wires are fastened to the bracket, from the extremities of which a stay is fitted with strain screws and stay-rod to the ground. Similarly, for high tension line crossing, the telecom alignment should cross the high tension alignment at right angles, below the power line for about 20 yards on each side. The power line should be enclosed in a cradle guard and the telecom alignment wires should be protected by a net guard and the minimum separation between the two guard wires should be 3 feet.

2.2 Calculation for Height of Mast

Let us take a wire suspended on two posts A and B, as shown in Fig. 2.4. The maximum vertical sag of the wire span in feet is shown as dip D. If L is length of the span in yards, then we know $D = \frac{3L^2}{8C}$ in feet, where C = ratio of the tension of the span to weight per unit length, i.e. $C = \frac{t}{w}$.

For wires used in India, C = 1500 yards for iron wire, 4500 for stranded steel wire, and 6000 for solid steel wire. It is 5000 for steel span wire with copper centre.

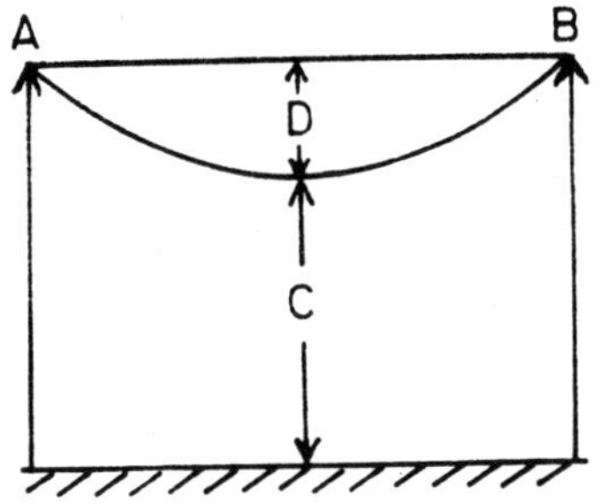

Fig. 2.4: A suspended wire in Catenary Form

$$\text{Thus } D = \frac{L^2}{4000} \quad \text{for iron wire}$$

$$= \frac{L^2}{12000} \quad \text{for stranded steel wire}$$

$$= \frac{L^2}{16000} \quad \text{for solid steel wire}$$

$$= \frac{L^2}{13333} \quad \text{for steel span wire with copper centre}$$

Case I: When the banks are on the same level on both sides of a river, the height of a mast is given by $M = H - h + D$,

where M = height of the mast in feet

H = headway above flood level in feet

h = height of bank above flood level in feet, and

D = dip in feet.

The vertical separation of wire is given by $V = \frac{1}{2}\sqrt{D}$ feet with a minimum of 2 feet, and the horizontal separation is given by $H = 0.4\sqrt{D}$ feet.

Case II: If the banks are at unequal levels, let M_H be the height of mast on higher bank and M_L that of the lower bank. Then $M_L = H - h + D$ and $M_H - M_L - D$, $|M_H = M_L - D|$ where H = headway above flood level in feet and h = height of lower bank above flood level in feet and d = difference of level of banks in feet.

Span terminal posts are erected on both sides of the masts as shown below in Fig. 2.5, so that the wires of the O.W. trunk line are first terminated on line terminals, then to the span terminals and then to the masts, so that if masts are carried away with high flood water, the line terminals remain safe with the rest of the alignment.

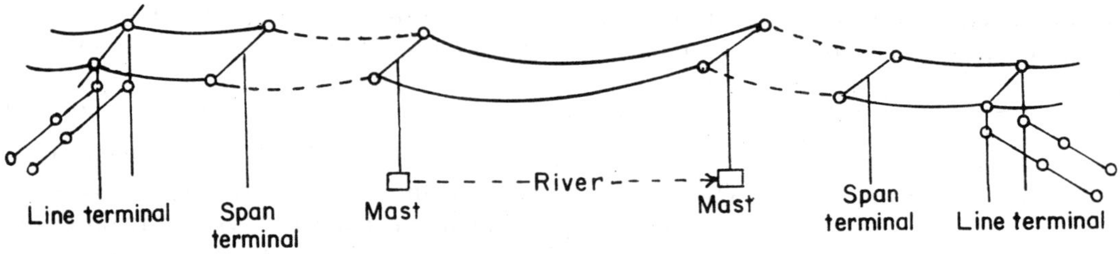

Fig. 2.5: Erection of Mast on Both Sides of a River.

For this purpose, light wires of iron, say 75 lbs per mile, is used between the mast and the span terminal. Continuous steel solid wires with any joint are used for mast span and the masts, as also the mast terminals and line terminals are all stayed away from the river on both sides. The masts may be (a) tubular masts, (b) lattice masts or (c) rail masts. They are erected on solid foundation, six feet square of concrete foundation for tubular masts, and 150 feet high. With a horizontal wire spacing of 3.5 feet, the maximum mast span is given by $S = 14.5\sqrt{c}$ yards. Necessary height of mast is given by a combination of A, B, C, D, E, F or even G posts, as required.

For working multi-channel carrier systems, the trunk lines are transposed at every alternate post of the alignment under *C*-8 or *C*-*J* schemes to get rid of inductive interference. 8-wire bracket with 16 inches limb separation and 8 inches pair separation with two pairs on each side of post, where the middle pair separation is 18" for the post, is used as shown in the chart below.

Section	Symbol	No. of span	Av. length	Min. length	Max length
Full section	T	256	8 miles	7.75 mile	8.25 miles
Half "	H	128	4 "		4.15 "
Quarter "	Q	64	2 "		2.10 "
1/8 "	E	32	1 "		1.05 "
1/16 "	X	16	½ "		0.53 "
Zero "	Z	8	¼ "		0.27 "

The railway track should be crossed at right angles using wires not heavier than 150 lb iron or 300 lb copper, keeping the headway at least 22 feet from the lowest bracket. At curvature of the railway track, the chord joining the alternate posts should not fall within 9 feet from the centre line of the track. Before construction, these specifications are to be approved by the railway authorities. Only tubular posts and not rail posts are to be erected in railway station areas. For erection of posts along the railway route, the minimum horizontal distance from the centre of the nearest track to the post should be equal to the height of the post plus 6 feet for narrow gauge (2′6") railway; it should be 6′6" for meter gauge railway and 7′ in case of broad gauge (5′ 6") railway. When a pair of rails have been laid and another pair is to be laid on the same side of the telegraph alignment along the railway route, a further distance of 14 feet is to be added to this. All the posts at railway crossings should be elongated and cross – stayed away from the railway track.

2.3 Details of Construction of Lines at Hills

To diminish the chance of contact between wires due to jungles and upheaval ground in hills and mountains, an additional separation of 6" for every 110 feet of span in excess of 220 feet is required. This is given by the equation $S = \frac{L \times 6}{110}$ inches, where S = separation in inches and L = length of span in feet. The dip of the wires erected in hills and mountains is adjusted by a dynamometer, as is done in plains. If there is a difference in levels between the supports, then the dip of wires below the lower support is given by the formula $x = \frac{L}{2} - \frac{dc}{3L}$ and $D = \frac{3x^2}{2c}$,

where L = length of the span in yards

d = difference of level of supports in feet

$c = \frac{t}{w}$ and

D = dip in feet.

The distance x from the lower support to the lowest point of the span wire will have a negative value if $\frac{dc}{3L}$ is greater than $\frac{L}{2}$, as happens at times when the lowest point of the span is not between the two supports and is beyond the lower support, as shown in Figs. 2.6 (a) and 2.6 (b) below.

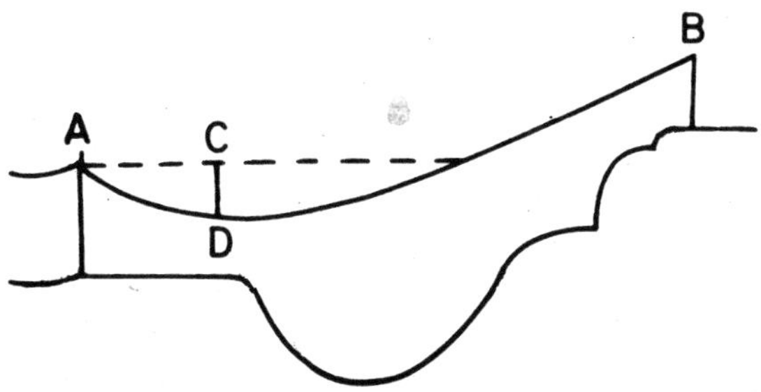

Fig. 2.6(a): When the Lowest Point of the Span is between the Two Supports *A* and *B*

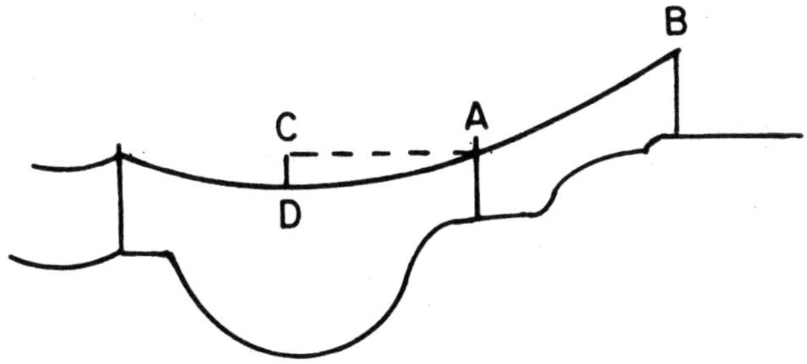

Fig. 2.6(b): When the Lowest Point of the Span is beyond the Two Supports *A* and *B*

In Fig. 2.6(a), x is positive and is the distance AC within the span AB. But in Fig. 2.6(b), x is negative and is the distance AC beyond the span AB. But in both cases the dip is the distance CD in feet.

2.4 Installation Tests

After the completion of construction of O.W. trunk lines, the following tests are taken:

(1) *Insulation test:* It should be done with battery voltage of not less than 100 volts. The insulation resistance should not be less than 1 mega ohm per mile in wet weather and 5 mega ohms per mile in dry weather, both for earth and cross insulation when measured by a Wheatstone Bridge or by meggar of 100 volts.

(2) *Conduction test:* The limb difference in conduction resistance should not exceed 2 ohms when measured by a Wheatstone Bridge.

(3) *Cross talk test:* The cross talk should not exceed – 40 db when measured by a cross talk meter shown below in Fig. 2.7.

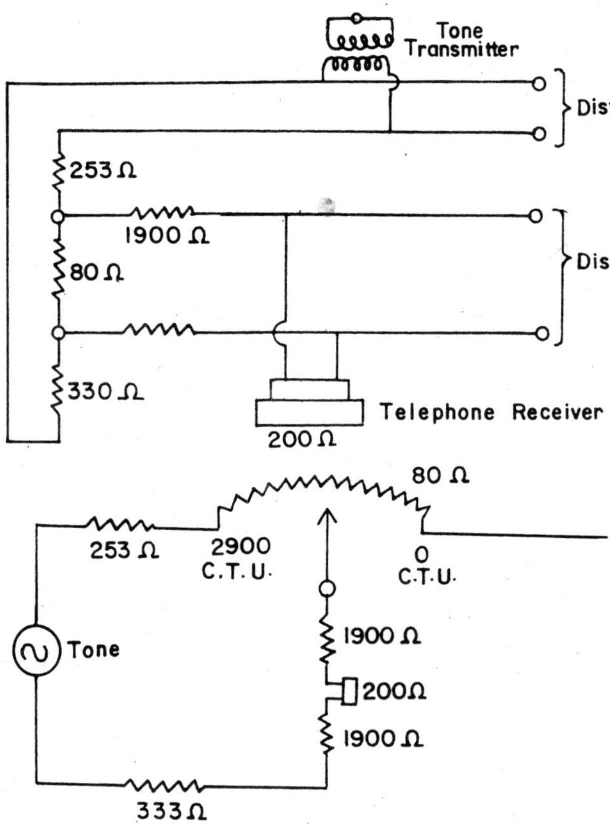

Fig. 2.7: Cross – talk Meter

(4) *Frequency attenuation test:* It is taken with frequencies of 300, 500, 800, 1000, 2000 and

2500 Hz and plotted on a graph, it should produce a curve in which the difference between any two consecutive readings are not greater than 0.5 db.

(5) *Impedance measurement:* It is taken with frequencies of 300 to 2700 Hz. With distant end terminated by characteristic impedance, it should show a curve with steady fall with rise in frequencies.

(6) *Noise level test:* It is taken with a volume indicator which should give a level of less than – 40 db when measured with a TMS at length end.

2.5 Construction of Cables

The laying of cables is done in two ways: (i) the laying of underground cable below the surface of earth, and (ii) the laying of under-water cable below the surface of water. Underground cable is required to be laid in congested and built-up areas in towns and cities. Underground symmetrical and coaxial cables are laid between cities for working long-distance cable carrier systems. Under-water cables are laid on the bed of the vast rivers, when no bridge is erected to link both sides and submarine cables are laid on the bed of the sea or ocean to connect cities and countries on both sides. On rivers where strong bridges link both sides, the O.W. alignments on both sides are linked by wires or overhead cables erected on the bridge – fixers. Occasionally overhead cables are laid on cable racks, rigidly fixed on iron brackets on overhead poles or on walls of buildings.

Telephone Cables: In towns and cities, local telephone exchanges work on local underground cables, 5 or 10 lbs per mile dry core lead-sheathed copper wires, varying from 20 pairs, 40 pairs, 60 pairs, 80 pairs, 100 pairs, 200 pairs, 1000 pairs + 5 per cent extra for each type. The next consignments are 20 lbs and 40 lbs per mile cable, mostly used for junctions between exchanges. 14/40 (14 pr of 40 lb) paper core quad twisted cables are used for leading in trunk lines from the terminal posts to the exchange Trunk Distribution Frame or Carrier Station R.S.D.F. respectively. For trunk cable lines, 70, 100, 150, 200, 300 lb copper wire cables are used. These are twin-twisted for 2-conductor pairs and when such pairs are together twisted for 4-wire multiple twist, the electric induction effects as also electromagnetic inductive effects are overcome. The other type of cable has one group of 4 conductors twisted together called quad – twist.

Telegraph Cables: In case of telegraph L.S. cables of 20 conductors, no such twisting methods are applied if they are of short lengths. But if they are of very long length, each wire is screened properly to decrease the inductive effect and when this screen is connected to earth, electric induction is decreased but electromagnetic induction remains as it is. Therefore, for telegraph working generally short lengths of cables are used and for long lengths, no earth circuit is used and thus the electromagnetic induction is got rid of by using a loop of two wires like telephone circuits. In general, the trunk cables are of composite type, in which the telegraph and telephone circuits are kept together by screening the telegraph wires and twisting the telephone wires.

In modern telegraph and telephone cables, each wire is separately screened and multipled, twined or quad-twisted. In each group there are 3 telephone and 1 telegraph or 2 telephone and 2 telegraph or 1 telephone and 3 telegraph or 4 telegraph circuits, which can work simultaneously.

Due to frequent theft of lead sheathed cable on account of high rise in price of lead and brass, the L.S. cable is replaced by (a) paper insulated or double Stalpeth cable, and (b) polythene insulated or PAP cable. These are generally laid underground direct at a depth of 3 to 5 feet, as necessary. The name Stalpeth has been derived by taking (st) from steel, (al) from aluminum and (peth) from polyethylene. Its transmission characteristics are the same as that of L.S. cable. In case of ordinary stalpeth, a polyethylene jacket is put round the cable pairs and for double stalpeth, an additional sheathing steel tape is put in the ordinary polyethylene jacket. A double stalpeth sheathed cable is shown in Fig. 2.8(a) below and polyethylene insulated or *PAP* cable is shown in Fig. 2.8(b) below.

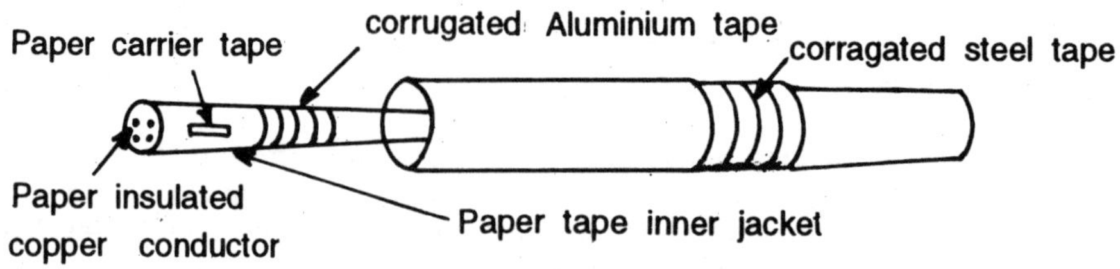

Fig. 2.8(a): Double Stalpeth Cable

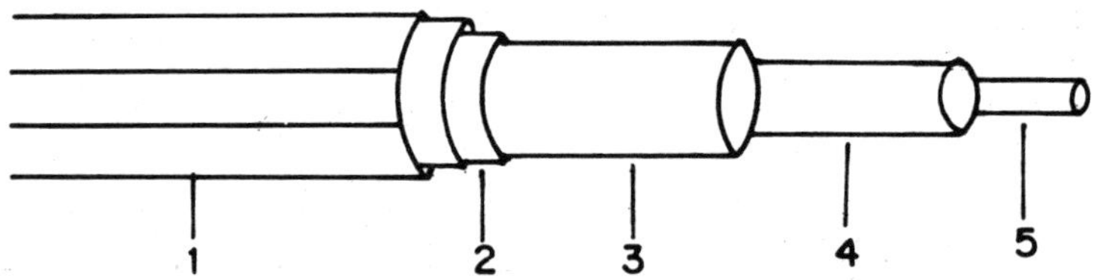

Fig. 2.8(b): PAP Cable

As given in the figure, stage 1 is polyethylene of high weight molecules, stage 2 is aluminium tape, 3 is polyethylene of high molecular weight, 4 is dry core tape, and 5 is polyethylene insulated copper conductor with colour code. The main copper conductor is kept insulated below these five layers . The *PAP* is nothing but a composite sheet which is set above the plastic insulated cable. It is used in place of Hepeth sheath of standard '*PIC*' cable. *PAP* is made of (*P*) for polyethylene, (*A*) for aluminium and (*P*) for polyethylene. A black polyethylene jacket solidly covers the cable core inside the sheath. Below this is a corrugated tape of aluminium wound round the inner polyethylene jacket, whose diameter is 0.65 inch, so that no vaporous material can penetrate the sheath whose dielectric strength is thereby increased.

Number scheme of polyethylene cable: Two conductors are twisted 6 times within 10 inches to form a pair. Thus 10 pairs are made to form 1 unit. Such units can be identified by seeing the colour of plastic binder. *A* and *B* conductors can be identified on seeing the colour of insulation. These are described below.

Table 1

Unit	1	2	3	4	5
Colour of plastic binder	B	O	G	Br	S

Table 2

Pair No.	Conductor A(mate)	Colour B	Pair No	Conductor B(mate)	Colour B
1.	W	B	6	R	B
2	W	O	7	R	O
3	W	Gr	8	R	Gr
4	W	Br	9	R	Br
5	W	S	10	R	S

Of the jointing materials, the main items are (i) ½" tape paper B marker, (ii) cotton tape 1", (iii) cement, (iv) 3/4" or 2" D.R. tape, (v) 4" aluminum B tape, (vi) 1" glass B tape, (vii) 3/4" fix on tape, (viii) primer, (ix) green line tape, (x) B sealing tape, (xi) polyethylene B tape, (xii) moslin 4," etc.

Cable jointing procedure: Polyethylene jacket is removed for 10 inches from the end where a joint is to be made and the Mylar tape wound round the conductor is removed up to the remaining polyethylene jacket. Then 4½ inches of insulation should be removed from the end of the conductor and sleeve is to be put on it. Each pair of wire is to be held by the thumb and fore finger of the left hand and two twists are to be made 4½" from the end. Then holding the pair 1 3/4" from the twist point and bending the ends, turns are made for about ½" to make them hard and then the remaining part of the wire is cut off, after keeping about 1½" to 1¾" from the twisted point. Then the twisted part is bent on the conductor and the sleeve is put on the same. In case of PAP cable, the conductors are wound 3" with B polyethylene and 1" with paper B tape twice. But in the case of Stalpeth cable, dry Moslin 36" long is wound round the joint and keeping about an inch on both sides, the two ends are bound by 1" cotton tape.

2.6 Underground Long-Distance Symmetrical Cable

When the railway electrification plan was taken up, it was observed that there cannot be any open wire trunk telephone alignment along the railway route as the inductive effect of 25 KV on 50 Hz used for the railways will have inductive interference on communication lines. So unless the telephone trunk route is shifted to a distance of at least 50 metres away from the railway route the inductive interference will render them unsuitable for communication purposes. So the telephone overhead lines were replaced by underground symmetrical cables, on which long-distance underground symmetrical cable carrier systems were worked with repeaters at a distance of about 12 miles or so. The specifications of the symmetrical cables are:

1. They are 4 quad, capacitance .042 μF and every conductor of 40 lb per mile.
2. Insulation resistance at 60^0 F and 5000 V = 4.27 MΩ per mile and DC resistance at 60^0 F = 21.93 ohm/mile.
3. Average attenuation at 50^0 F and 252 KHz = 4.27 db/mile and at 50^0 F and 60 KHz = 1.95 db/mile.
4. Far end cross-talk at 252 KHz and 60 KHz = 71 db.
5. Near end cross-talk at 60 KHz = 56.5 db.
6. Impedance at 60 KHz = 170 ohm.
7. Each drum of 4 carrier quad + 40 or 20 VF quad is of length 500 yards of cable.

Balancing of Symmetrical Cable: For balancing, 42 cable drums are required for a repeater section of 12 miles, in which 8 drums give a full-sling length of 4000 yards for balancing. For a full repeater section of 12 miles, we require five full slings and a part sling. The construction of a sling of 4000 yards length is shown in Fig. 2.9 below.

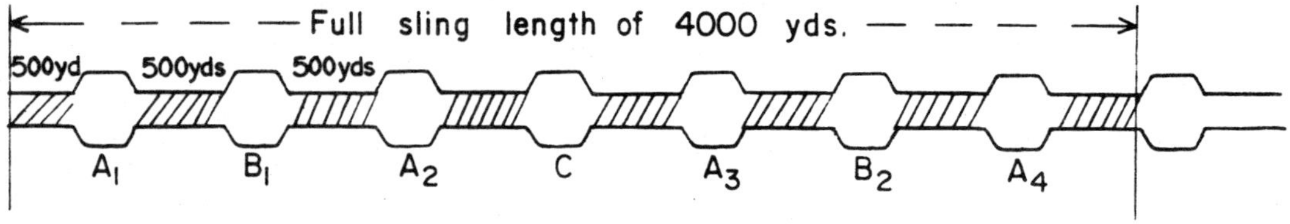

Fig. 2.9: Construction of a Full-Sling Length of 400 yds

The first two drum lengths are jointed by A_1 joint. Each drum gives 1/8th of a sling length and two drums give a quarter sling jointed by a B_1 joint for each such two drums jointed by a A_2 joint to give a half-sling length. In case of the other half of the sling, two such drums jointed by A_3 and A_4 joints on each side are jointed by a B_2 joint. The full-sling length is made by a C joint jointing these two half slings as shown. In case of A joints, one pair is jointed straight and the other pair crossed to attain the electrical neutrality. In one quad, the highest capacitance section is jointed to the lowest capacitance of the next quad to decrease the capacitance unbalance. Thus by changing the position of links in a pair, the side-to-side capacitance unbalance is brought to a minimum by taking a test with 800 Hz tone. The conductor positions are interchanged to give a satisfactory standard of cross-talk for poling underground symmetrical cable, which is enclosed in an aluminium sheath and then armoured.

The VF quads are loaded with 44 millihenry loading coils at distances of 2000 yds. The characteristic impedance of these loading coils = 1120 ohm with an attenuation of .4 db/mile. The impedance and capacity balancing is done in a special procedure as described above and the conductors are poled for transposition. For working in carrier frequencies, there are three layers in which the quads are divided. The four inner central quads consist of the first layer. This is paper insulated. The second layer consists of 13 paper insulated VF quads and 4 'Pet' insulated VF quads. The third layer consists of 23 VF quads with paper insulation.

2.7 Long Distance Underground Coaxial Cable

In India, the coaxial cable used is of 2 cored 375 type.

Specifications: The central conductor consists of a solid copper wire of .104 inch (2.66 mm) diameter, which is kept fixed at the central position of an annular copper plate of thickness .01 inch (.25mm) forming a hollow conductor around the solid one. The inner diameter of the hollow conductor is .375 inch (9.5 mm). The central solid conductor passes through central holes in polyethylene disc insulators of size .075 inch (1.9 mm) each placed at distances of 1.3 inches apart in the hollow tube conductor. Thus the solid inner copper wire is the first conductor and the hollow copper tube surrounding the first conductor is the second conductor. The centre of these two conductors being the same, the cable as a whole is called the coaxial cable.

A coaxial cable thus formed is twice wound by a mild steel tape of .0005 inch (.13 mm) thickness and then twice wound with paper tape. Thus the losses caused by electric and magnetic inductive effects are nullified and the radiation effects are brought to the minimum. Though the core of a coaxial cable is not balanced, the steps taken at the time of construction of the cable sufficiently perform the balancing work and the cable remains completely free from undesirable noise. Earthing of the cable at limited distances also decreases the coupling effects to some extent. At the time of jointing, some polyethylene discs are replaced by polytetra – fluoraethelene (PTFE) discs.

The diagram of a 375 type two cored coaxial tube is shown below in Fig. 2.10.

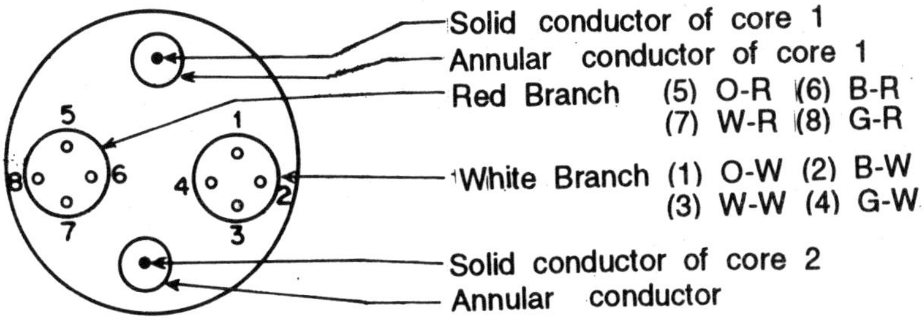

Fig. 2.10: Sectional Diagram of 2-cored Coaxial Cable.

Installation tests: On completion of laying of coaxial cables, the electrical transmission characteristics are determined by the following tests:

1. Dielectric strength test: When the central solid conductor is subjected to 3 KV for 2 minutes with the outer tubular conductor earthed and it is found that there is no breakdown of insulation, then it is to be taken that the dielectric strength is correct. This test is carried out with a DC high voltage tester.
2. Insulation test: When tested with 500 volt meggar in the range of 100000 meg ohm, the insulation test should show:
   i) insulation between C and $T \not< 8000$ meg ohm/km section,
   ii) insulation between T and $E <$ 1000 meg ohm per km section,
   iii) the DC resistance measured by Houston Bridge for 20° centigrade, for 'C' should show 5.1 ohm ± 4% per mile (3.17 ohm ± 4% per km) and for T it should show 3.8 ohm ± 4% per mile (2.36 ohm ± 4% per km) and difference of these two $(C - C_2) > 0.2$ ohm per repeater section,
   iv) attenuation per mile = $3.75 \sqrt{F} + .002\ F$db per mile, where F = frequency in mega Hertz,
   v) characteristic impedance is 2.5 mega Hertz = 75 ohm.
3. Near end cross-talk (NEXT) test: Between the two coaxial cores from 60 KHz to 10 mega Hertz < 160 db or $80+73 \sqrt{F}$ db, where F = frequency in mega Hertz.
4. Far end crosstalk (FEXT) test: The reading should not be less than 160 db or $73+73\sqrt{F}$ db.

Joints are to be made within the group VF quad to VF quad. In the first quad, straight joints are to be made always. No. 2 quad should always be jointed to No. 3 quad. No 2 and 4 quads are to be respectively jointed to No 2 and 4 quads and doped. Similarly, for No. 5 and 7 quad for No. 6 and 8 quads.

2.8 Underwater and Submarine Cables

When it is decided to lay an underwater cable in a big river or to lay a submarine cable in a sea or ocean, the most important work is to select the location of the cable hut on the bank of the river or on the shore of the sea. For this purpose, the bank should be of hard earth of clay and not built on sandy ground. Instead of being abruptly high, the bank or shore should be of gradual slope to the river or seabed. The cable hut should be located at a far distant place from the steamer or ship station or harbour, since due to the pull of the anchor, the cable lying in the river bed can get attached to the anchor, which can damage the armour of the cable, and in extreme cases the cable is likely to be torn off.

After the soil test is over, the examiner should determine the type of cable to be put under water. Laying a submarine cable is very difficult and very expensive; it is also entwined with national interest and with the future demands of the inhabitants of the cities and countries on both side of the sea or ocean. Along with number of conductors of the submarine cable, it is also determined whether the submarine telegraph or telephone cable should contain the repeaters in it, immersed in the deep sea.

After determining the location of the cable hut, pillar box and terminal post, the laying work should begin in dry season, when the water level remains low. A canal is dug at least 5 to 6 feet deep in a zigzag way from the cable hut terminal to the water level. If the span of water for laying underwater cable is a marshy land, then big boats or wooden rafts are used for carrying the underwater cable to lay it on the bottom of the marshy land. But if the cable is to be laid in a big river, the same is done in one or two steam launches. If a submarine cable is to be laid in a sea or ocean, then a special marine ship built for the purpose is required. The speed at which the cable is to be laid is determined by a Brake dynamometer. Before laying of one drum of cable is over, the joint made with the second drum of cable should be completed, so that drum after drum of cable may be laid in continuation. When the other end of the marshy land, river or sea is reached, then the wooden raft, or boat or launch or special ship is to be tightly anchored first and the full end of the cable is to be taken to the cable hut by taking the same with the hands of the workers and keeping a sufficient length in excess beyond the cable hut. The cable is then slowly to be laid in the zigzag canal dug earlier, taking sufficient care to see that there is no unusual twist in the cable, which is then covered with earth. Pillars are laid on both sides of the cable route and on the high platform on both sides of the cable hut; two masts of wooden rods 30 feet high are erected. These masts are painted alternately with white and black paint for a distance of one foot for each colour, so that a sea-going ship can see the same from a distance and take sufficient care not to throw the anchor near that place. Moreover, buoys are placed on the water on the sea to give an indication of the cable route. These buoys are also painted similarly as the pillars, and lights are exhibited at night for necessary indication of the cable route.

Necessary installation tests are taken after the laying of the cable and terminating the same in the cable hut commutator through protective devices and joined to the aerial trunk lines through matching transformers joined in the lead-in cables. The records of these installation tests are kept hung in the cable hut or sign written in a tin or wooden board indicating the cable route, cable length with its conductor resistance, capacitance, etc. A copy of this record is also kept with the testing master and engineer concerned.

In case of a cable fault, the same is located by taking insulation test, electrification test, capacitance test, etc. by a Nalder test set to determine the distance to fault. An experienced engineer with a cable repairing party is sent to the probable distance to fault either in a steam launch or a cable ship. With the help of a 'graphel', i.e. a multi-mouthed anchor, the faulty cable is pulled up and kept flat on the floor of the cable ship or launch. If the fault is found in the part of cable already pulled up, then the cable is cut on both sides of the affect point at a suitable distance on both sides and a piece of new cable of identical type is jointed on both sides which are properly sealed, taped and armoured and the cable is retested before it is again pulled down and laid on the bed of the river on sea. But if the fault is not found in the portion of the cable pulled up, then it is relaid on the bed of the river or sea and the next portion of the cable is pulled up and tested. This procedure is continued till the fault is found and repaired, retested and the cable relaid.

In our country, we have got underwater cable laid across the Ganges linking overhead trunk lines of north and south Bihar near Patna, from Police Line cable hut to Sambalpur cable hut, with a length of about 2300 yds or so. Similar underwater cables have also been laid in other vast rivers where no bridge is available to join the overhead trunk routes on both sides. The submarine cable laid in the Bay of Bengal is from Dhanuskoti in India to Talaimaner in Sri Lanka.

Fig. 2.11 (a): Anglo-Dutch Paper-core Submarine Cable Cross-section

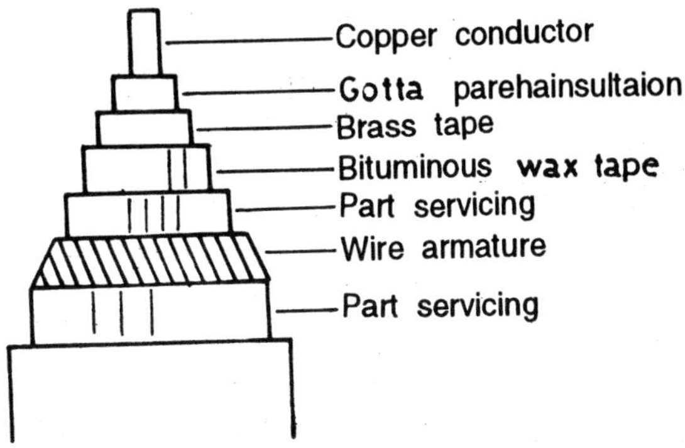

Fig. 2.11 (b): Ideal Unloaded Deep-sea Telegraph Cable Head.

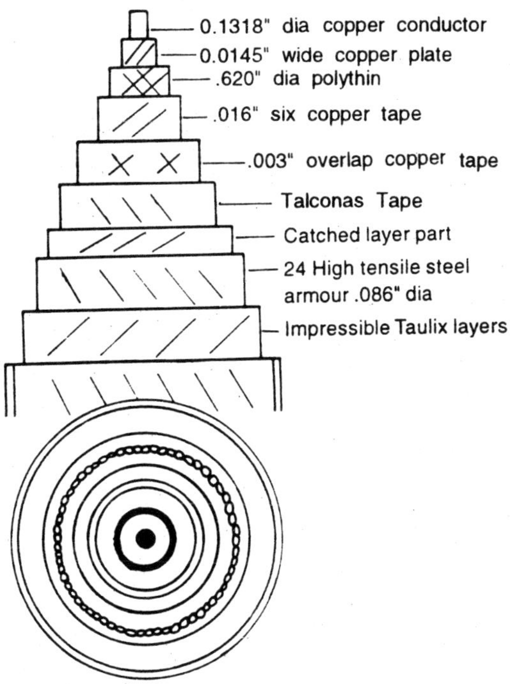

Fig. 2.11 (c): Trans-Atlantic Telephone Cable (used below 700 fathom in sea water, 1 fathom = 6 ft)

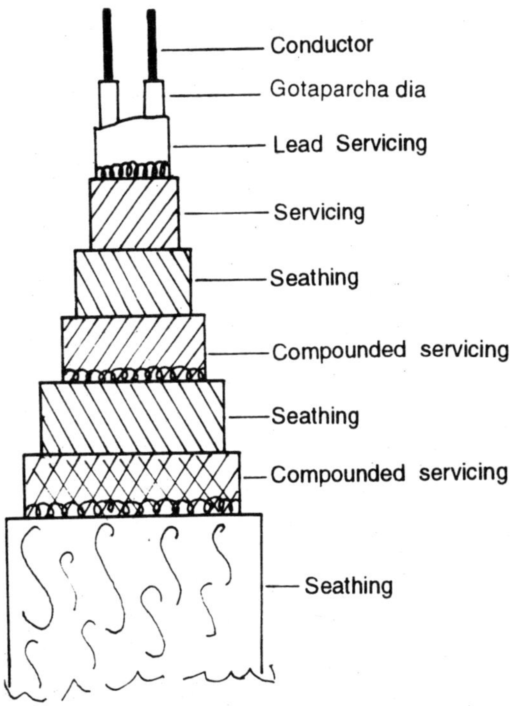

Fig. 2.11 (d): Telegraph Cable Heads for Impregnable and Armoured Submarine Cable

Fig. 2.11: Various types of Submarine Cable

It was 'Oliver Heanyside' who, after prolonged research, found out by the end of the nineteenth century that if lumped or continuous inductance are used in the cable, the speed of telegraph working in the cable is increased. Though unloaded cable was first laid in 1891 from St. Margaret to Sun Gate, coil loaded solid dielectric cable was first laid in 1910 from Abbots' cliff to Grisnez. A little earlier, continuous loaded solid dielectric cable was laid from Elsinore to Helsingbry, and air-space and paper-insulated lead sheathed cable was laid from Grttsiel to Borkaw. The Anglo-Dutch Cable is the oldest British paper-core cable that was laid in 1924 from Aldebergh to Dombarg. These submarine type cables with their manufacturing technique, etc. are diagrammatically represented in the following pages.

The maintenance of underwater or submarine cable is done by keeping dry air cylinder pressure constant inside the cable as indicated by the air pressure indicator. A continuous flow of dry air or gas is sent inside the cable so that a constant pressure is maintained inside the cable for which any outside gas, air, water, clay, etc. cannot enter into the cable and decrease the insulation of the wire conductors inside the cable. The air or gas pressure is constantly observed on the pressure indicator visually and an alarm bell also rings if the same decreases beyond a limit, when the air or gas cylinder is replaced. But if this decrease of gas or air pressure still persists inside a cable, it indicates a leakage giving earth fault, the same is located by a Nalder testing set for underwater cable. There is a Shrader valve fitted at an interval of every 500 yards on the route in the case of underground coaxial cable. By using a portable manometer on the route, the air or gas pressure can be continuously noted and from a graph drawn with distance on one side and air pressure on the other, the point of leakage, i.e. the irregularity point giving the distance to fault, can be found out. This is shown in Fig. 2.12

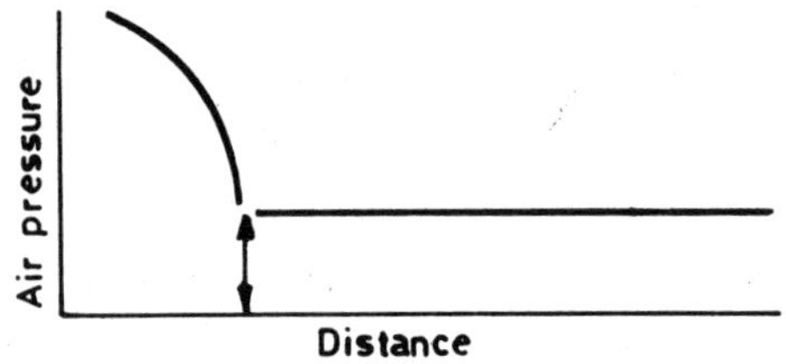

Fig. 2.12: Distance versus Air Pressure Curve

The normal dry air pressure is taken as .6 Kg/sq cm. sent from a 150 Kg/sq cm. dry air cylinder. If the air pressure falls to 50 Kg/sq cm., an alarm bell goes on ringing till the cylinder is replaced by a new one.

2.9 Testing of Open Wire Lines

Open wire lines are tested by Wheatstone Bridge for conduction, insulation and contact between the conductors. The principle of Wheatstone Bridge is described below in Fig. 2.13.

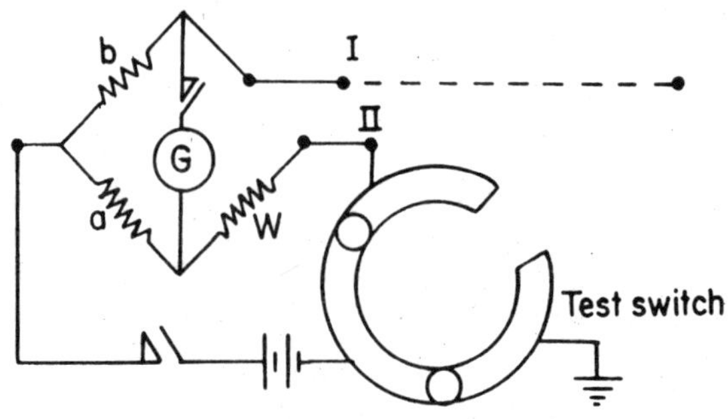

Fig. 2.13 (a): Insulation Resistance

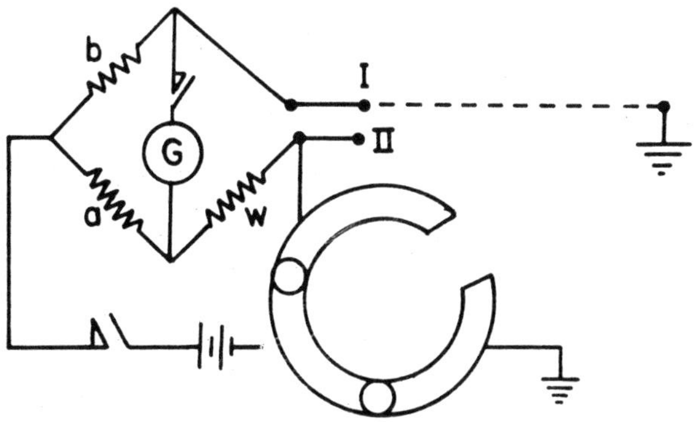

Fig. 2.13 (b): Conduction Resistance

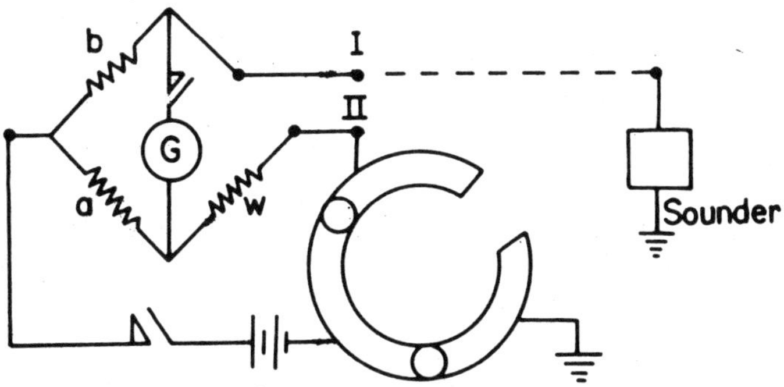

Fig. 2.13 (c): Circuit Resistance

Fig. 2.13: Resistance Measurement with Wheatstone Bridge

The first and the second arms of the bridge are of resistance a and b. The resistance x of the conductor to be measured, placed on the third arm, is determined by balancing the same with known resistance w placed in the fourth arm when no current will pass through the galvanometer. This is done by putting the battery switch first and then the galvo switched by making the battery and earth connections by means of a test switch with bridge terminals as shown, so that in the balanced condition.

$$\frac{b}{a} = \frac{x}{\mathrm{w}} \quad or \quad x = \frac{b}{a}\,\mathrm{w}.$$

With this principle, the insulation resistance, conduction resistance, and circuit resistance are measured.

Fault localisation: Three types of faults occur in O.W. lines. They are earth, contact and break faults.

The earth faults are localised in case of dead earth by measuring the conduction resistance to fault and converting the same to distance to fault with reference to the gauge book for this alignment available with the testing staff. In case of partial earth, the same is localised by sectional tests given by the line staff on the route, when no other line is available on that route. But when a good line is available, the earth faults, dead or partial, are localised by Varley test as shown in Fig. 2.14.

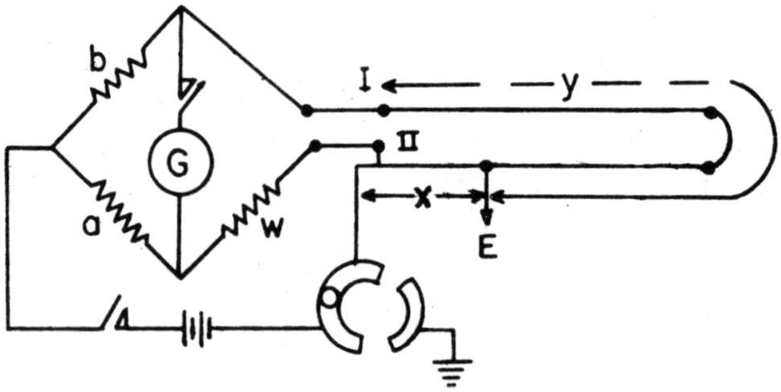

Fig. 2.14 (a): Varley Loop Test

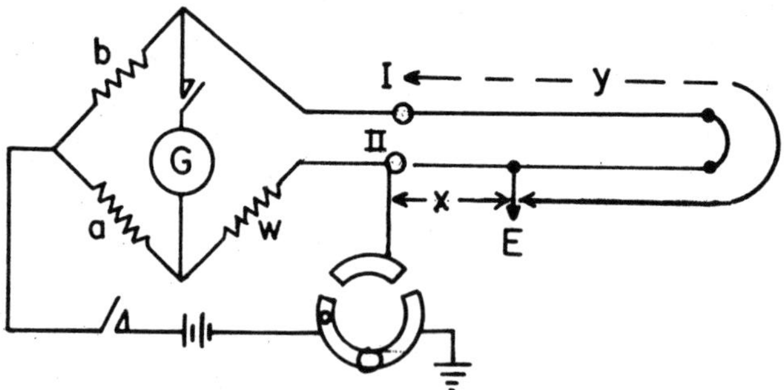

Fig. 2.14 (b): Varley Localisation Test

Fig. 2.14: Varley Tests

To begin with, the insulation and conduction tests for the faulty or good wire are taken. If the insulation test result for the good wire tallies with the insulation test result in the gauge book, the same wire may be used for localisation test, otherwise another wire with good insulation has to be selected. Then, the distant end is asked to give a loop on the good and faulty wires; at the testing end, the good wire is joined with terminal I and the faulty wire with terminal II. The resistance r of the loop is measured with arms *a* and *b* kept equal so that $r = x + y$ (1)

where,

x is the resistance along the faulty wire from the test point to the point of fault, and

y is the resistance of the remaining part of the loop. The testing master then disconnects his battery from terminal II and connects it direct to earth as shown in Fig. (b) and takes the localisation test in which at balance,

$$w + x = y \tag{2}$$

From (1) and (2), we get $x = \dfrac{r - w}{2}$, i.e. resistance to fault $= \dfrac{r - w}{2}$ which can be converted into distance to fault in comparison with the gauge book, after correcting for the temperature which is taken as 60^0F in India. For this purpose, formula $R_t = R_0\,(1 + \alpha t)$ is applied, where

R_t = resistance of wire at temperature t^0 C at the time of test,

R_0 = resistance at 0° C,

t = rise of temperature, and

α = temperature coefficient of the material of the wire.

3-Wire test for contact fault localisation: In case of a contact fault occurring between two wires, the same may also be localised by Varley loop test if a third good wire is available by taking loop-localisation test as shown in Fig. 2.15 below.

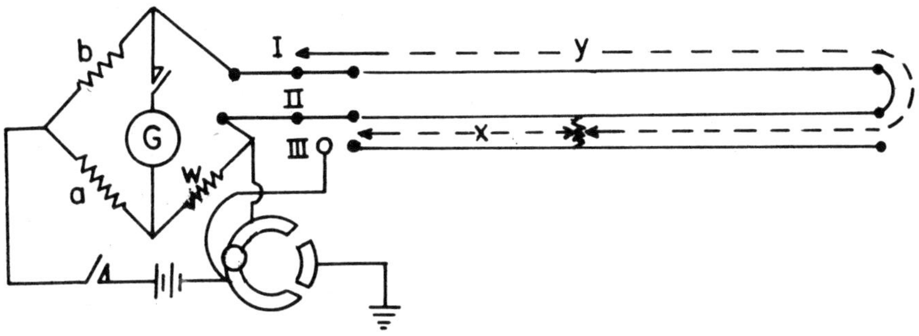

Fig. 2.15 (a): Loop test

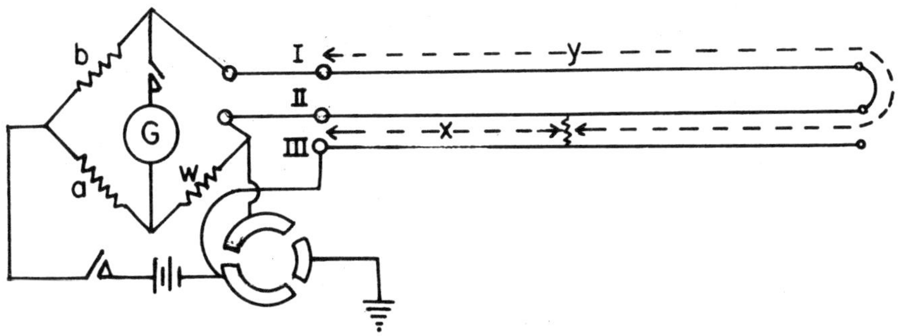

Fig. 2.15 (b): Localisation Test

Fig. 2.15: 3-Wire Test for Contact Fault Localisation

In the loop test, the good wire is connected to terminal I of the bridge and the faulty wire connected to terminal II, these are looped at the distant end, as shown in Fig. (a). Then keeping the other faulty wire insulated, balance is made by adjusting the third arm resistance keeping $a = b$. Then if x is the resistance to fault and y is the resistance of the remaining part of the loop, then at balance

$$r = x + y \tag{1}$$

In the localisation test, the other faulty wire is connected to terminal III of the bridge as shown in Fig. (b). Then at balance, when $a = b$, we get

$$w + x = y \tag{2}$$

From (1) and (2), we get, $x = \dfrac{r+w}{2}$,

i.e. resistance to fault $= \dfrac{r + w}{2}$.

2 – *wire test*: If a third good wire is not available, then the distant end is asked to put earth on one wire and keep the other insulated. Then conduction and localisation tests are taken as shown by Figs. 2.16(a) and (b).

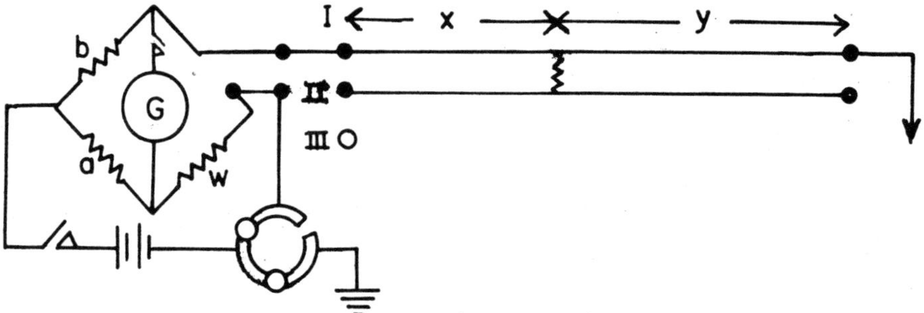

Fig. 2.16 (a): Conduction Test

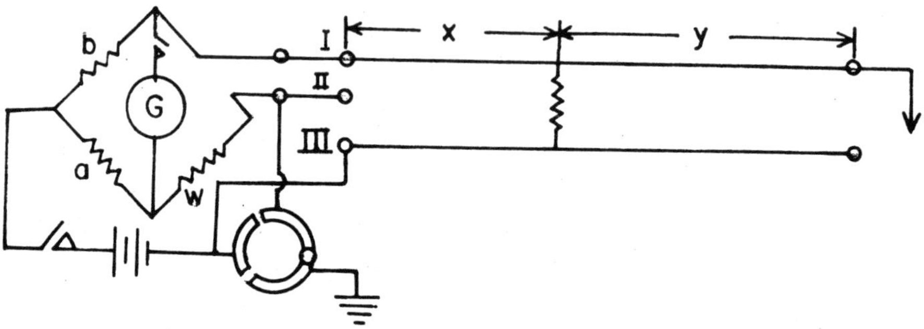

Fig. 2.16 (b): Localisation Test

Fig. 2.16: Two-wire Test for Contact Fault Localisation

In the conduction test, calling the resistance to fault as x and the resistance to the remainder of the wire under test plus earths as y, and putting on terminal I the wire whose distant end is earthed, we have, at balance,

$$r = x + y \tag{1}$$

In the localisation test, the second wire is connected to terminal III of the bridge as shown in Fig. (b). Then at balance, when $a = b$, we get,

$$w + y = x \tag{2}$$

From Eq. (1) and (2) we get, $x = \frac{r+w}{2}$,

i.e. resistance to fault $= \frac{r + w}{2}$.

2 -*wire test*: If a third good wire is not available, then the distant end is asked to put earth on one wire and keep the oher insulated. Then conduction and localisation tests are taken as shown by Figs. 2. 16(a) and (b).

If no balance is obtained in the localisation test, then the bridge terminal I is connected to earth and terminal II to the line and the bridge is set as for localisation in 3-wire contact test, vide

Fig. 2.15(b), when at balance the resistance to fault $= \dfrac{r - w}{2}$.

Conversion of resistance to fault into distance to fault: The line resistance book or the gauge book of the testing station contains the following particulars for each wire separately:

(a) Length of various sections.
(b) Resistance from one end to other at 80^0 F.
(c) Total distance to the end of section.
(d) Resistance per mile of the section at 80^0 F.

Before converting the resistance to fault into distance to fault, the corrected resistance to fault is calculated at 80^0 F $= x\dfrac{R}{r}$, where

(a) in case of single wire, R = total line resistance as given by the line resistance book, and r = conduction resistance;
(b) in case of 2-wire contact test, R = total resistance plus the earths as given by line resistance book, and r = conduction resistance;
(c) in case of loop test and 3-wire contact test, R = sum of resistances of two wires forming the loop as given by the line resistance book, and r = measured loop resistance. R/r is called the conduction ratio or the correcting factor, which is either above or below 1 as the temperature is below or above 80^0 F.

2.10 Cable Testing

(a) For local cable, tests are generally taken by a meggar or bridge meggar or a telefault locator for break, contact or earth fault, and faults are localised by sectional tests from D.P. boxes, etc.
(b) For long distance cable or underwater or submarine cable, the same faults are localised by Nalder testing set fitted with a moving coil galvanometer, with a lamp and scale, kept insulated by three insulators on table, a universal shunt, a cable testing key, a standard 10,000 ohm resistance, a standard 1 μF or sometimes 1/3 μF condenser and a B.P.O. pattern Wheatstone bridge, as shown in Fig. 2.17(a) below.

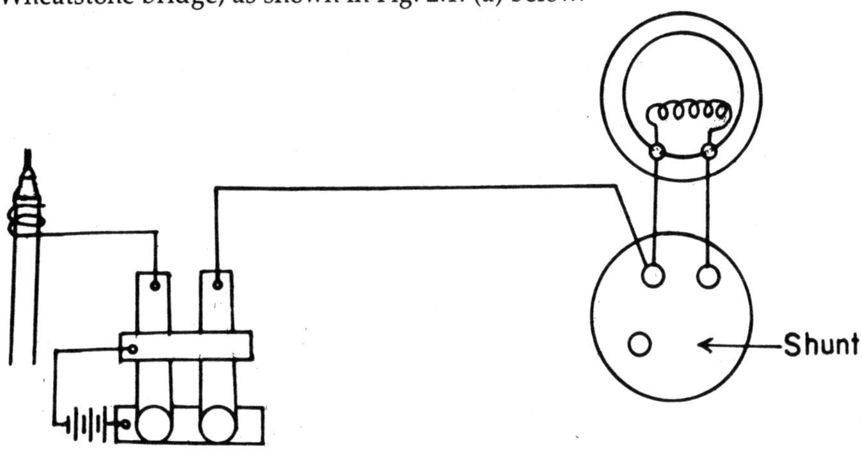

Fig. 2.17(a): Insulation of Testing Set

The levelling is adjusted by the levelling scales till the moving coil is at its null-point. The insulation of the testing set is taken by connecting the whole of the insulation testing battery to the key contacts (positive to front contact and negative to rest contact) and left hand key lever to earth with L terminal kept insulated. Then on pressing the first key, there should be no deflection for any change of shunt. The same should be the case with other key pressed. If there is a deflection of more than two divisions, the insulation of the leads, battery and the keys, etc. should be checked till the deflection is limited. Even when the L terminal of the shunt is connected to earth and the left hand key lever is insulated, a repeat test should show limited deflection.

The resistance constant of the galvanometer is found out by finding that resistance which in series with the galvanometer (shunt at 1) and the whole of the testing battery would produce a deflection of one scale division, as shown in Fig. 2.17(b).

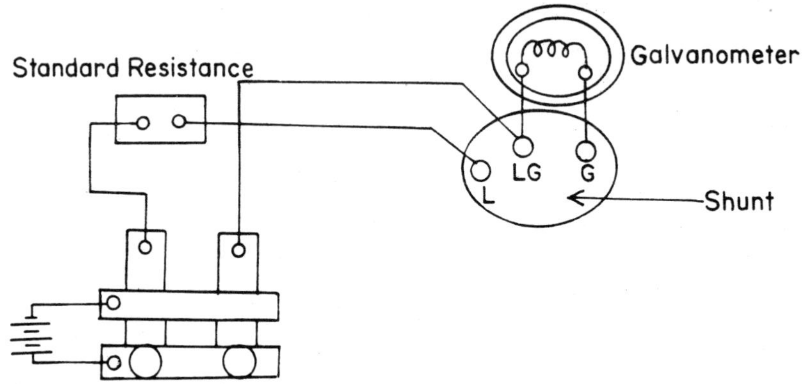

Fig. 2.17(b): Determination of Resistant Constant of a Galvanometer

Thus if D is the deflection found with a single cell through 10,000 ohm (= .01 mega ohm) with shunt S, then as the deflection is inversely proportional to resistance, we find that for one scale division and the shunt at 1, the resistance for 1 cell would be $D \times S \times .01$ mega ohm. The ratio of the whole testing battery to 1 cell is given by E/e. So the constant of the galvanometer for the whole battery is given by

$$K = \frac{E}{e} \times D \times s \times .01 \text{ mega ohm.}$$

Thus if a cell through 10,000 ohm and $S = 1000$ produces a mean deflection of 120, then the mean x shunt is 120,000. If for 1 cell with 1 μF and $S = 10$, producing a mean deflection of 60 gives a mean x shunt = 600 and if whole battery with 1 μF and $S = 1000$ producing a mean deflection of 50 gives a mean x shunt = 50,000.

$$\text{Then } K = \frac{5000}{600} \times 120000 \times .01 \quad \text{mega ohm}$$

$$= 100{,}000 \text{ mega ohm}$$

For electrification, insulation and capacity test, the connections are made as shown in Fig. 2.18.

To test for electrification, the cable is allowed to discharge for a few minutes, the galvo shunt is set to zero, then with a stop watch when seconds' hand reaches 60, the right hand battery key is pressed connecting the positive pole of the battery to cable, the negative being joined through the left hand key to earth. After about 10 sec, the shunt is removed to $\frac{1}{1000}$ or so, till a good deflection is obtained and the reciprocal of its number, i.e. the power shunt is noted. Then at the end of first minute after the key is pressed, the deflection is noted for about 10 min. at the end of each successive minute. Then the shunt is put back to zero and the battery key is released. The cable is completely discharged for about 15 min. by joining it direct to earth by a piece of wire. This earth is removed after 15 min. and the test is repeated by pressing the left hand key, connecting positive pole to earth, the negative being connected via the right hand key to the cable. The deflections are noted at the end of each successive minute for 10 min. If the cable is of multi-wire type, say, wires *A*, *B*, *C* and *D*, then the positive electrification tests are taken for *A* when *B*, *C* and *D* are earthed; for *B*, when *A*, *C* and *D* are earthed, and so on. Similarly, readings for negative electrification tests are taken for each wire for 10 min. at the end of each successive minute where the remaining three are earthed. Thus the eight tests take about 1¼ hr, otherwise it would have taken double the time.

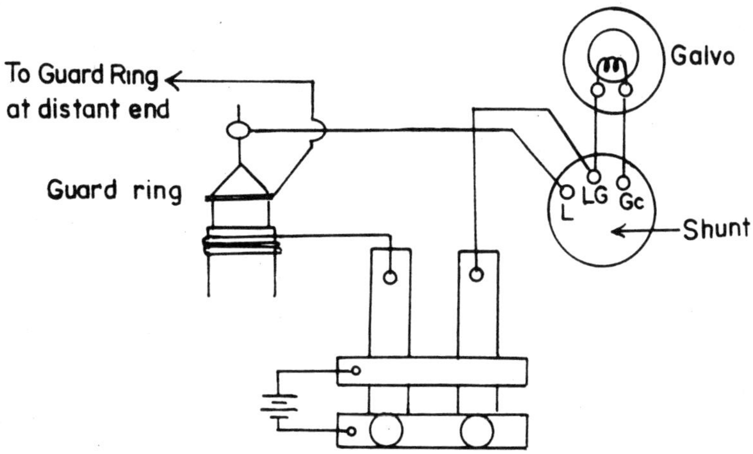

Fig. 2.18: Electrification, Insulation and Capacity Test

Then electrification curves are drawn by plotting time in seconds and deflection in degrees as shown below in Fig. 2.19.

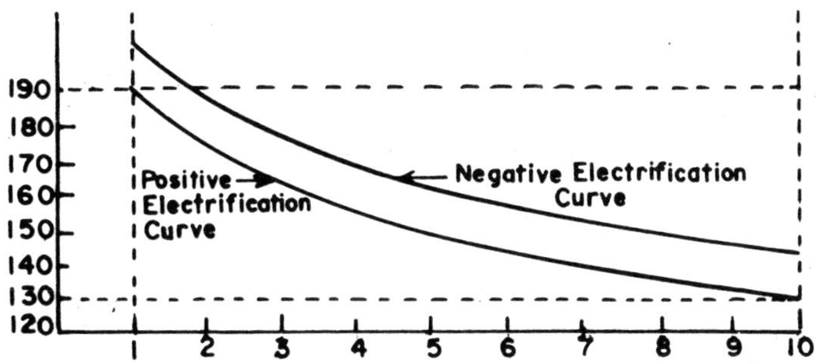

Fig. 2.19: Electrification Curve

If the curve shows any irregularity, it indicates that a fault is developing even if the insulation of the cable is high.

To test for insulation resistance, the same is calculated from the first positive and first negative readings obtained in the last test, as shown in the following example. Suppose in a set of tests, the following results were obtained:

Insulation of testing set and leading wire ($S = 1$) for two readings $+7$, -7 and mean 7 .....(1); insulation for testing set, leading wire and the cable ($S = 1$) for two readings -178 and $+182$ with a mean 180....(2), when $K = 100{,}000$ mega ohm. Subtracting (1) from (2), we get deflection due to cable only $d = 180 - 7 = 173$. Absolute insulation resistance of cable $= K/d = \frac{100{,}000}{173}$ or 578 mega ohm.

To test for capacity of a cable, the capacity constant of a galvanometer is first determined. The capacity constant of a galvanometer is that capacity which when charged through the galvo (shunt = 1) by a testing battery, produces a throw of one scale division. For this purpose, in place of 10,000 ohm standard resistance, as in Fig. 2.17(b), a standard condenser is inserted. Then with a battery of 5 to 10 dry cells the mean of three negative throws are taken, noting the value of the shunt given. If this deflection is taken as d, the deflection with the galvo with shunt 1 would be d x s, where S = multiplying power of the shunt. Hence if c is the capacity of the standard condenser used, then to produce 1 scale division, a capacity of $\frac{C}{d \times S}$ would be required. The throws of the capacity constant are taken immediately after the throws to find the resistant constant $\left(\frac{E}{e}\right)$ and before the electrification test.

After taking the negative electrification test, the insulation testing battery is replaced by the same cells as used in the capacity constant test. Allowing some time for the cable to discharge, the shunt is set to $\frac{1}{1000}$. Then the left hand key is depressed and the throw is noted. If it is very small, the shunt is reduced to $\frac{1}{30}$ or further. If the deflection is off the scale, the spot is adjusted for a false zero to minimise the observation error. After adjusting the shunt and the scale of the moving coil reflecting galvo, the left hand key is kept pressed for 15 sec and the permanent deflection d is noted. The left hand key is then released and the cable allowed to discharge. Then three throws are taken one after another using left hand key (negative to cable) allowing time for the cable to discharge between each throw. If the mean of these three throws is taken as D, then the capacity throw of the cable $= D - d$.

As an example, if the capacity throw of the leading wire ($S = 1$) gives a mean reading of 338, then mean x shunt = 338. If the capacity throw of the leading wire+cable ($S = 10$) gives a mean reading of 336, then mean x shunt = 336x10 = 3360, and if the permanent deflection for the same ($S = 10$) = 2, then mean x shunt = 2x10 = 20. Now if the capacity constant of the galvanometer is given by

$$K = \frac{1}{5800}, \text{ then the throw of the cable only}$$

$$= 3360 - 20 - 338 = 3002.$$

$$\therefore \quad \text{The capacity of the cable only} = \frac{3002}{5800} = 0.5175\ \mu\text{F}.$$

Determination of conduction resistance of a cable is done after the electrification and capacity tests are completed.

To test for conductor resistance, the insulation of the Wheatstone bridge is first ascertained by making connections as shown in Fig. 2.20 below.

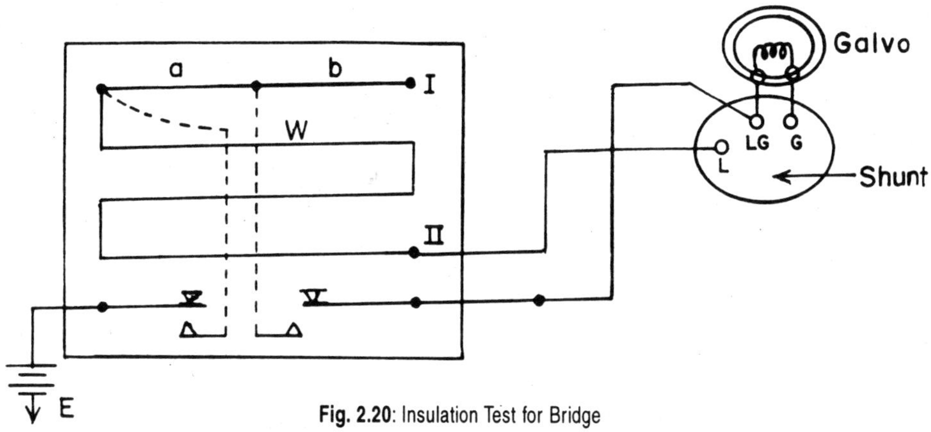

Fig. 2.20: Insulation Test for Bridge

By making *a*, *b* and *w* arms all of 1000 ohm, the keys are pressed, first for battery and then for galvo. If there is any deflection at any stud of the shunt, then it indicates a faulty insulation of the bridge and its cause should be traced and removed. If there is no deflection, it shows that the bridge is fully insulated and the conduction test for the cable may be taken by making connections as shown in Fig. 2.21 for one cable conductor only.

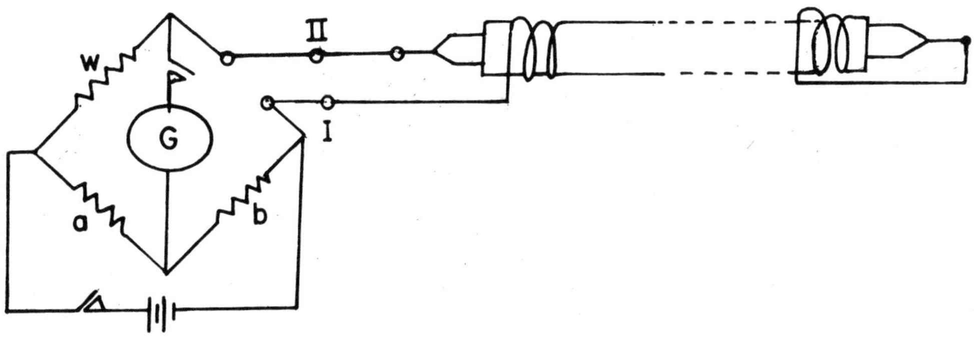

Fig. 2.21: Conduction Resistance Test (when no return wire is available)

When an insulated return wire is not available, the general wires are used as a return wire in place of ordinary earths. As the resistances to be measured are usually small, the ratio arms should be unequal, the ratio $\frac{b}{a}$ being very small. In balancing, the tester should first obtain balance approximately with $S = 1000$ and finally with $S = 1$ by altering the shunt. For conduction test, a battery of three dry cells is taken. Here at balance, we have, $x = \frac{b}{a}$ W, where the values of x includes the resistance of the return circuit.

When a second wire is available, the resistance of each is obtained by looping them at the distant end and applying earth at the loop. Then loop and localisation tests are taken by making connections as shown in Fig. 2.22(a) for loop test, and 2.22(b) for localisation test.

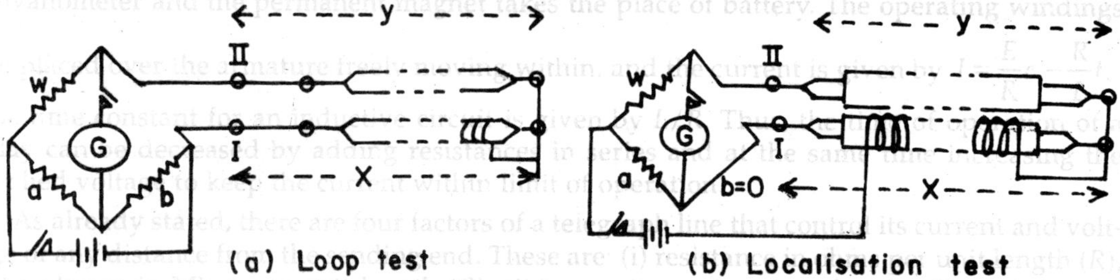

Fig. 2.22: Conduction Resistance when Two Cable Wires are Available

In the loop test, on balance, we get,

$$x + y = \frac{b}{a} W = \gamma \text{ (say)} \tag{1}$$

In the localisation test, on balance,

$$wx = ay \tag{2}$$

From (1) $x = \gamma - y$ and from (2)

$$R = \frac{a}{W} y \therefore \gamma - y = \frac{a}{W} y \text{ or, } \gamma W - yW = ay \therefore y = \frac{\gamma}{a+W} W$$

Therefore, from (1) and (2) we get $x = \gamma\left(\frac{a}{a+W}\right)$ and $y = \gamma\left(\frac{W}{a+W}\right)$

Note 1: To reduce the absolute measured insulation to per mile value, it is multiplied by the length of the cable in yards and divided by 1760. Thus, if the insulation of a cable 3000 yards long = 590 mega ohm, then its average insulation = $\frac{590 \times 3000}{1760}$ mega ohm per mile.

$$= 1005 \text{ mega ohm/ mile.}$$

Note 2: To reduce the absolute capacity of a cable to per mile value, multiply it by 1760 and divide by the length of cable in yards. Thus if the capacity of a cable 3000 yard long is .515 μF, the average capacity = $\frac{.515 \times 1760}{3000}$

or = .3021 μF

Note 3: To reduce the absolute conduction resistance of a cable to per mile value, multiply by 1760 and divide by the length of the cable in yards. Thus if the conduction resistance of a cable 3000 yd long is 8.118 ohm, then the average conduction

$$= \frac{8.118 \times 1760}{3000} = 4.754 \text{ ohm/mile.}$$

NB: A nautical mile = 2029 yds.

Temperature correction: In India, 60^0 F is taken as the normal standard temperature for all cables. So all the measured resistances should be corrected to this temperature. The temperature of the underwater and submarine cable is measured by immersing a thermometer tied to a piece of string into deep water and taking its reading on taking it out of water after 3 min.

Thus, if the temperature of a G.P. Cable is 79^0 F, its insulation resistance per mile = 1146 mega ohm and the conduction resistance per mile = 4.56 ohm, then its insulation resistance per mile at 60^0 F = 1146x3.24 or 3713 mega ohm. From the correcting table, it is found that the coefficient for G.P. at 79^0 F = 3.24 and the coefficient for copper at 79^0 F = .954.

Therefore conduction resistance per mile at 60^0 F

= 4.56x.954 = 4.34 ohm.

2.11 Coaxial Cable Testing

For the purpose of jointing and testing, a length of about 2000 yard of coaxial cable consisting of four drum lengths of 500 yard each is taken as a basic unit, called a sling, five of which generally cover a repeater section of 5.8 mile, i.e. 10,208 yard. The first two drums are jointed by a B_1 joint to produce a piece of coaxial cable of half sling length. The other half of the sling is similarly made by jointing the third and fourth drum by a B_2 joint. The two halves of the sling are jointed together by a C joint. So one set of tests are taken during laying and another after laying.

Preliminary tests during laying of coaxial cable: The following tests are taken during the laying of coaxial cable:

(a) High volt DC test
(b) Capacity unbalance test
(c) End-to-end tests of a sling length
(d) Pressure tests
(e) Pulse echo test
(f) Frequency-attenuation Test on coaxial cable core, and
(g) NEXT and FEXT tests.

(a) *High Volt DC Test*: It applies 3000 V DC between the central and the annular conductor for 2 min and is taken at every stage of cable laying. A block schematic of high voltage DC. Test set is shown in Fig. 2.23.

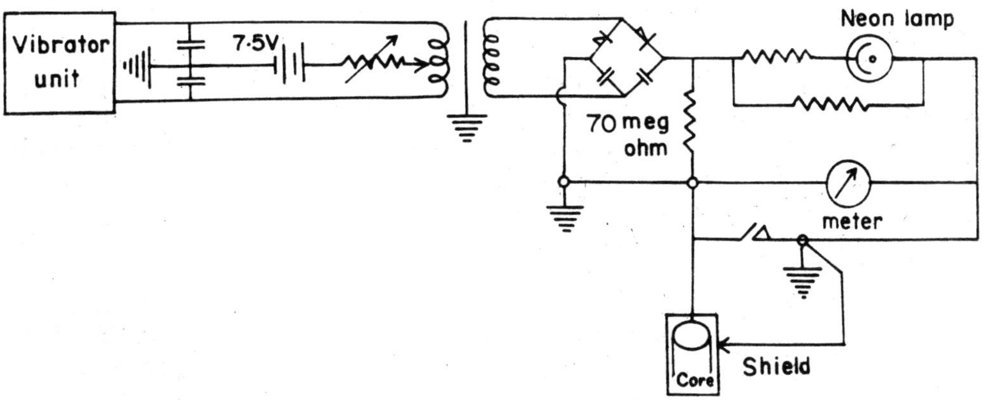

Fig. 2.23: Block Schematic of a High Volt DC Test Set

The set works with five 1.5 V dry cells in series and when the push button is operated, 7.5 V DC is applied to the vibrator circuit which supplies an AC voltage of 220/2200 V across the secondary winding of the vibrator transformer. This in turn is applied to a rectifier bridge of voltage doubler type and a smoothing circuit, delivering an output of 700 to 5000 V DC continuously variable. The leakage, if any, is indicated by the neon lamp flashing and a sudden downward fluctuation of the meter. The extent of the low insulation can be roughly assessed by noting the voltage at which the neon lamp starts flashing. This starts when the leakage current exceeds 150 μA. As the capacity of the cable is high, a long time is taken in charging and discharging of cable. Therefore, even when the set is switched off after the tests are over, a dangerously high voltage can exist on the cable. This can be avoided by including a relay which applies a short circuit across the cable, the moment the set is switched off.

The high voltage DC test is taken on each individual drum length of the cable, and after every jointing operation before plumbing.

(b) *Capacity Unbalance Test*: The various interpair and interwire unbalances of capacitances, connected with a quad for measurement of side-to-side capacity imbalance are as shown in Fig. 2.24.

By analyzing the above network side-to-side unbalance $[(w - x) - (z - y)]$ should be reduced to minimum and also side-to-earth capacitance unbalance $(a - b)$ and $(c - d)$ should be reduced to a minimum to get minimum crosstalk. The mean direction from mutual capacity should also be minimum. Thus the principle of balancing will be to reduce $[(w - x) - (z - y)]$, $(a - b)$ and $(c - d)$ and the deviation from mutual capacity minimum as far as possible, by making suitable pair and wire crossings.

At each B_1 and B_2 point, the side-to-side capacitance unbalance is measured for carrier quads-while side-to-side and side-to-earth unbalances are measured for V.F. quads. For this purpose, a capacity unbalance set with an auxiliary condenser box is used. The schematic set for measurement of side-to-earth (S/E) capacity unbalance is diagrammatically shown in Fig. 2.25(a).

The measurement is made by using a differential and variable condenser. Capacity is added or subtracted from side circuits so as to get minimum tone in one pair when the tone is sent on the other pair.

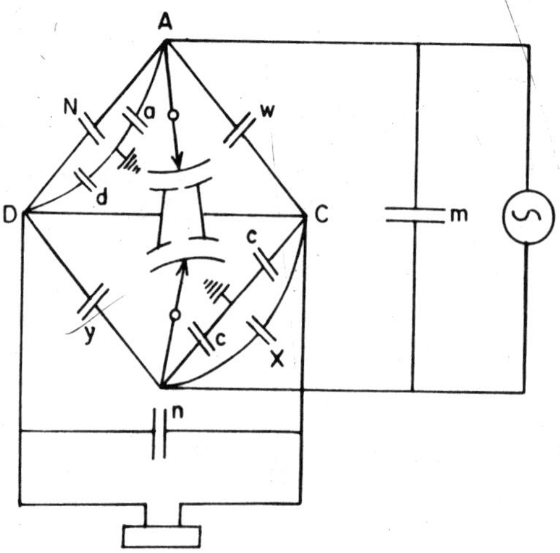

Fig. 2.24: Schematic of Set for Measurement of Side-to-Side (S/S) Capacity Unbalance with the Quad under Test

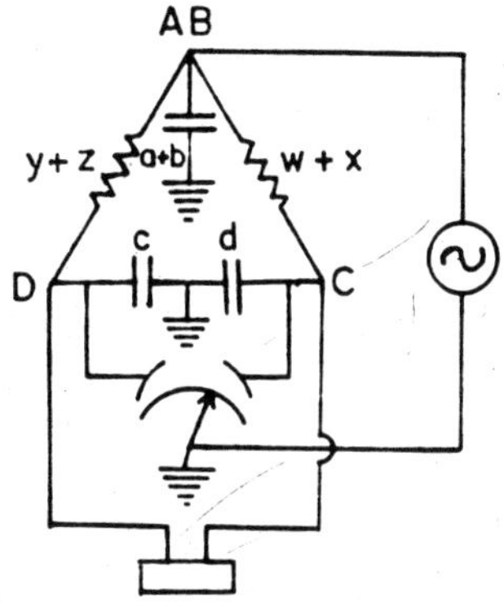

Fig. 2.25(a): Scheme for Measurement of Side-to-Earth (*S/E*) Capacity Unbalance with the Pair under Test

The side-to-earth capacity unbalance is not taken on C.F. quads as more rigid control is exercised during manufacturing them in the case of V.F. quads. In the case of V.F. quads, these unbalances are reduced to a minimum by building up an unbalance in B_1 (say) to be cancelled by an equal figure on the other half of the sling. But in case of carrier quads, no build-up in unbalance is permitted as they invariably result in high imbalance at carrier frequencies. At joint *C*, the resistance unbalance side-to-side, capacity unbalance and mutual capacity deviations are measured for carrier quads. For V.F. quads, resistance unbalance, side to side capacity unbalance and side-to-earth capacity unbalance are taken. Of these, the resistance unbalance is taken first as it contributes to differences in straight and reverse readings of cross-talk at carrier frequencies. The mutual capacity duration, i.e. capacity across one pair as compared with that across another contributes to differences in impedances. Side-to-side capacity unbalance of pairs contributes at carrier frequencies generally to conductance unbalance, which is to be kept low. Hence for preparing a jointing schedule, resistance unbalances are canceled first, and then both mutual capacity deviations and side-to-side deviations are considered jointly.

(c) *End-to-end tests of a sling length*: Tests on coaxial core are: (i) high volt DC test, (ii) insulation, (iii) DC resistance (copper), and (iv) pulse echo test from both ends. Tests on interslice quads — carrier quads (polythene) are: (i) side-to-side capacity unbalance, (ii) mutual electrostatic capacity (M.E.C), (iii) insulation, and (iv) DC resistance and identification (ring through). Tests on V.F. quads (paper) are: (i) side-to-side capacity unbalance, (ii) side to earth capacity unbalance, (iii) insulation, and (iv) DC resistance and identification (ring through).

The results obtained through these are as follows. In case of carrier quads, side-to-side capacity unbalance is found to be about 5 – 10 pF (pico farad), the maximum reading up to 30 – 40 pF. The resistance unbalance is about 0.16 ohm, the maximum reading is up to 0.3 ohm. The maximum percentage deviation from mean mutual electrostatic capacity (M.E.C.) is of the order of 2 per cent, while the limit is of 3 per cent in rare cases.

In case of V.F. quads, the average S/S and S/E capacity unbalance is of the order of 30 - 40 pF in both the cases and the maximum is up to 100 pF and 200 pF respectively. The resistance is about 0.04 ohm, the maximum being 0.1 ohm with a maximum limit of 0.25 ohm.

(d) *Pressure tests*: The cable is then tested for any leakage of inert gas or dry compressed air filled into the cable at a constant pressure by using a reducing valve. The gas is filled at a pressure of 24 lbs per sq inch and is allowed to stabilise for 24 hr. Then the pressure readings are taken at fixed time of the day for two days to avoid any change due to temperature. If any leakage is observed, necessary action is taken to eliminate the same before proceeding further.

(e) *Pulse echo test*: When the basic link has been pressure tested, a pulse echo tester is used to localise the spot, magnitude, and sign of each irregularity, either in jointing or in cable, by testing with frequencies up to 10 MHz/sec. The block schematic of a pulse echo test set is shown in Fig. 2.25(b) below.

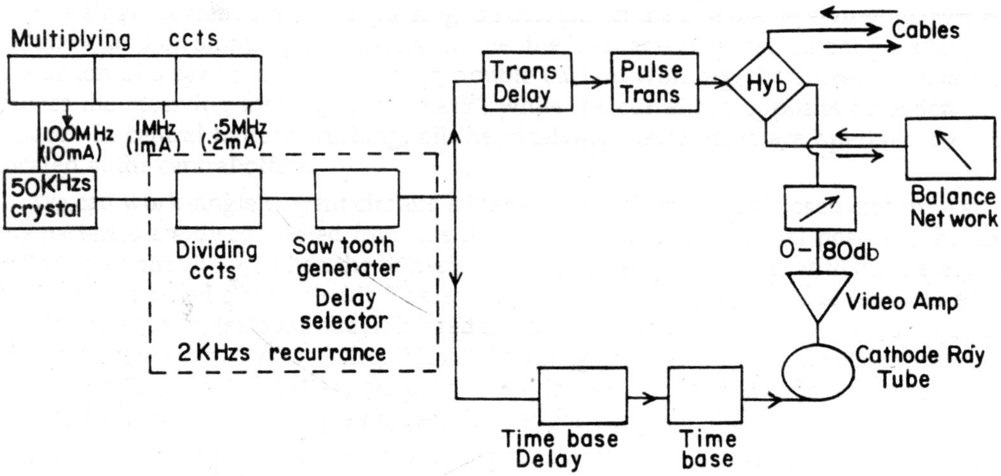

Fig. 2.25(b): Block Schematic of Pulse Echo Test Set

It transmits a high frequency pulse on the coaxial cable allowing the resultant echoes formed due to impedance irregularities along the length of the cable to be displayed on a cathode ray tube. With this method the end impedance of the cable with reference to a standard core at 2.5 MHz can also be measured.

More details with regard to localisation of the spot, magnitude, and sign of each irregularity are given below.

Position: The output from 50 KHz crystal oscillator is divided by a multivibrator first into five and the output is again divided by another multivibrator by five to give a frequency of 2 KHz, which is used to operate a linear single shot saw toothed generator. The output from this is applied to two delay selector circuits, of which one supplies the pulse for triggering the time base and the other for triggering the transmitting pulse. Three frequencies 100 KHz, 1 MHz and 5 MHz are provided by the calibrator calibrating the time bases of 10 μ sec, 1 μ sec and 0.2 μ sec, respectively.

These transmitter pulses are applied to the coaxial cable under test through a hybrid coil with a balancing network to match the cable. Due to irregularities in the cable, the pulse gets reflected and is received in the hybrid and thence fed to variable attenuator (0 - 80 db) and a video amplifier finally puts the output to the cathode ray tube. The reflected pulses control the y - axis of the vertical deflection of cathode ray tube while the x - axis on the horizontal motion is controlled by the time base. The coaxial cable is connected to the hybrid by a length of flexible coaxial lead and the balancing network is also connected to the hybrid by an electrically identical length of flexible coaxial cable to minimise the mismatch due to test leads.

The differences in micro seconds between the initial pulse from the reference point and the reflected pulse to the same point represent the distance to the irregularity from the sending end. A calibrated control for delaying the time lease provided in the set is read in micro seconds. It is also directly calibrated in yard/km and the distance to irregularity can be directly read.

Magnitude: A variable attenuator is used to measure the ratio of transmitted pulse to received pulse for any impedance irregularity. For this purpose, the attenuation is noted for a vertical deflection of a cathode ray tube for a transmitted pulse, say 70 db for 1 cm height. The attenuation can be varied till the received pulse gives the same deflection as the transmitted pulse. Then the magnitude of the irregularity in ohms, giving a reflection Kdb is given by

$K = 20 \log \frac{Z_a + Z_b}{Z_a + Z_b}$, where Z_a and Z_b are impedances before and after the mismatch. For 375 type of coaxial cable, a mismatch of .05, .15, .25 and 1.5 ohm correspond to reflections of 70, 60, 55 and 40 db, respectively.

Sign of irregularity: If the reflection is in the same sense as the original pulse, the change results in an increase of impedance and if the reflection is in opposite direction, the change is to a lower impedance.

(f) *Frequency-attenuation test on coaxial cable core*: At the distant repeater station, the two cores are looped by a flexible coaxial cord. At the testing end, the signal generator is connected to core 1 and the heterodyne detector amplifier to core 2 as shown in Fig. 2.26.

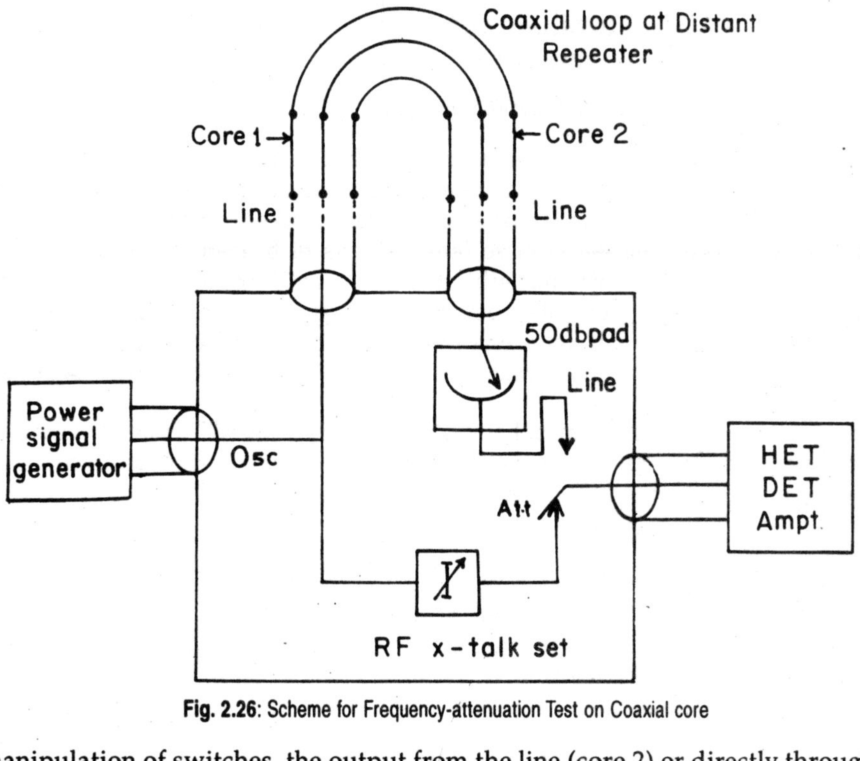

Fig. 2.26: Scheme for Frequency-attenuation Test on Coaxial core

By manipulation of switches, the output from the line (core 2) or directly through the pads may be connected to heterodyne detector amplifier. Thus the two tones can be heard, one the oscillating tone after getting attenuated through the line, and the other, the direct oscillating tone after getting attenuated through the pads adjusted to the known value. When the two tones are equal as compared audibly or on a meter, the test is repeated for frequencies from 60 KHz to 10,000 KHz, for which the impedance of attenuation pads should be same as that of line and is within a margin of ± 10 per cent variation. The attenuator and the X-talk set is so designed that the line and attenuator pads are connected in parallel to the oscillator and when the Het. Amplifier is connected to the return line, the attenuation side remains open-circuited. A 50 db pad is permanently connected to avoid any change in oscillator output. The attenuation in db per mile of the coaxial core is given by the formula = $3.75 \sqrt{f} + .003f)$, where f = frequency in MHz. The loop attenuation at 10 MHz for a repeater section comes to about 130 - 140 db. The measured attenuation values are corrected to a temperature of 10^0 C. The resistance of the coaxial core is measured and the variation in temperature is determined from its value. The D.C. resistance coefficient of temperature is .41 per cent/°C and the attenuation correction coefficient is .0021/°C.

(g) *NEXT and FEXT tests*: These are used for near-end cross talk and Far-end cross talk. For the near-end cross talk, the instrument used and the circuit arrangement of this test is the same as that of frequency-attenuator test, except the difference that at the distant station both the cores are terminated with 75 ohm. The method used is also similar to the frequency-alternator test. The pads are introduced successively with the tone heard in both positions, until the tone received in core 2 and through the pads is equal. The arrangement is shown in Fig. 2.27.

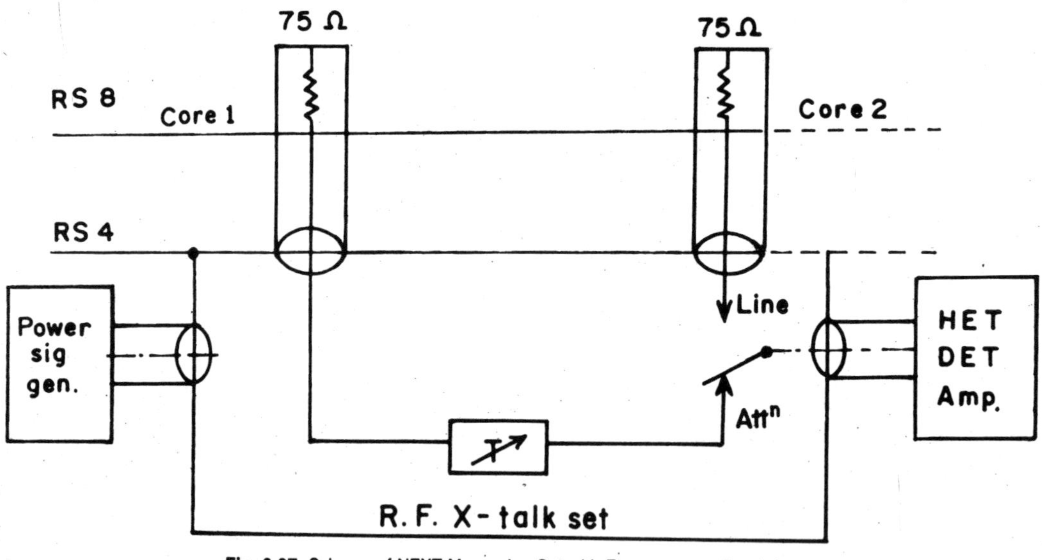

Fig. 2.27: Scheme of NEXT Measuring Set with Frequency on Coaxial cores

The pads introduced give the value of attenuation in db. After about 600 to 700 KHz, the limit is recorded, i.e. 160 db. Then the heterodyne detector amplifier noise level is compared to the received level of X-talk.

For the Far-end cross-talk, the instruments used are the same as used for NEXT, except that the signal generator in this test is kept at the far end, from where the tone is sent on one core, while the other core is terminated with 75, as shown in Fig. 2.28 below.

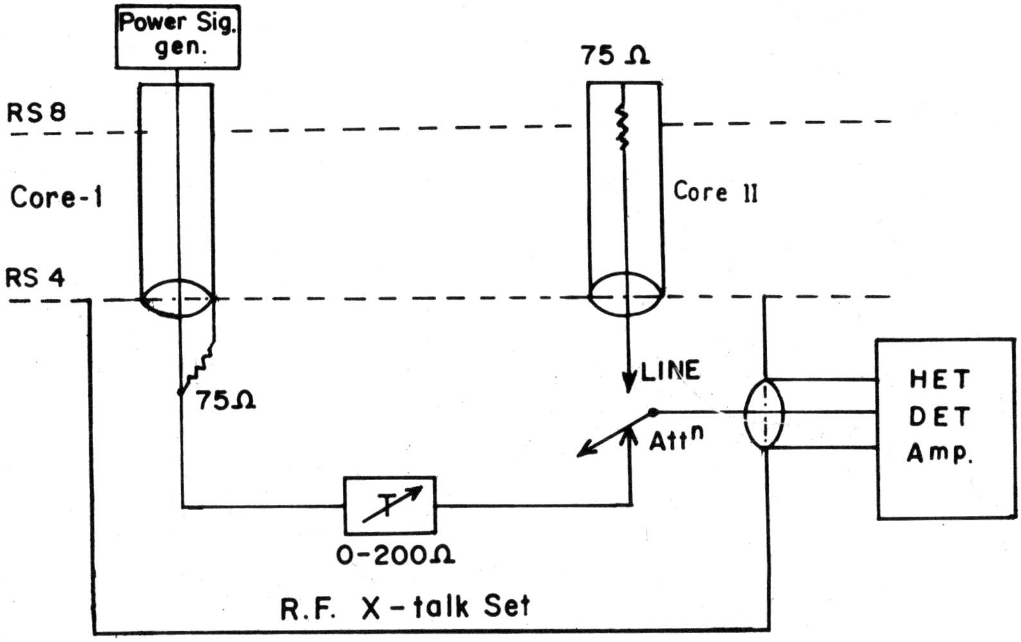

Fig. 2.28: Scheme of FEXT with Frequency on Coaxial Cores

The attenuation set is kept in FEXT position and the core 1 in the oscillator position, while core 2 is connected in line position. Thus we can get the direct tone from the oscillator after getting attenuated through the adjustable attenuation pad in attenuation position of the switch and the X-talk induced in core 2 in the 'line' position of the switch. The two tones can be made equal by suitable adjustment of pads in the X-talk set. The FEXT in db will be equal to the line attenuation at the frequency plus attenuation pads introduced. The test is repeated for various frequencies, till it reads more than 160 db for successive readings. After the construction of a complete repeater section of coaxial cable, one of the above tests is taken for the complete section again and the reading noted.

3

Direct Current Telegraphy

For a D.C. circuit, we know that the resistance of a homogeneous material is its tendency to oppose flow of current through it. It is found to vary directly with its length and inversely with its cross-section and, it depends upon its temperature. Thus for a particular temperature, $R \alpha \frac{l}{a}$, where R = resistance, l = length, and a = cross sectional area, i.e. $R = \rho \frac{l}{a}$, where ρ is a constant called the specific resistance or receptivity, i.e. resistance between the opposite faces of a unit cube of the material. We know that within the range of temperature varying from 0^0 C to 100^0 C, $R_t = R_0 (1 + \alpha t)$ for most of the electric conductors, where R_t = resistance at temperature t°C, R_0 = resistance at 0^0 C, t = rise of temperature above 0^0C, and α is constant, called the temperature coefficient of the material. The specific resistances in micro ohms per cube and temperature coefficient of various electric conductors used in telecommunication circuits is given below.

Name of material	Sp. resistance	Temp. coeff.	Place where used as a conductor
Silver	1.55	0.004	Little used owing to cost.
Copper	1.62	0.0043	Mostly used as wires, cables, switches, etc.
Aluminium	2.65	0.0043	For lightness and cheapness
Iron	10.0	0.006	Resistance rises with current rise
Platinum	10.9	0.0035	For contacts
Lead	20.0	0.004	For accumulator containers, grids and connecting legs
Mercury	94.0	0.00075	Basis of international standard
Manganin	42.0	0.0000	Used as resistance coils
Carbon	4000-7000	- 0.005	Used as resistance material.

3.1 Component Parts

(a) *Resistors*: These are moulded carbon, painted or encased in ceramic. For small size, the maximum voltage rating is 150 V and for medium size, the maximum voltage rating is 250 V. The other types are carbon film, metal oxide and wire-wound. A colour code is given on the body of resistor with rings from one end to indicate the nominal value of resistance as given in the table below. The first two rings indicate digits, the third indicates multipliers and the fourth ring gold or silver indicates tolerance of ± 5% or 10%, respectively. If no fourth ring is used, we take tolerance as ± 20%. If rings are orange/white yellow/gold, the code indicates the value of the resistor as 390 K ± 5%, brown/black/red bands indicate 1K ± 20%.

Colour	Digit	Multiplier
Black	0	x 1
Brown	1	x 10
Red	2	x 100
Orange	3	x 1000
Yellow	4	x 10,000
Green	5	x 100,000
Violet	7	
Grey	8	
White	9	
Gold		

From Ohm's law, we know that the resistance of a conductor in

$$\text{ohms} = \frac{\text{potential difference in volts}}{\text{rate of flow of current in amperes}}$$

The equivalent resistance of a series of resistances $R = R_1 + R_2 + R_3$ etc., where R is the equivalent resistance and R_1, R_2, R_3 are the resistances in series. Similarly, if the resistances are in parallel, then their equivalent resistance is given by $\frac{1}{R} = \frac{1}{R_1} + \frac{1}{R_2} + \frac{1}{R_3} \cdots$.

Now considering the current carrying capacity of a circular wire of length l, diameter d, receptivity P, and emissivity K, i.e. power in watts dissipated per unit area per unit temperature difference above the surroundings, we find that the rate of heat production = rate of heat dissipation, i.e. $I^2R = K(\pi\ d.l)\theta$, where I = current in amperes θ = steady temperature,

$$\text{or} \qquad I^2\left(\frac{lP}{\pi d_2/4}\right) = k(\pi dl)\theta$$

$$\text{or} \qquad I^2\left(\frac{\pi^2 k\theta}{4l\rho}\right)d^3 = \left(\frac{\pi^2 k\theta}{4\rho}\right)d^3$$

or $I = ad^{3/2}$ where a is a constant $\doteq \left(\dfrac{\pi^2 k\theta}{4\rho}\right)$ at a particular steady temperature.

We also know that an electrostatic force Q between two charges of electricity q_1 and q_2 in absolute units at a distance d cm apart in a medium of dielectric constant k is given by $Q = \dfrac{q_1 q_2}{kd^2}$. Thus if E is the intensity, i.e. force exerted on a unit charge at a point d cm from a charge of q units, $\in = \dfrac{q \times 1}{kd^2} = \dfrac{q}{kd^2}$.

For a spherical surface of radius r having a charge of Q units at its centre, the tubes of flux per unit area, i.e. the flux-density $D = \dfrac{Q}{4\pi r^2}$

But electric force at a distance r cm from a charge Q is given by $\in = \dfrac{Q}{kr^2}$

Therefore from the above two equations, we get, for any point in an electrostatic field, $\in = \dfrac{4\,\pi\,D}{k}$

(b) *Capacitors*: We also know that for an insulated conductor, the potential acquired is directly proportional to the quantity of electricity supplied proportional to the quantity of electricity supplied, i.e. $V \propto Q$ or $Q = VC$, where C is a constant called the capacity of the condenser. Thus $C = Q/V$, where Q is in coulombs, V is in volts and C is in farads. For a plate condenser made of two parallel plates of area A cm each and placed d cm apart, separated by a dielectric of specific inductive capacity K, we know, $V = Ed$, when A is large in respect of D. But $\in = \dfrac{4\pi D}{k}$ and $D = Q/A$, since each unit arises from a unit charge.

Therefore $V = \dfrac{4\pi}{K} \cdot \dfrac{Q}{A} \cdot d$

But $C = Q/V = Q \Big/ \dfrac{4\pi Q}{KA} \cdot d = \dfrac{KA}{4\pi d}$ absolute *e.s.u.*

Again in case of concentric condensers for a coaxial cable of length l cm of outer hollow conductor of radius r_2 and inner solid conductor of radius r_1 cm with a dielectric constant K for the medium, we know that the capacity $C = \dfrac{Kl}{2\log\dfrac{r_1}{r_2}}$ absolute *e.s.u.*

$$= \frac{0.039K}{\log 10\dfrac{r_1}{r_2}} \mu K \quad \text{per mile.}$$

The equivalent capacity of a number of condensers of capacities C_1, C_2, etc. in series is given by $= \frac{1}{C_1} + \frac{1}{C_2} + \frac{1}{C_3}$ etc., and when joined in parallel $C = C_1 + C_2 +$ etc.

(c) *Electromagnetic induction and operation of relays*: The magnetic lines of force in a magnetic field of a straight wire can be taken as concentric lines of force with wire as centre and the direction of the magnetic force can be found by a right handed cork-screw turned to move along the conductor, when the directions of rotation gives the direction of magnetic force round the conductor. If the straight wire is made to form a circular loop on a solenoid or a torroidal coil, the lines of force pass in one direction giving the relative directions of current and magnetic force as shown below in Fig. 3.1.

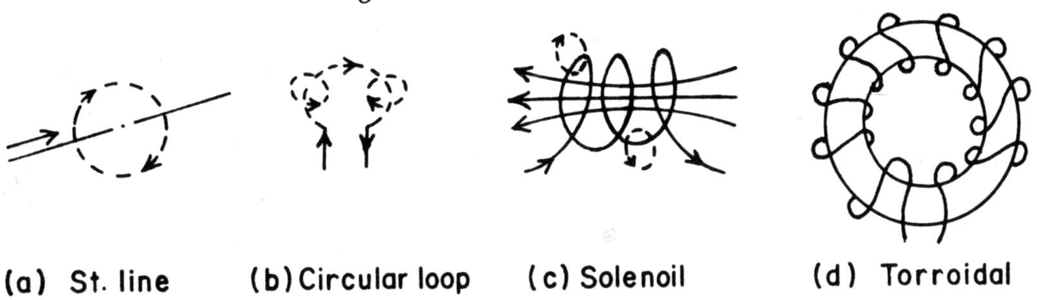

Fig. 3.1: Magnetic Field due to:
(a) Straight line, (b) Circular loop (c) Solenoid (d) Torroidal.

In case of straight conductors, we know that the field strength $H = \frac{2I}{r}$ units, where I is the current in absolute units and r is the shortest distance of the conductor from the point of consideration.

Now if I is in amperes, then $H = \frac{2I}{10r}$ c.g.s. units. In case of solenoids, the field strength $H = \frac{4\pi NI}{l}$ c.g.s. units, where N = number of turns, I = current in absolute units and l = length of coil in cm. If I is in amperes, $H = \frac{4\pi NI}{10l}$ c.g.s. units.

In case of magnetisation of iron, we know that iron greatly intensifies the density of flux lines in magnetic circuits. Thus, with iron as a magnetic material for a field strength of H c.g.s. units, the flux density will be increased to B flux lines per sq cm and the ratio of B to H is called the permeability (μ) and is given by $\mu = B/H$.

Now, since $B = H$, the flux density within the solenoid is $4 \pi NI \mu / 10 l$ and the total flux ϕ across any cross-section is given by $\phi = \frac{4\pi NI\mu a}{10l} = \frac{\frac{4\pi NI}{10}}{\frac{l}{\mu a}}$

Comparing this with an electric circuit equation $I = E/R$, we find that the magnetic flux ϕ is analogous to current I, $\frac{4\pi NI}{10}$ being a measure of the work done in moving unit pole once round the magnetic circuit. It is analogous to emf E, which is the work done in moving a unit charge round the electric circuit. This is why $4\pi\, NI/10$ is called the magnetomotive force (m.m.f), NI being the ampere turns, while $l/\mu a$ is analogous to electrical resistance and is called magnetic resistance or reluctance (S). Thus from an analogous comparison with $R = \rho l/a$, we can consider $\frac{1}{\mu}$ as the specific resistance offered by the material to the passage of a magnetic flux. μ being the reciprocal of specific reluctance may be considered as the conductivity of the material for magnetic flux.

In contrast to the above analogy, there are three main points of difference also. There is no flow of current in a magnetic circuit, as in an electric circuit. In a magnetic circuit, energy is expended only when a flux is created, while in an electric circuit, energy is expended continuously when the current flows. In an electric conductor the specific resistance is independent of current strength, while in a magnetic material, the permeability varies greatly in accordance with the flux density. Further we know from Faraday's laws of electromagnetic induction that the induced emf = rate of change of linkage, i.e. $E = -\frac{d(n\phi)}{dt}$, where e = induced emf and $d\,(n\phi)$ is a very small change of linkage in a small time dt seconds and the negative sign shows that the emf produced opposes the emf, causing the current to flow. As number n is a constant, denoting number of turns of linkings, $d(n\phi) = n(d\phi)$. Thus $e = -\frac{n(d\phi)}{dt}$.

(d) *Inductors*: When permeability remains constant, the flux density is directly proportional to the magnetising force, i.e. $\frac{d\phi}{dt} \alpha \frac{di}{dt}$ where di is a small change of current in a small time dt seconds, producing a change of flux $d\phi$. Now since $e \,\alpha -\frac{d\phi}{dt}$, $e \,\alpha -\frac{di}{dt} = -L\frac{di}{dt}$, where L is a constant called inductance. If e is in volts, i is in amperes, L will have a value 10^{-9} of its absolute value in the equation $e = -L\frac{di}{dt}$. L in practical units is now called henrys, when a current changing at the rate of one ampere per second induces an emf of one volt.

In case of a solenoid, we know

m.m.f. $F = \frac{4\pi NI}{10}$ and

reluctance $S = \frac{l}{\mu a}$

Then $$\phi = \frac{mmf}{reluctance} = \frac{F}{S} = \frac{4\pi NI}{10} \Big/ \frac{l}{\mu a} = \frac{4\pi NI\mu a}{10 l}$$

Now if a coil has N turns and if ϕ lines of flux is produced by a current I amperes in time T seconds in the coil with inductance L henries, then the change of linkage produced by a change of current from 0 to I is $N\phi$ and the induced emf in volts is $LI/I = \dfrac{-N\phi}{I \times 10^8}$ or $L = \dfrac{N\phi}{I \times 10^8}$.

$\dfrac{N\phi}{I}$ may be called the linkage per ampere and thus inductance $= \dfrac{\text{linkage per ampere}}{10^8}$

Thus in a solenoid, the linkage produced

$$N\phi = \frac{4\pi N^2 I\mu a}{10l} \text{ and linkage per ampere}$$

$$= \frac{4\pi N^2 \mu a}{10l} \therefore \text{Inductance} = \frac{4\pi N^2 \mu a}{10^9 l} \text{ henrys.}$$

Now a telephone or telegraph relay acts like an electromagnet operated by a tractive force $= B^2A/8\pi$ dynes, where

B = flux density in the air gap (gauss),

A = area of magnet face (sq cm).

When the total reluctance of the magnetic circuit is S, $B = Q/A = 4\pi NI/10SA$ and the tractive force = $(1.257\ NI/SA)^2 \times A/8\pi$ dyne

$= 0.06284\ N^2I^2/S^2A$ dynes

$= 6.4\ N^2 I^2/10^5S^2A$ gms.

(e) *Tractive force of a relay*: Thus the tractive force of a relay varies directly as the square of the number of turns and current and inversely as the reluctance, and the area of the pole face. But due to loss of a part of flux at the air gap, a leakage factor$= \dfrac{\text{total flux generated}}{\text{useful flux}}$ is introduced to make good the leakage effect of fringing or spreading of flux between pole face and armature.

Thus, for a relay coil the flux change when the current changes from 0 to 1 ampere = $4\pi N/10S$. If change is in 1 sec, the induced emf in one turn of the coil = $4\pi N/10S$ electromotive units. Thus for N turns, the total emf induced = $4\pi N^2/10^9S$ volts

($\therefore$ 10^8 e.m.u. = 1 volt).

$\therefore$ Inductance L of a relay coil = $4\pi N^2/10^9S$. henries.

Now due to presence of an inductance in a relay coil its current does not rise in accordance with Ohm's law, as the increasing flux produced by the rate of change of current rise cuts the coil turns and produces a back e.m.f. in the opposite direction of the voltage applied. On account of this, the current rises until it reaches a steady value when the relay operates, as shown in the Fig. 3.2.

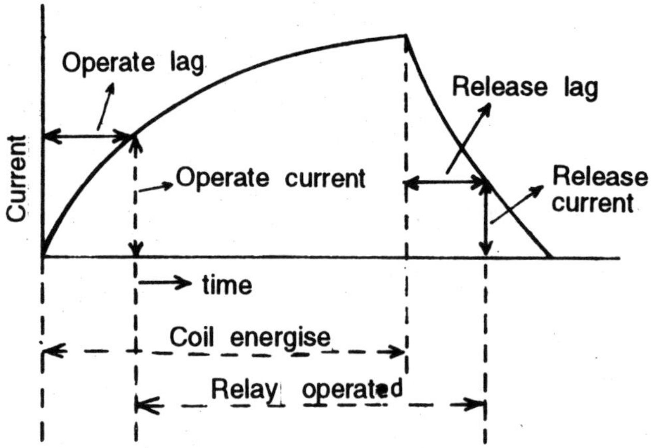

Fig. 3.2: Rise and Fall of Current in an Induction Coil (Relay coil)

The time delay between the switching of a circuit and the relay operation is called the operate lag. If relay circuit is broken, the decrease of current induces an e.m.f. in the same direction of the applied voltage, prolonging the operating current and causing the release lag as the time delay between the switching off of the circuit and the actual release of the relay.

(f) *Telegraph Relays*: The telegraph relay consists of soft iron core, over which is wound an insulated coil of copper wire turns through which the current for operating the relay passes, making the coil an electromagnet. The magnetic flux in the electromagnet attracts the armature irrespective of the direction of the flow of current in the relay coil in the case of non-polarised telegraph relays. But in the case of polarised telegraph relays, the operation of the armature depends not only on the magnitude of the current flowing but also on the direction of flow of current in the coil.

In non-polarised telegraph relays, the relay has two differentially wound coils with half of it on two soft iron cores of electromagnet, as shown in Fig. 3.3.

The electric and magnetic characteristics of each coil being equal, the windings can be used either in opposition for differential working or in series or parallel aiding. There are two soft iron armatures *A* mounted on a pivoted brass spindle *B*. Each armature is split in halves at the spindle by a non-magnetic separator to prevent the possible formation of a closed magnetic circuit with electromagnets. There is a light contact tongue *T* at the upper end of the spindle. This tongue plays between two adjustable contact screws *M* and *S* and it is electrically connected to the relay base. It is normally held over the contact screw *S* by the spring *C* adjusted by a screw *D*. When a current passes through the coil, the armature plays between the poles of the electromagnet as a magnetic circuit is formed round the iron cores, across the four air-gaps and the armature in series. Its principle of operation is the same as that of a telephone relay explained earlier.

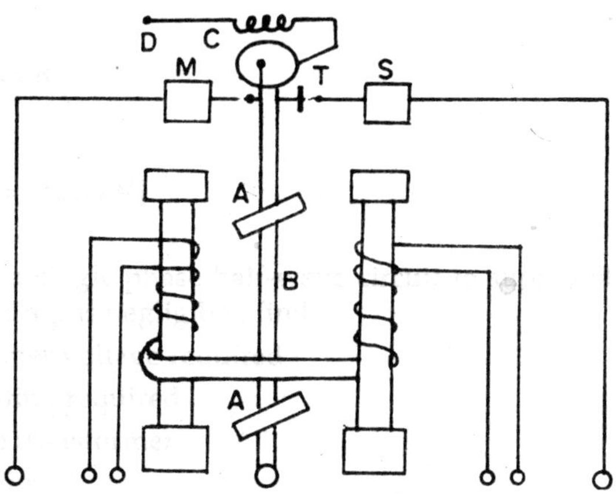

Fig. 3.3 Non-polarised Telegraph Relay

In polarised telegraph relay, the operation of the armature depends on the magnitude and sense of the current in the operating windings. *A* permanent magnet supplies the fluxes from the polarising magnet, as shown in bold type lines in the figure. When a current passes in the operating winding, the magnitude and nature of the flux distribution is controlled by the magnitude and sense of the operating current, making the relay equally sensitive to operate in either direction. The four basic types of polarised relays are shown in Fig. 3.4.

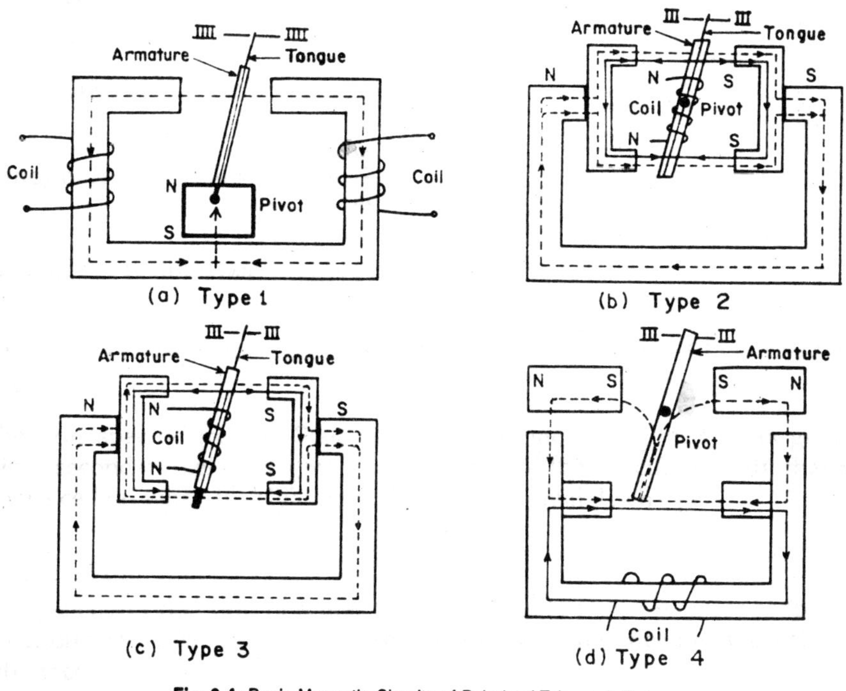

Fig. 3.4: Basic Magnetic Circuits of Polarised Telegraph Relays

In Type 1 shown in Fig. (a) the whole permanent magnet flux is passed through the armature and then it is divided into two paths through the pole pieces and the yoke. The windings are placed on the yokes. In types 2 and 3 as shown in Figs. (b) and (c), the magnetic reluctances are analogous to the resistances in a Wheatstone bridge, where the armature takes the place of galvanometer and the permanent magnet takes the place of battery. The operating windings are placed over the armature freely moving within, and the current is given by $I = \frac{E}{R}e - \frac{R}{L}t$. The time constant for an inductive circuit is given by L/R. Thus, the time of operation of a relay can be decreased by adding resistances in series and at the same time increasing the applied voltage to keep the current within limit of operation.

As already stated, there are four factors of a telegraph line that control its current and voltage of any distance from the sending end. These are: (i) resistance in ohms per unit length (R), (ii) leakance in Mhos per unit length (G), (iii) capacitance in micro farads (C), and (iv) the inductance in henries (L).

On open wire lines with high insulation resistance, capacitance does not play an important role in controlling the speed of signalling, but on lines with considerable length of underground cable or with low insulation, the speed of signalling is affected, as longer times are taken by the line to charge and discharge.

At ordinary speed, the inductive effect of a line does not affect the speed of signalling, but at high speed it is helpful, as it neutralises the capacitance of the line to a certain extent and improves the signalling. The ratio of the absolute insulation of a line to its conductor resistance is an important factor in signalling speed in telegraphy. Thus in a single current working, the current in one direction brings the receiver tongue to mark, and the stoppage of the current brings the tongue of the receiver to space by the mechanical force of the spring attached to the tongue. The marking force depends on the incoming current, which depends on the ratio of the insulation resistance of the line to its circuit resistance. With double current working, the varying insulation of the line varies the strength of both, marking and spacing signals. Thus with double current working, the ratio of the insulation resistance to the conductor resistance of the circuit may be very much less than for single current working. In case of single current working for simplex circuits, the absolute insulation of a line must be at least three times the circuit resistance, while double current simplex circuits can be worked on a line whose absolute insulation resistance is two-thirds of its actual circuit resistance. Also, double current duplex circuits can be worked on a line whose absolute resistance is greater than three-fourths of its circuit resistance.

As the effect of inductive interferences from adjacent lines on long lines create disturbances such as mutilation of letters or figures, there shall always be current on the line, i.e. the circuits should always work as closed circuits, either on single current simplex or on double current duplex. Thus it is found that double current working is more advantageous than single current working on lines with high capacitance, and on lines with low insulation, as compared with their circuit resistance and on lines subject to inductive interferences.

3.2 DC Telegraph Instruments

(a) *Single current key*: It is used in transmitting signals in simplex circuits. The three terminals are mounted on the base board as a switch. With the switch on the 'send position' the centre terminal is joined to the terminal towards the back end of the base, keeping the terminal nearest to the front end insulated and with the switch on the 'receive position' the centre terminal is joined to the terminal nearest to the front end of the base, keeping the terminal towards the back and insulated, as shown in Fig. 3.5.

Symbolic representation

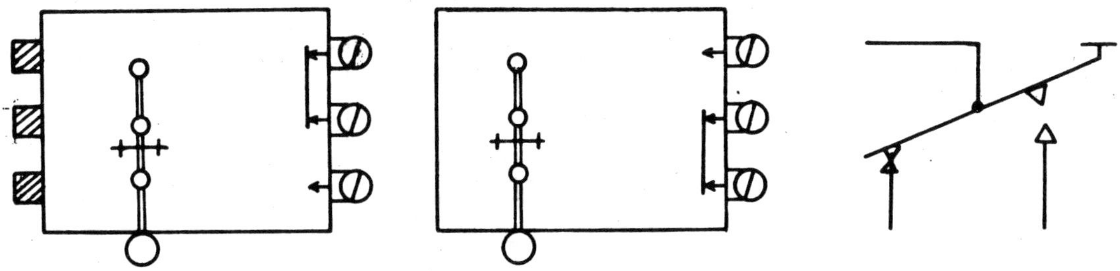

(a) Switch to send position (b) Switch to receive position (c) Single current key

Fig. 3.5: Single Current Key in Simplex Circuits

(b) *Duburn Sounders*: A Duburn sounder is a non-polarised receiving instrument, in which the two coils of the electromagnets are wound with a single insulated wire. The lower ends of the two cores are joined by a soft iron yoke and the upper ends project over the brass plate and the face of armature on opposite sides.

When there is no current in the coils, the armature remains held against the stop contact and when a current passes through the coil, the armature T is attracted to the stop contact 2. One coil is connected to terminals L and J and the other to terminals J and E, so that both the coils are in series between the terminal L and E, as shown in Fig. 3.6.

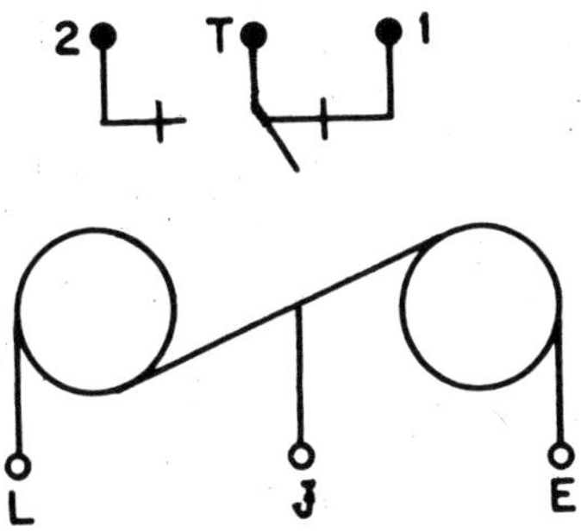

Fig. 3.6: Duburn Sounder

In case of BPO sounders, the electromagnet consists of two soft iron cores whose lower ends are joined to the yoke as shown in symbolic representation, Fig. 3.7 below. There is a bobin in each of the cores of the coils and an armature is fixed to the lever of the instrument, so arranged that it comes above the upper ends of the cores.

(d) *Polarised Sounder*: In a polarised sounder, there are two equal windings as shown in Fig. 3.8 below. The lower ends of the cores are fixed to the brass base and the U-shaped permanent magnet is fixed to the underside of the brass base in such a manner that the poles come close to the cores but do not touch them. Thus there is always a magnetic pull on the armature due to the permanent magnet, which is counter-balanced by the tension of the spring. One winding is connected to the terminals *U* and *D* while the other is joined to terminals *U* and *D*. The two terminals *U* and *D* are joined together so that the two windings are in series.

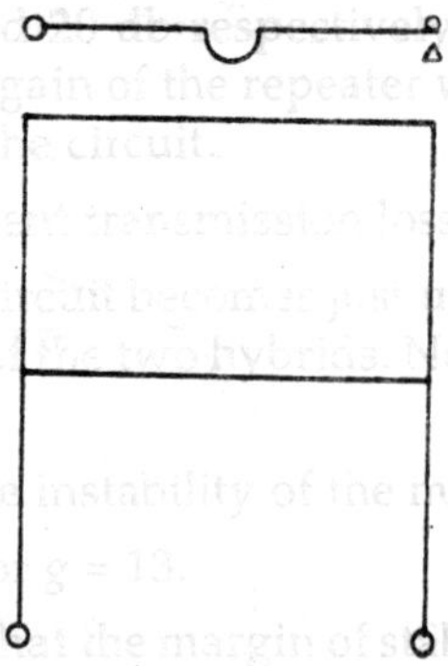

Fig. 3.7: B.P.O. Sounder

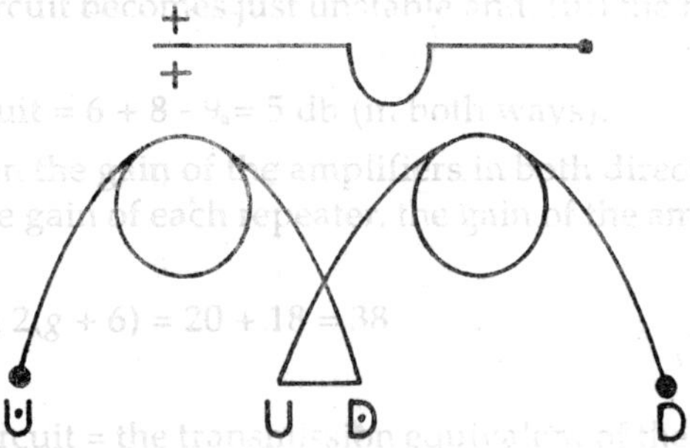

Fig. 3.8: Polarised Sounder

Like the symbolic representation of sounders, the non-polarised and polarised relays can also be symbolically represented. These are shown below in Fig. 3.9

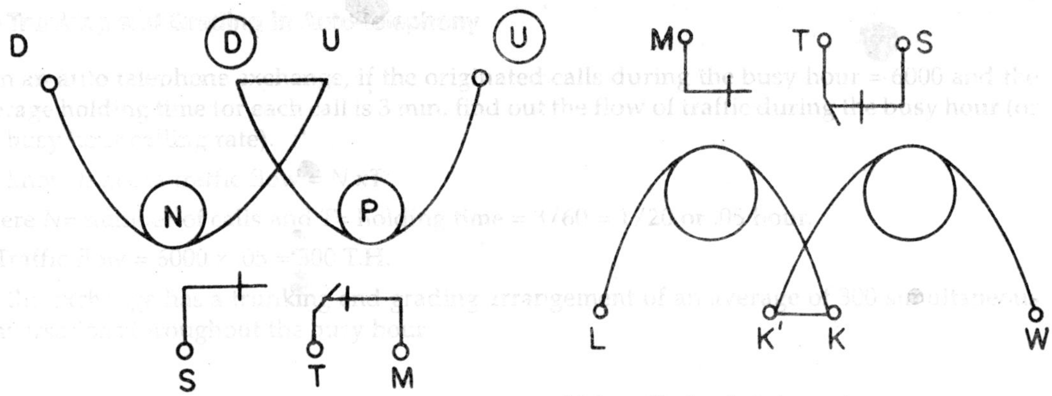

(a) Non-polarised relay (b) Polarised relay

Fig. 3.9: Symbolic Representation of Polarised and Non-polarised Relays

(e) *Switches*: These are required for short-circuiting and insulating the lines. They consist of two pieces of brass, each with a terminal screw outwardly fitted on a piece of wooden base, so that a brass plug when inserted between them short-circuits the two pieces and when the same is taken away, the two pieces become insulated, as shown below with different types of switches.

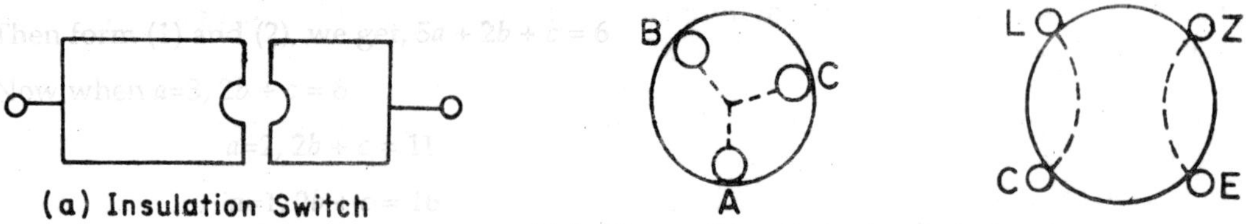

(a) Insulation Switch (Two Terminal)

(b) Three-Way Switch (Three Terminal)

(c) Current Reversion Switch (Four Terminal)

Fig. 3.10: Different types of Switches

(f) *Double-current Key*: Here the key works like a battery-reverse and consists of a double lever playing between the contacts, so connected that when the key is at rest the battery is connected to the terminals in one direction, and when the key is pressed, the battery is connected in the reverse direction. Each lever is provided with a steel-pivot. The two levers are mechanically fixed to each other but are electrically insulated from each other, one being longer than the other, as shown below in Fig. 3.10(d) & (e).

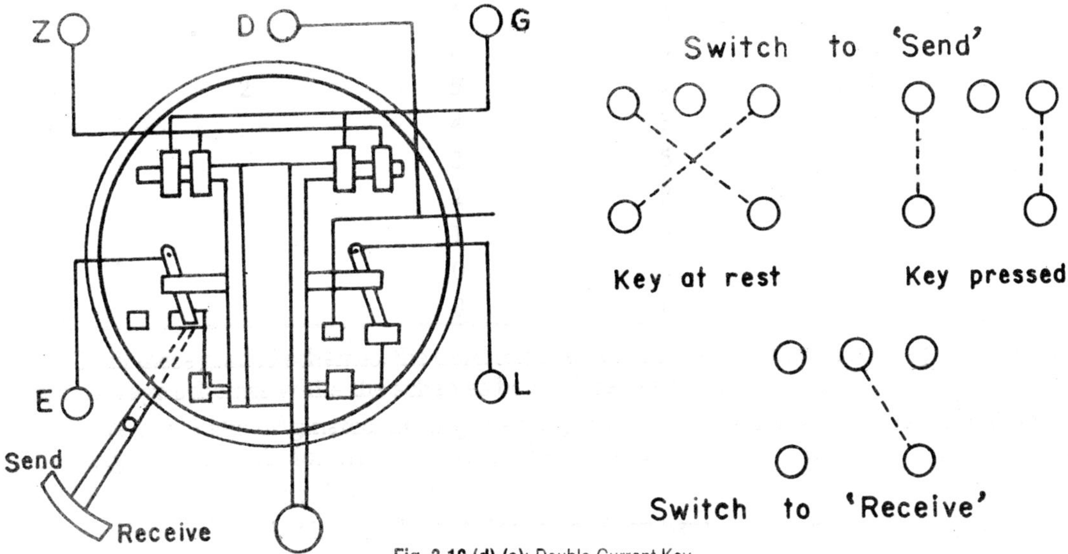

Fig. 3.10 (d)-(e): Double Current Key

Two springs, called the rest springs, with platinum contact at their upper side are arranged to make contact with lower contacts on the levers when key is at rest. Two other springs, called the marking springs, having platinum on the underside are arranged to make contact with the upper contacts on the lever when the key is pressed. The rest springs of the left hand or *E* lever and the marking springs of the right hand or *L* liver are joined to terminal *C* while the rest spring of *L* lever and marking spring of *E* lever are joined to terminal *Z*. If positive to *C* and negative to *Z* is the battery connection, then, at rest, *E* lever is connected to positive pole and *L* lever to the negative, and when the key is pressed, *L* lever is connected to positive pole and the *E* lever to the negative.

Thus, the switch is a two-pole two-way type, one pole connecting to *L* terminal and the other to *E* terminal. With the handle of the switch on send position, *L* terminal is joined through to line lever, and *E* terminal is joined through to *E* lever while the *R* terminal is insulated. With the handle of the switch on the receive position, *L* terminal is joined through to *R* terminal, *E* terminal is insulated. The line and *E* levers are disconnected from *L* and *E* terminals as shown in symbolic representation.

(g) *Power plants required for telegraph offices*: For small telegraph offices, the primary cells of 3-pint size, as a special form of Daniell cells or Carsac cell of the same size, as a special form of Leclanche cells, are used for wet cells. For dry cells, two sizes of Obach cells are in common use — *N* and *P* size. For big telegraph offices, secondary cells are used. They require regular charge and discharge with necessary charging equipments. These will be described in the chapter on power plants and measuring instruments, in detail.

3.3 Direct Current Simplex Telegraph Circuits

These are simplex, duplex and multiplex. In case of simplex, messages can be sent only in one direction at a time on the line. In case of duplex, messages can be sent in both the directions simultaneously on the same line. In the case of multiplex, more than two messages can be sent and received on the same line simultaneously. In every case, the circuits can be worked on single current, double current or on central battery system.

In a single current circuit, on passing the current on the line the receiving instrument goes to 'mark' and on stopping the current on the line, the receiving instrument goes to 'space' by the action of a spring or vice versa. In double current working, the current in one direction causes the receiving instrument to 'mark', while the current in opposite direction puts it to 'space'. In central battery working, all the receiving instruments are operated by a battery located at the central office.

We can work single current circuits either on open circuit or on closed circuit basis. When the batteries are disconnected from the lines with keys in the normal position, the circuit is called open circuit and when the batteries are connected to the lines, with all the keys at rest, the circuit is called closed circuit. These circuits are used to meet various requirements. Thus single current simplex open circuit is used mainly when there are only two stations on the line. Single current simplex closed circuit is used on long lines where a number of offices are working on the line. Central battery simplex circuit is used for several offices working in a small area and double current simplex system is used for very long lines.

Single current simplex open circuit: A diagram for single current simplex open circuit is given in Fig. 3.11.

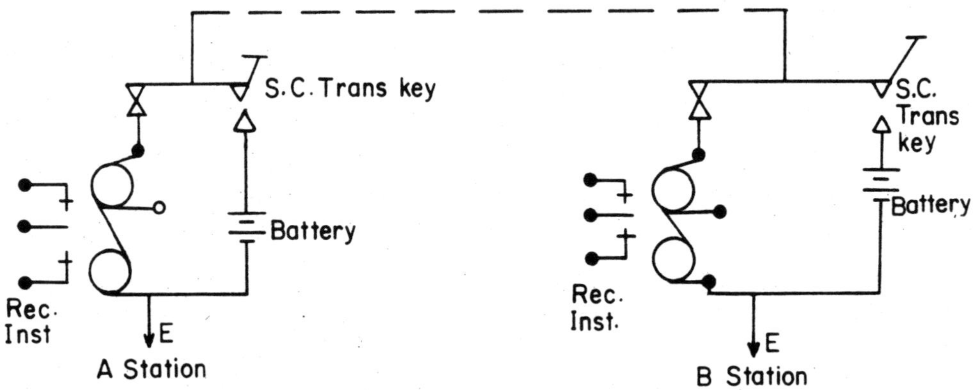

Fig. 3.11: Single Current Simplex Open Circuit

At each station there is a single current key to send the code signals with electric current from the battery through the make contact of the key when pressed. There is also a receiving instrument joined to reproduce the signals received from the distant station, as shown. The line is connected to the lever of the sending key and one terminal of the receiving instrument is connected to the rest stop of the key lever and the other end through earth to the negative terminal of the battery.

With the key at rest, the receiving instrument is joined to the line, while the batteries at both the stations remain out of circuit. When a telegraphist at station *A* depresses his key, his battery is disconnected from his receiving instrument and is connected to the line through which the current passes and energises the receiving instrument at the distant station *B* to give a mark signal so long as the *A* station key is kept pressed. When the *A* station key is released, *B* station receiving instrument is de-energised giving a space signal. Thus by depressing and relaxing the send key, the coded signals can be sent from one station to the other at one time. But when one station is sending, the other station cannot send at the same time, as both the receiving instruments will remain disconnected at that time.

Single current simplex closed circuit: A diagram for single current simplex closed circuit is given in Fig. 3.12.

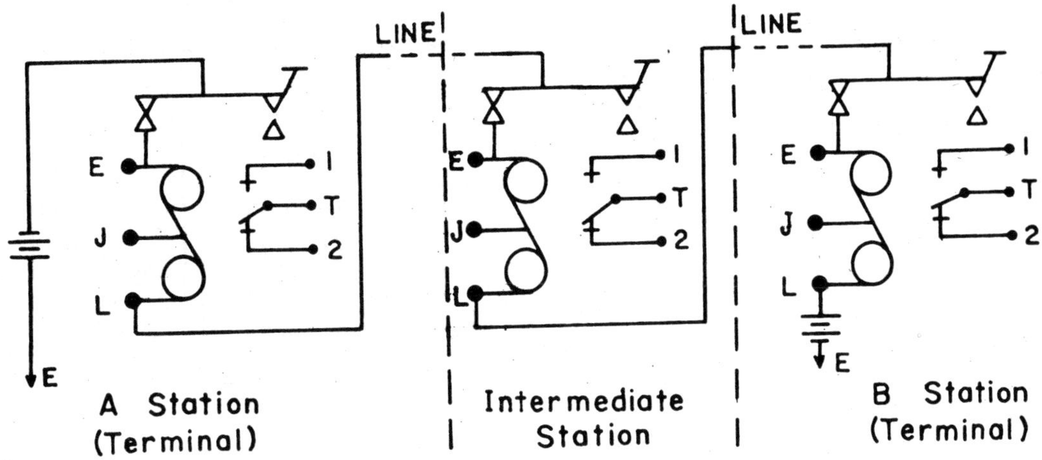

Fig. 3.12 : Single Current Simplex Closed Circuit

The batteries are connected in both the terminal stations only and no battery is used in the intermediate stations which work on the power supplied by the batteries of the terminal stations. The connections of the batteries in the terminal stations are made in such a way that when send key is not pressed in any terminal or intermediate station, the current set up is always in the same direction. This is called an idle condition, when current passes on the line and keeps the armatures of the sounders attracted. When any key is pressed breaking the flow of current, the sounders become de-energised and relax the armature. Thus the operation of the armatures are just the reverse of that in the open circuit.

Central Battery Simplex Circuit: Here, the battery is located only at the central station and each station is provided with a key, a polarised sounder and a 2 μF condenser, and the line is connected to the centre of the key and to one terminal of the condenser, as shown below in Fig. 3.13.

The other terminal of the condenser is connected to the U terminal of the polarised sounder and the front contact of the key is earthed. At the central station where the battery is located, the negative terminal of the battery is connected through the feed resistance to the line and the positive terminal is connected to the earth. When all the keys are at rest, no current passes to the line but all the condensers are charged through the coils of the polarised sounder and through earth.

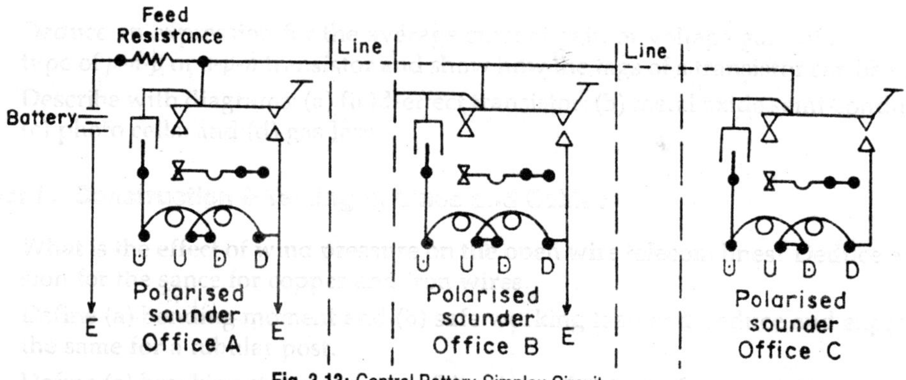

Fig. 3.13: Central Battery Simplex Circuit

When station *A* presses the key, the line is earthed and the condenser *A* discharges through the sounder *A* in the *U* to *D* direction which is the marking direction. Then the condenser at Station *B* discharges through the sounder *B*, the line between stations *A* and *B* and earth. The sounder *B* then comes to marking position. Similarly, the discharge of condenser at station *C* through the sounder C, line between stations *A* and C and earth brings the sounder C to marking position.

On release of the key at station *A*, the earth is taken off the line, whereby the line with all the condensers are charged, the charging current passing from the positive pole to the negative pole of the battery via earth, *D* terminal to D terminal of all the sounders, condensers, line and feed resistance. Hence the charging current being in the spacing direction, all the sounders will come to spacing. Thus all the sounders will respond when station *A* signals. Similarly all the sounders will respond when any other station sends signal and thus communication can be established.

Double Current Simplex Circuit: Generally, single current simplex circuits are worked for short lines, because on long lines whose insulation becomes low, simple current simplex circuits are not workable. But on long lines, double current simplex circuits work very well with polarised sounders used as receiving instruments, with marking and spacing current in opposite directions as shown in Fig. 3.14.

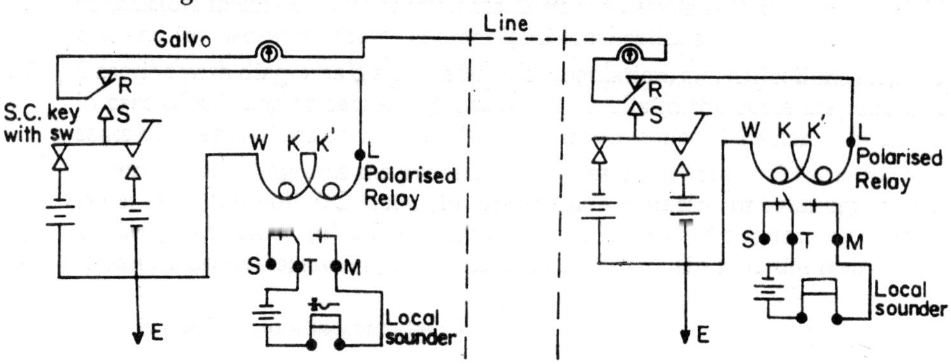

Fig. 3.14: Double Current Simplex for Two Stations

The reversal of current is produced by a single lever key with switch and a double battery at each station. The same may also be produced by a double lever key and a single battery. The line is connected through a single current galvanometer to the lever of the key switch. Generally, the switch is on receive position, the line is connected to the polarised relay terminal *L* and the key is kept insulated. When the switch is on send position, the line is connected to the centre of the key and thence to the positive or negative pole of the double battery according as the key is pressed or is at rest.

When there are more than two stations, all the stations are connected, as shown below in Fig. 3.15.

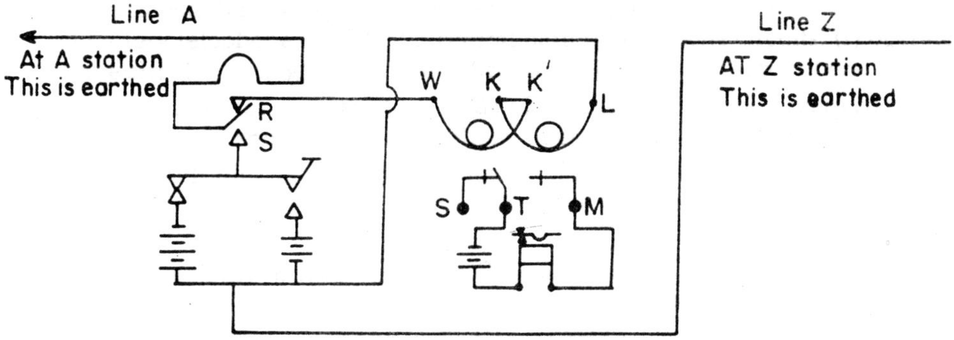

Fig. 3.15: DC Simplex with more than Two Stations

For DC simplex working, a polarised sounder may be used in place of polarised relay and local sounder. At the terminal stations *A* and *Z* the wires marked line *A* and line *Z*, respectively are connected to earth.

DC simplex with individual battery: In the above diagram, the battery at Z station may be common to several circuits, i.e. a universal battery and the batteries at intermediate stations and at terminal *A* station will be individual, i.e. associated with only that particular circuit. Where individual batteries are used, a DC key with single battery is used for two stations, as shown in Fig. 3.16 (a), and for more than two stations in Fig. 3.16 (b).

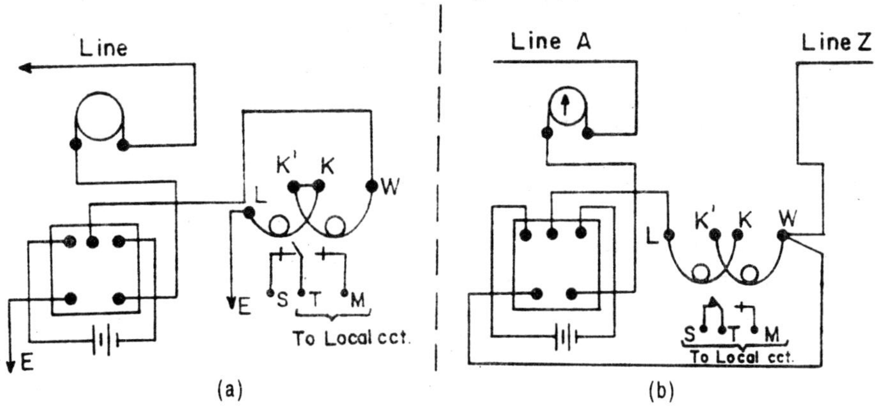

Fig. 3.16: DC Simplex with Individual Batteries

Concentrator Working: In a big telegraph office where a large number of local lines remain idle for a long time and do not require to be manned for all times, they are brought to a switchboard called central battery concentrator, where an operator joins these lines as and when required to a spare set, as shown in Fig. 3.17.

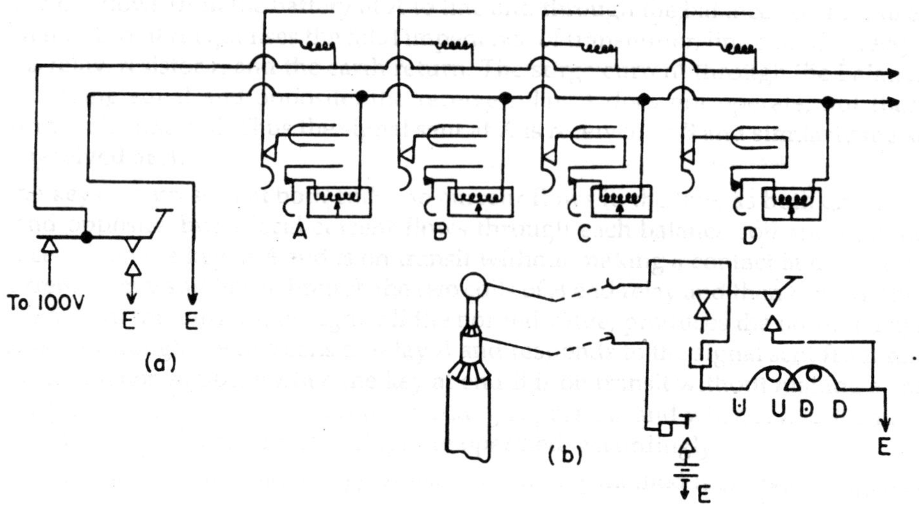

Fig. 3.17: C.B. Simplex Concentrator

In the concentrator, each line is connected through its feed resistance to the battery and to a short outer spring on a jack in front of the switch board. The jack has four springs of which the two outer springs push the two inner springs. When an office calls, the operator sees the particular indicator and then plug is inserted by the operator into that jack, when the tip and sleeve are connected to the positive and negative poles respectively. The sleeve via a push button and the tip via earth E and the inner springs remain disconnected. Thus, signals are exchanged between the two stations, at the end of which the terminal set telegraphist presses a button giving a signal to the concentrator operator to disconnect.

Though the concentrator is used to work only C.B. simplex circuits, this can be used to work any type of telegraph circuit with necessary modification.

3.4 Direct Current Duplex Telegraph Circuits

In duplex telegraph circuits, two messages can be simultaneously transmitted in opposite directions from both the terminals on a single circuit and received simultaneously at the other end. For this, the receiving instrument is to remain unaffected by the movement of the transmitting key at the home station by the proper balance of the circuit elements and their resistances and capacitances.

Duplex telegraphy can be worked on (a) single current differential with batteries in opposition or aiding position at both stations, or on (b) double current differential duplex with batteries in opposition or aiding at both ends.

The differential winding of a relay coil means winding the coil with two twisted wires, so that the two windings are exactly similar in resistance and magnetic effect, as shown below. Here, equal currents flow in opposite directions through two equal differential relay windings. Since the current produces equal and opposite magneto-motive process in magnetic circuit of the relay, their resultant is zero and the armature remains in contact where it was touching before, i.e. at M, as shown in Fig. 3.18.

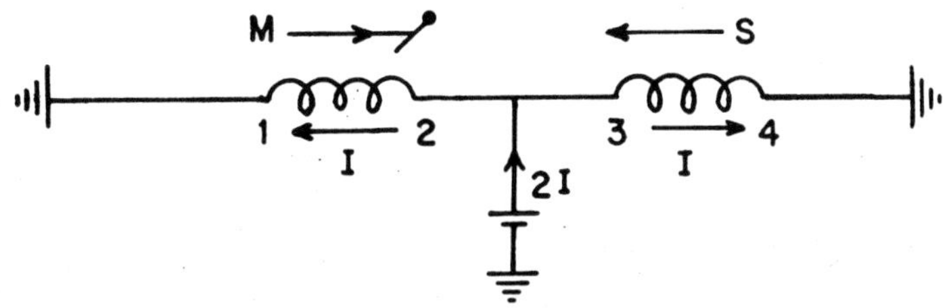

Fig. 3.18: Differential Relay Winding

In B.P.O. practice, the current passing through the coil from 2 to 1 or 4 to 3 operates the relay tongue to the marking contact and that passing in the opposite direction operates the relay tongue to the spacing contact.

Differential duplex can work either in opposition method or in combination method. But in both methods, the following transmission conditions are required to be fulfilled: (a) when transmitting tongue at each end is resting on mark contact, the receive relay at each end must be held on mark contact; (b) when the transmitting tongue at end A is operated to space, the receive relay tongue at end *B* must move to space, whilst the receive relay at end *A* must be a mark; and (c) when both the transmitting tongues at *A* and *B* end are on space, both the receive relays at *A* and *B* must be held at space.

(i) *Single current differential duplex-opposition method*: This method is shown diagrammatically in Fig. 3.19.

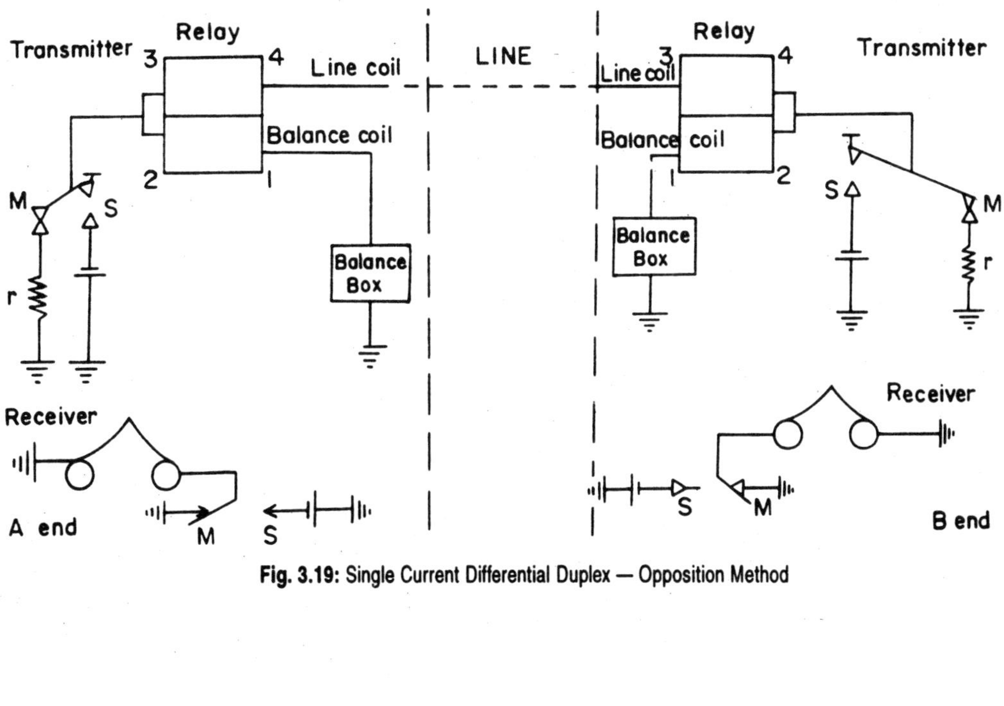

Fig. 3.19: Single Current Differential Duplex — Opposition Method

The line and the receiving equipment at the other end is balanced by the resistors and capacitors in the balance box and resistors r is equal to the internal resistance of the transmitting battery at each end.

When the transmitting keys at both ends are at rest on its mark contact, no current flows in the circuit and the receiving relay contacts remain on mark at both ends. When the key at A is on space, current flows from the battery at A to line and through the balance box at A to earth. Now the balance box at A equalises the total impedance of transmitter, line coil of A relay, line, line coil of B relay, resistor r, and the earth return. The surge current through the balance coil and line coil being equal and opposite, the receive relay A does not operate, but that at B operates through its line coil. Thus the signal sent at A is received at B and similarly the signal sent at B is received at A.

When the keys are pressed at both ends, no current flows in the line, as both end batteries are equal and opposite, but a local current flows through each balance coil and operate the relay to space. When the key at A end is on transit without making a contact and B end key is pressed, current from B end flows through the two coils of A end relay and then to earth through the balance box. But this current, though half the normal value, produces the normal magnetic effect on passing through the two coils of relay A and responds to the signal sent from relay B. The same phenomenon happens when the key at end B is on transit without making a contact and end A key is kept pressed. Thus duplex working is possible and when A and B end transmit simultaneously, the distant station relays are operated accordingly.

(ii) Double current differential duplex — opposition method. This method is shown in Fig. 3.20.

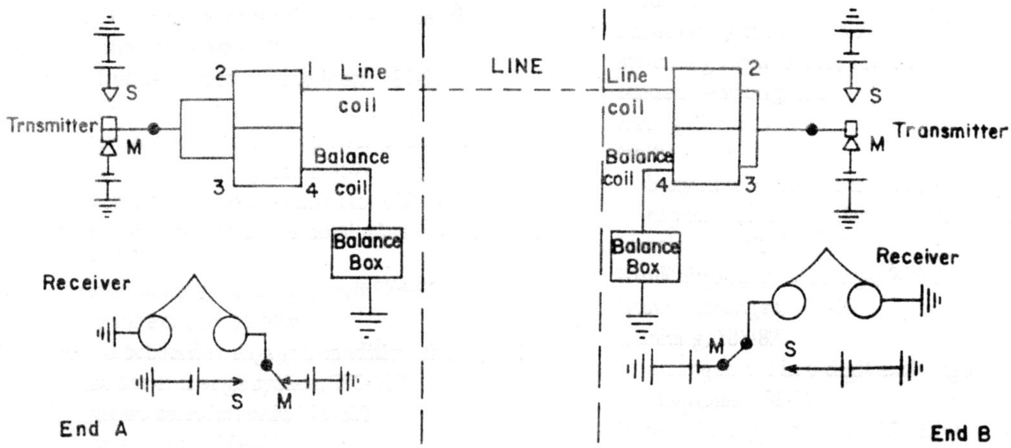

Fig. 3.20: Double Current Differential Duplex—Opposition Method.

When both the transmitters are in the mark position, as shown, no current flows in the line due to equal batteries at each station in opposition, but a marking current flows in both the balancing circuits holding the tongue at mark at both ends. But when the transmitter at A is in space position, the current in the balance circuit at A is reversed. It tends to bring the receive relay to space contact. But the batteries at A and B now being in series — aiding through line circuit, and the line current being twice of that in balance circuit, as the balance circuit has the same resistance as the line and distant receiving equipment, the line current on the line coil at A operates the receive relay at A to mark. The current in the line winding of relay B is in spacing direction and is twice that of the balance coil, hence the receive relay at *B* goes to spacing contact. Similarly, when *B* sends spacing signal to *A*, the receive relay at *A* goes to spacing contact. When *A* and *B* send simultaneously space signals to the line, no current flows, but the space current flow in each balance circuit operating their receiving relays to space contact.

If the transmitter at *B* is at space and the transmitter tongue at *A* is in transit to mark contact, the space current flowing from *B* through the line coil and balance coil of relay *A* in series works like a normal balance current (though its value is half normal balance current), and the response of receive relay at *A* to the signal sent from *B* is not disturbed. Under this condition at *B*, the current in the line coil being half of that in the balance coil, the latter continues to maintain the relay to space, until the transmitter at *A* reaches its mark contact and at this time the relay at *B* operates to mark also. Thus, the correct transmission of signals is maintained during transit condition also.

iii) Single current differential duplex — combination method. This method is diagrammatically shown below in Fig. 3.21.

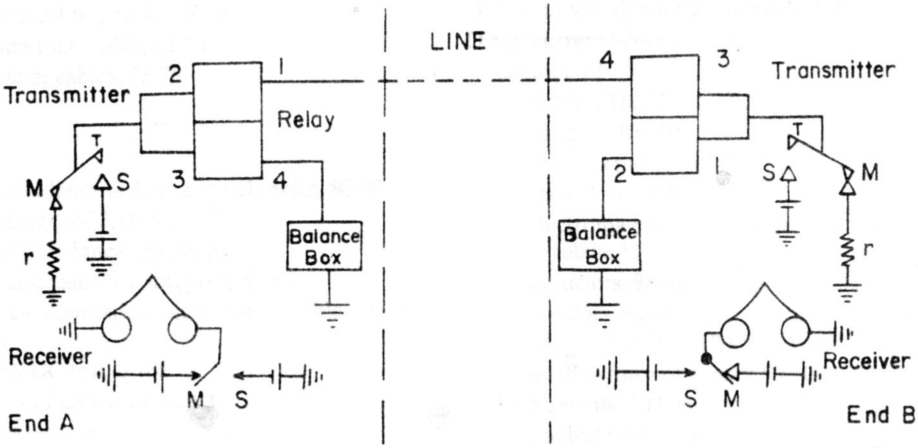

Fig. 3.21: Single Current Differential Duplex-combination Method

Here, the transmitters at *A* and *B* are connected for single current operation. At *A* when transmitting key is pressed, negative battery is transmitted to the line but at *B* when the transmitted key is pressed, positive battery is transmitted to the line.

When the transmitting keys at *A* and *B* are at rest, no current flows in the line or the balance circuits and the receive relay tongues rest on the marking contacts under the control of marking bias of the relay. When both the transmitters are operated to spacing contact, the batteries at *A* and *B* stations are series-aiding, the line current being double of that in the balance coil, each receive relay operates to space contact. When the transmitter at *A* is on space contact and that at *B* is on mark contact, the current in the line coil and balance coil of receive relay *A* will be equal and opposite. The tongue of relay *A* will be operated to mark by marking bias of the relay. At station *B* the line current flowing through the line coil to earth through resistor *r*, operates the relay to space. Thus the signals are identically received at the other end when they are transmitted separately or simultaneously from both ends.

The Balance Box: The operation of the receive relay in a duplex circuit depends on the correct balancing of the line impedance by the balance circuit, so that the instantaneous values of the two currents are equal. Otherwise the transmitted signal will disturb the receiving relay of the same station and thereby the duplex working will be upset. For this the basic network for balance circuit is shown below in Fig. 3.22(a).

Generally, the resistor R_1 equals the resistance of short lines and condenser C_1 simulates the capacitance of the line and the receiving equipment at the other end. But the resistor R_2 delays the charge and discharge time of capacitor C_1. But if the line is longer, the time taken for the current to reach the steady value is longer and thus the time to charge and discharge the condenser in the balance circuit becomes longer. For this purpose, resistor R_3 and capacitor C_2 are added to the balance circuit for suitably balancing the longer lines as shown in Fig. 3.22(b).

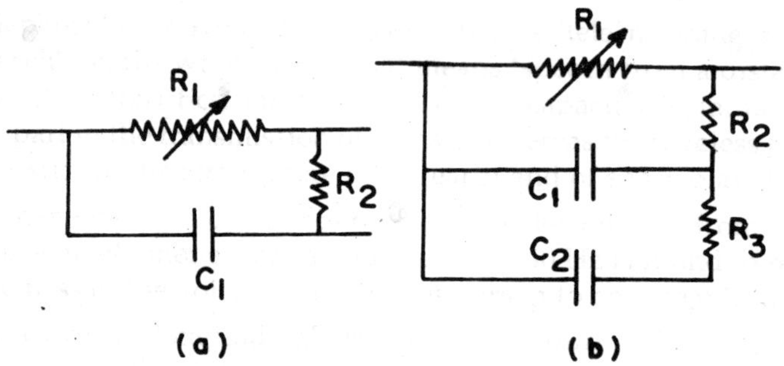

Fig. 3.22: Balance Circuit Networks

The operation of a bridge duplex telegraph circuit is on the balancing done by the principle of Wheatstone bridge as shown below in Fig. 3.23.

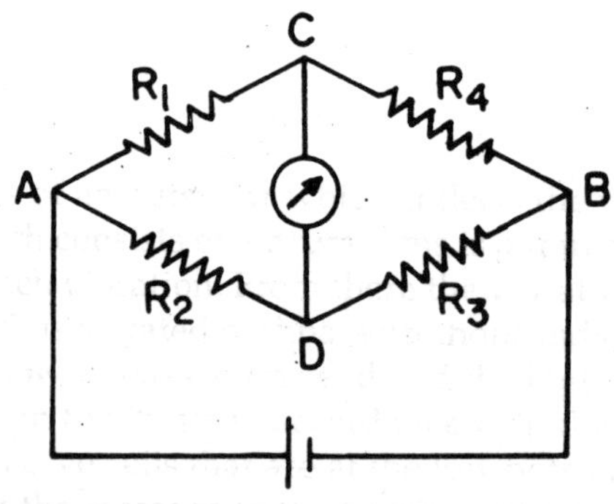

Fig. 3.23: Wheatstone Bridge Principle

The bridge is balanced when $R_1:R_2 = R_3:R_4$, i.e. when $R_1 \times R_4 = R_3 \times R_2$, then only the bridge is balanced and points C and D will be at the same potential and no current will pass through the galvanometer. *Bridge duplex*: A bridge duplex telegraph circuit is shown below in Fig. 3.24.

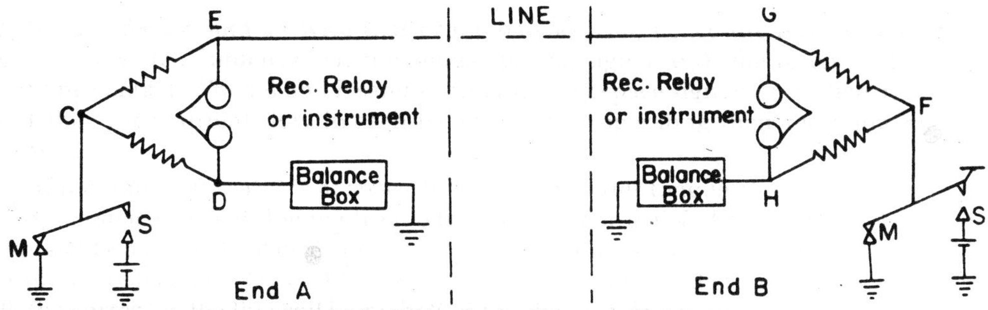

Fig. 3.24: Bridge Duplex Telegraph Circuit.

The resistance arms *CE* and *CD* are made equal to the resistance arms *FG* and *FH*. The line and the distant receiving instrument are connected, as shown, across *E* and *D* at *A*, and *G* and *H* at *B* end are balanced in the same way as was done in the case of differential duplex.

At rest position of the transmitters on mark contact, no current flows. But when key at *A* is operated to space, current is equally divided at *C* so that voltages at *E* and *D* are equal and no current can flow in the receive instrument at *A*. The current through *CE* passes through the line to *G* at the other end *B*, where a part of it flows through *GF* to earth and the other part through the receive relay to balance box and *HF* in parallel to earth. Then, the receive relay at *B* end is operated to space contact. Similarly, the operation of *B* station key to space operates the receive relay *A* and to space without disturbing the receive relay at *B* end. If both end keys are operated to space, no current can flow in the line as point *E* and *G* are at the same voltage in opposite direction, but a part of the current at each end flowing through the receive relay and balance box to earth operates the receive relay at both ends to space.

When the *A* end key is in transit and *B* end sends, say, space signal, the current from *B* dividing at *E*, a part flows through *ED*, operating *A* end receive relay to space. The other part flows through *EC* and *CD* in series, and unites with the first part at *D* and flows via duplex balance box to earth, keeping relay *A* operated to space.

Thus, duplex work is possible in all conditions of transmission from both ends.

4

Manual Telephone Exchange Working

The word 'telephony' is derived from the Greek words, 'tele' meaning 'at a distance' and 'phone' meaning 'speech'. So the word 'telephony' means 'speech from a distant place'. For line telephony, the line characteristics, namely its resistance, capacitance, inductance and leakance play an important part in the transmission of signals, whereas for wireless telephony, the transmission characteristics of the atmosphere of the earth and of space play the most vital role.

Now a human voice is a compilation of frequencies ranging from 75 Hz to 10,000 Hz. But for practical use in telephone a band width of 300-340 Hz is found to represent the pitch, volume and exactness of the voice of a particular person to be recognised for commercial use.

The sound wave is a longitudinal wave for which during the transmission of sound waves in a material medium, the molecules of that medium do not bodily move away from the source but merely oscillate about their mean position, passing some energy to the adjacent molecule.

We know, if v is the velocity of propagation of sound wave and if f is its frequency, then the time taken to travel one complete cycle = $\frac{1}{f}$ sec and the wave length, i.e. the distance travelled in $\frac{1}{f}$ sec = $\lambda = \frac{V}{f}$.

The human ear consists of an outer ear which collects the vibrating waves of sound which travel to the inner ear which consists of a drum. This is a thin membrane containing three tiny bones which move with each vibration. From there the vibrations travel to a tube which is like the shell of a snail in which is situated cochba with thousands of nerve cells arranged like the keys of a xylophone with lower notes at one end and the higher notes at the other. The tube is filled with fluid which begin to vibrate in accordance with the vibrations of sound waves and the fluid passes on to the nerve cells that are at the low note end. They are very sensitive and pass on the intelligence of the message to the hearing centre of the brain, from where we understand the sensibility of the message.

Transmission of speech: Now the speech is transmitted through a circuit by a telephone transmitter converting the mechanical vibrations into electrical waves, which impinge on the diaphragm of a receiver at the other end of the speech circuit, where the electric waves are again converted into mechanical vibrations in the air and these vibrations impinge on the ear drum of the subscriber at the other end. The called subscriber at the other end, on hearing the speech understands the intelligence of the speech and replies on the same circuit through his transmitter furnishing the necessary particulars required by the calling subscriber who receives.

4.1 Relays, Indicators, Transformers and Telephone Circuits

Telephone Relays: In the signalling and switching circuit of telephony, the most important item is the relay, which is an electromagnet. When electric current passes through its coil of wire, the core becomes magnetised and attracts an armature to which is fitted a movable contact and the other contact is fitted to a fixed position, so that the contacts may either make or break when the armature is attracted. When electric current is stopped, the reverse action takes place. Thus a telephone relay virtually works like a switch from a distance. Its principle of construction is shown below in Fig. 4.1.

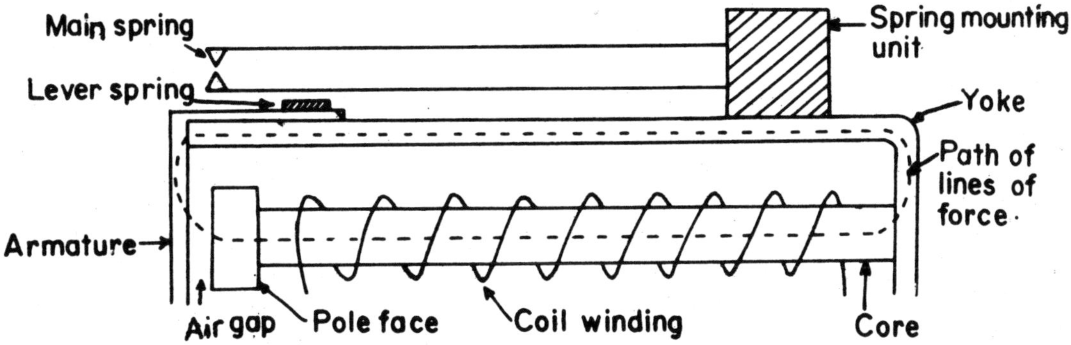

Fig. 4.1: Elements of a telephone relay

The component parts consist of (a) core of soft iron rod with a flat end called the pole face, (b) a coil of insulated wire wound round the core, (c) an L-shaped yoke of soft iron attached to one end of the core, (d) An L-shaped soft iron armature pivoted at the front end of the yoke just above the pole face with an air gap, (e) a set of springs carrying necessary contacts for make or break. These are fitted above the yoke on a spring mounting unit, so that their other ends fitted with contacts are controlled by the movement of the armature. These contacts may be either make, break, change over and make before break, etc.

A relay is said to be fast operated if no special feature is introduced to affect its natural speed of operation and release. But its speed of operation and release is affected by the insertion of a thick copper ring called 'slug' over one end of the core, as the magnetic effects of eddy current depend on the position of the slug on the core. It is observed that while a heel-end slug has little effect on operation, the operate lag of a relay with armature end slug is greater than of a relay without slug, and that the greater spring tension will produce greater operate lag. Thus the relay with heel-end slug becomes slow to release, whilst the relay with armature-end slug becomes slow to operate and slow to release.

Telephone indicators: These are also some electromagnetic devices through whose coil when current is passed, the armature in the form of a flag or disc is operated to give a visual indication. These indicators may be either (a) drop indicators which give a continuous signal when operated for a moment or (b) eyeball or flag indicators which give visual signal as long as the operation key is on. These are shown below in Fig. 4.2.

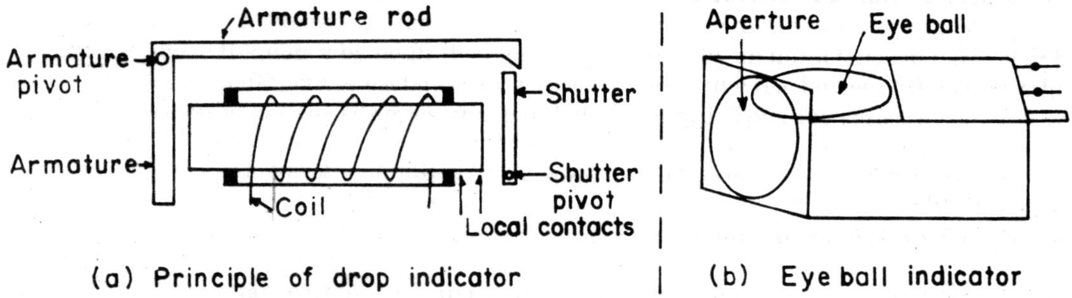

Fig. 4.2: Telephone Indicators

In case of drop indicators, the shutter attached to the armature rod drops down as soon as a pulse of current passes through the coil attracting the armature, which tilts the armature rod rigidly fixed to it and lets loose the shutter which falls due to gravity. The operator has to manually put back the shutter linking it with the projection of the armature rod. But in case of eyeball indicators, the eyeball type of armature revolves on the carriage and fills the aperture, as long as the current passes through the electromagnet, thus drawing the attention of the operator and as soon as the current ceases, the eyeball indicator restores its position under gravity. But in each type, there is an arrangement made by local contacts of the indicator on operation to give some audible alarm to expedite the attention of the operator.

Transformers: With the help of a transformer, we can electromagnetically couple two or more AC circuits together and by suitable adjustment, the alternating voltage of one circuit can be transformed to a desired value in the other circuit. The transformer consists of two coils of wires wound round a common core. These windings are known as the primary and secondary.

We know that if a current passes through a conductor, it produces a magnetic field round the conductor and the intensity of the field at any instant is proportional to the current passing. Again we know that if a conductor is placed in a varying magnetic field, an emf is generated in the conductor and the value of this emf is proportional to the rate of charge of the magnetic field. Thus if I is the alternating current passing through the primary coil, a magnetic field is induced in the core, varying as the current varies. If e volts per turn of coil is induced, the total induced voltage in the primary = $e\,N_1 = E_1$, say where N_1 = number of turns in the primary. Similarly the total voltage induced in the secondary = $e\,N_2 = E_2$, say where N_2 number of turns in the secondary.

$$\frac{E_1}{N_1} = \frac{E_2}{N_2}\ or\ \frac{E_1}{E_1} = \frac{N_1}{N_2} \qquad (1)$$

Thus the ratio of primary and secondary voltage is equal to the ratio of the number of primary and secondary turns. Now if there is no inherent loss in the transformer, the input power in the primary must be equal to the output power in the secondary, i.e. $I_1E_1 = I_2E_2$, where I_2 is the alternating current passing through the secondary, i.e. $E_1/E_2 = I_2/I_1$. But $E_1/E_2 = N_1/N_2$ $\therefore$ $N_1/N_2 = I_2/I_1$. Thus the ratio of primary and secondary current is inversely equal to their turns ratio.

The transformers are designed to suit the frequency or range of frequency for which they are to be used. Thus the power transformers are designed for 50 Hz and the transformers required to be used for telephone circuits are designed to allow for the passage of various ranges of frequencies with the least loss of power.

There are five main sources of power loss: (i) shunt loss due to magnetising current, (ii) hysteresis loss due to iron core, (iii) eddy current loss in iron core, (iv) series loss due to incomplete linking flux with each turn of primary and secondary windings, and (v) ohm resistance loss of the windings. A better performance is obtained by increasing the inductance of primary windings at low frequency; reducing the iron and copper losses at the middle frequencies, and reducing the leakage flux at high frequencies.

A transformer to be used for the transmission of speech is therefore constructed with a core material of (a) high permeability to get high primary induction, (b) high specific resistance to get eddy current reduced, and (c) low hysteresis loss. Three types of cores are therefore used, namely, ring type, core type and shell type, as shown below in Fig. 4.3.

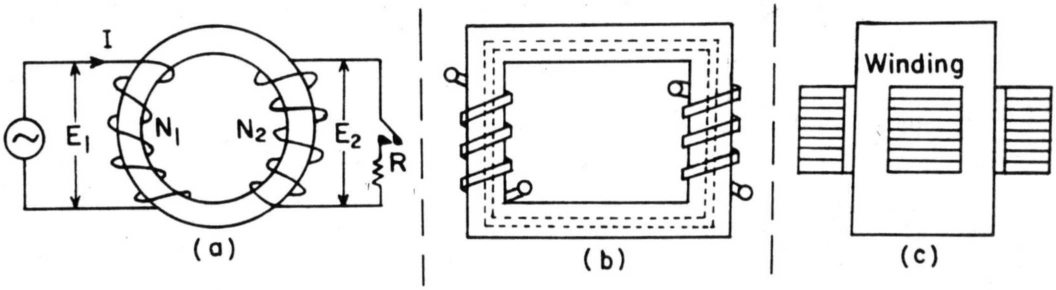

Fig. 4.3: Telephone Transformer
(a) 'Ring Type' Transformer Core (b) 'Core Type' Transformer Core (c) 'Shell Type' Transformer Core

4.2 Different Telephone Systems

In manual telephony, there are mainly two types of telephone exchanges, namely the magneto and the central battery (CB) system. In magneto systems the signalling is done by a low frequency alternating current produced by a magneto generator, while the current for sub's transmitter is supplied by a local primary battery at the sub's place. Due to these difficulties, the magneto system is seldom used now and as such has become more or less obsolete. So we shall describe mainly the C.B. system in detail. In automatic telephony, there are mainly three types, namely (i) the Strowger, (ii) the cross-bar, and (iii) the electronic systems. But as the Strowger and cross-bar systems work on space division switching network and are decimal systems, working on classical amplitude and frequency modulation, we shall describe these after the C.B. system in this volume. The electronic system is based on binary or two number (0 and 1), system, depending on Boolean Algebra and working on stored programme controlled time-division switching networks with digital integrated circuits for processing of signals using pulse code modulation (PCM) and delta modulation (DM) for speech band compression, and the same has been described in Volume II to avoid unwieldy size of Volume I.

The level of speech from a subscriber to another in an exchange is generally kept at -8 db to -20 db level for commercial purpose.

Thus a change of 1 db = a power ratio 5/4 nearly

Thus a change of 2 db = a power ratio 3/2 nearly

Thus a change of 3 db = a power ratio 2/1 nearly

Thus a change of 10 db = a power ratio 10/1 nearly

Thus a change of 20 db = a power ratio 100/1 nearly

The chief components of a telephone circuit are: a transmitter, a receiver, and line between them.

Principle of operation of a telephone system: The telephone systems in which a local battery is used at the subscriber's premises is called a local battery system, but when a bigger capacity battery is used at the telephone exchange and no local battery is used at the subs' place, the system is called a central battery system.

Local battery telephone circuit: We know that for the maximum transfer of power, it is necessary that the sending end impedance is equal and opposite to the receiving end impedance. Thus a maximum energy passes to the line when the impedance of the line and the distant receiver together is equal to the internal impedance of the transmitter. These two impedances can be matched by using a transformer called the induction coil for coupling the transmitter and the line at each end to obtain a local battery telephone circuit, as shown below in Fig. 4.4 and 4.5.

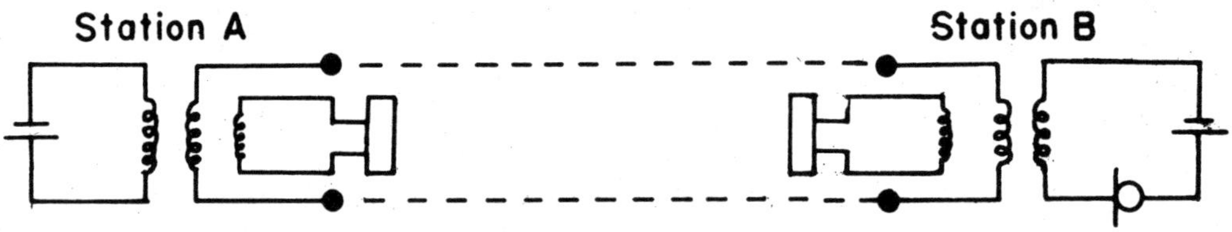

Fig. 4.4: Both way local battery telephone circuit with 3 winding induction coils

At *A* station when a subscriber speaks into his transmitter, the vibrations produced in the air by his speech waves strike the diaphragm which begins to vibrate accordingly. These vibrations vary the pressure on the granules causing corresponding variations of the resistance of the carbon granules and consequent variations in the current. This varying alternating current in the primary of the induction coil produces a varying flux in the core, which induces an alternating emf in the secondary coil and this induced emf flows through the line to the induction coil of the distant end receiver at B station. Here an alternating flux is produced in the pole pieces of the receiver. Then in conjunction with the static field of the permanent magnet it gives a variable tractive force on the receiver diaphragm, producing corresponding vibrations in the air which then impinging on the ear drum produce the sensation of original sound waves transmitted from station *A*. Similar sequences of operations take place when a subscriber speaks from station *B* to station *A*. The above principles are also applied to a modern local battery telephone circuit as shown in Fig. 4.5 where a three-winding induction coil is used in the form of an auto-transformer for matching the impedance at each end between the transmitter and line, and between the line and receiver.

Fig. 4.5: Both way local battery telephone circuit with 3 winding indication coil connected as an auto-transformer

The principle of auto transformer is shown in Fig. 4.6 below, where the alternating current flowing in winding N_1 produces an alternating flux in the transformer core of the auto-transformer,

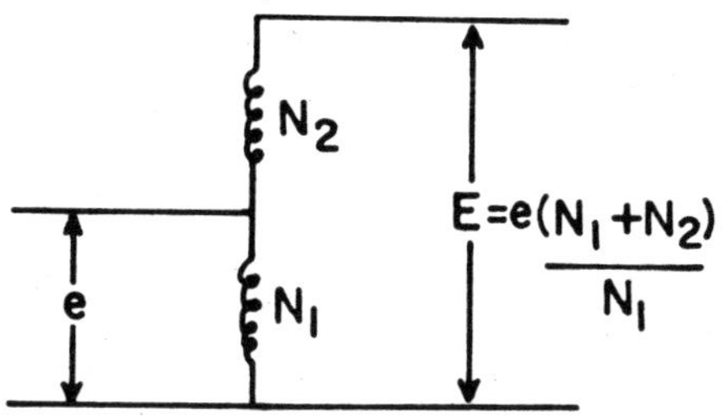

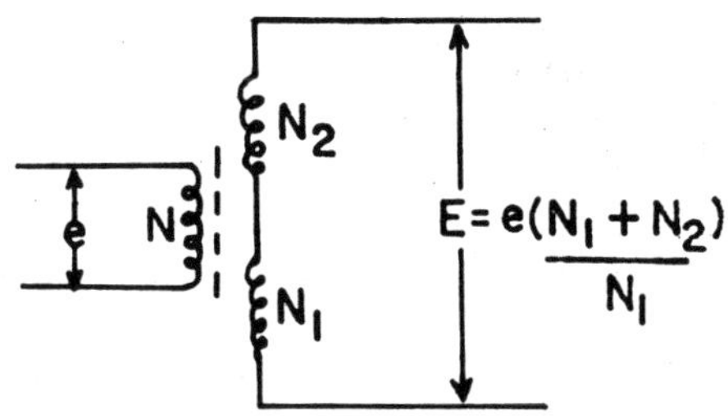

Fig. 4.6: Principle of Auto-Transformer **Fig. 4.7** Equivalent Transformer

which cuts the turns of both the windings N_1 and N_2 to produce an emf which is equal in magnitude to that which would be produced in a separate secondary winding of $N_1 + N_2$ turns.

The equivalent of this auto-transformer can thus be represented by a simple transformer as shown in Fig. 4.7 above.

Central battery telephone circuit: The local battery system is practically obsolete now due to high expenditure of maintenance of primary cells used at the sub's place. This system has been replaced by the central battery system, where one battery of secondary cells is provided at the telephone exchange instead of primary cells at sub's place as shown below in Fig. 4.8.

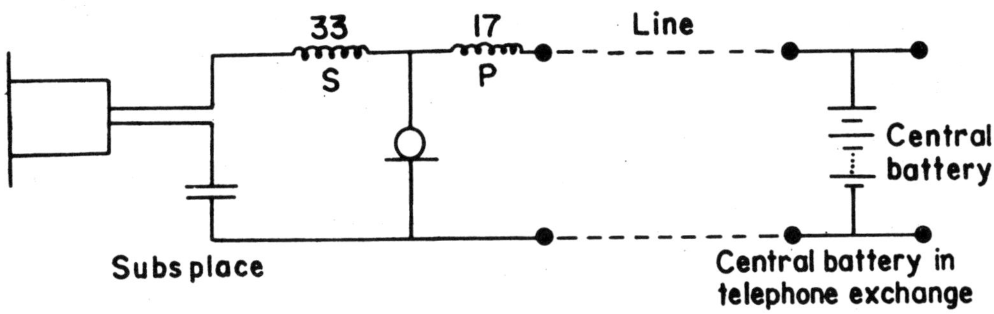

Fig. 4.8: Central Battery Telephone Circuit

The speech voltages and current in a C.B. telephone circuit with their principle of operation are shown below in Fig. 4.9.

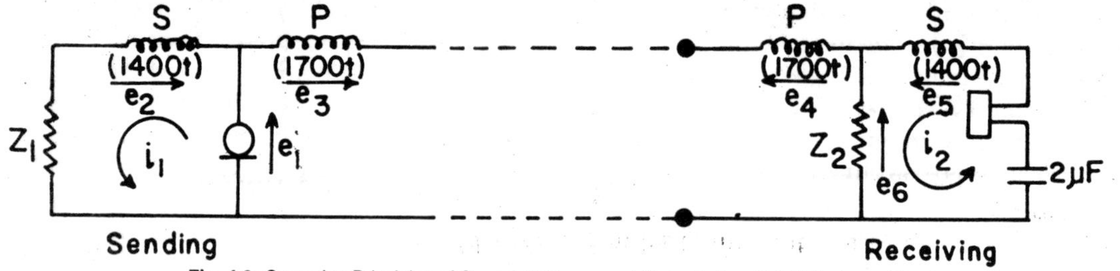

Fig. 4.9: Operating Principles of Speech Voltages and Currents in a C.B. Telephone Circuit

Z_1 = impedance of receiver and condenser

Z_2 = impedance of transmitter

The transmitter voltage e_1 gives a current i in the local circuit consisting of secondary of the induction coil, the receiver and the condenser producing a p.d. (e_2) across the secondary winding of the induction coil. The passage of current through the secondary coil induces an emf e_3 in the primary coil. Thus a voltage ($e_1 + e_3$) is applied to the line, which gives a voltage to the line greater than that of the transmitter alone and the induction coil behaves like a step-up transformer.

In the reception circuit, the greater part of the speech currents pass through the transmitter Z_2 as the receiver circuit is of higher impedance. The voltage e_4 passing through the primary coil induces a voltage e_5 in the secondary and produces a current i_2 in the receiver circuit. The passage of current in the transmitter Z_2 produces a p.d. e_6 in the receiver circuit which opposes the induced emf in the secondary coil. Thus the induction coil behaves as a step-down transformer matching the receiving circuit impedance to that of the line.

Thus the induction coil behaves opposite in the transmitting and receiving condition, for which their turns ratio for primary and secondary windings must be a compromise to give reasonable sending and receiving efficiencies.

Acoustic shocks in the receiver: There is a surge voltage applied to the receiver of any telephone producing a click in the ear. Sometimes when the surge voltage is very high, it is likely to cause acoustic shocks which are very harmful to the ear drum and may damage it. To eliminate this trouble, a device consisting of two rectifiers, connected in parallel with opposite polarity, is connected across a telephone receiver, so that the excessive voltage in either direction is diverted through the rectifier, allowing only a small portion to pass through the receiver, as shown in Fig. 4.10 and 4.11 below.

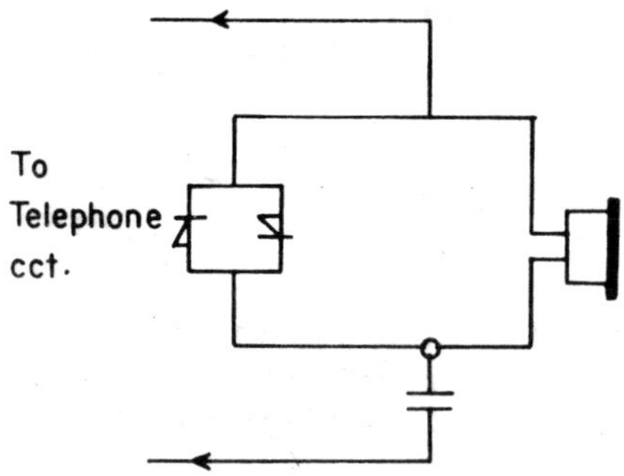

Fig. 4.10: Use of Rectifiers to Prevent Circui Accoustic Shocks,

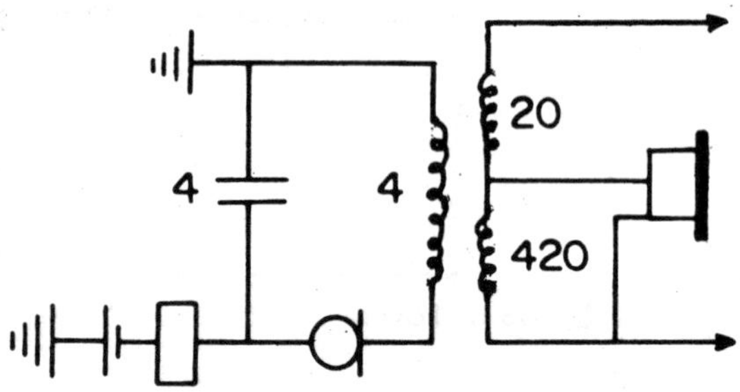

Fig. 4.11: Operator's Instrument with Side tone Suppression

Special arrangements are now made on all operators' telephone circuit to guard against this accoustic shocks by use of such combination of rectifiers across the receiver.

Apart from the receivers in general, the anti-side tone circuit is particularly used in operators' telephone set as shown in Fig. 4.11 above. The primary coil of 4 ohm winding of the induction coil is connected to the transmitter, to which is applied the current from the exchange battery through a retard coil to reduce the current to its proper value and the speed current passes through the 4 μF condenser as it cannot pass through the battery due to high impedance produced by the retard coil.

The turn ratio of the secondary winding is so arranged that the secondary coil with two windings of 20 ohm and 440 ohm is joined in series to the tip and ring of the cord circuit, plug and the receiver is connected across any of these windings. When the operator plugs into the jack of a line, the voltage drop across the receiver is neutralised by the equal and opposite p.d. from 440 ohm winding.

4.3 Transmission Bridge

When two subscribers are to be joined for communication through a telephone exchange with the help of an operator's cord circuit, it is required to provide a through path for speech from one sub to the other through the exchange cord circuit, while at the same time the two subs line should remain separate for the purpose of signalling. The circuit by which this can be done is called the transmission bridge, which may be either (a) an electrostatic coupling in which the two lines are connected for speech by large condenses or (b) an electromagnetic coupling where a transformer is introduced between the two lines. The first type is called the Stone transmission bridge and the second type is called the Hayes transmission bridge. These are shown below in Fig. 4.12 (a) and (b), respectively.

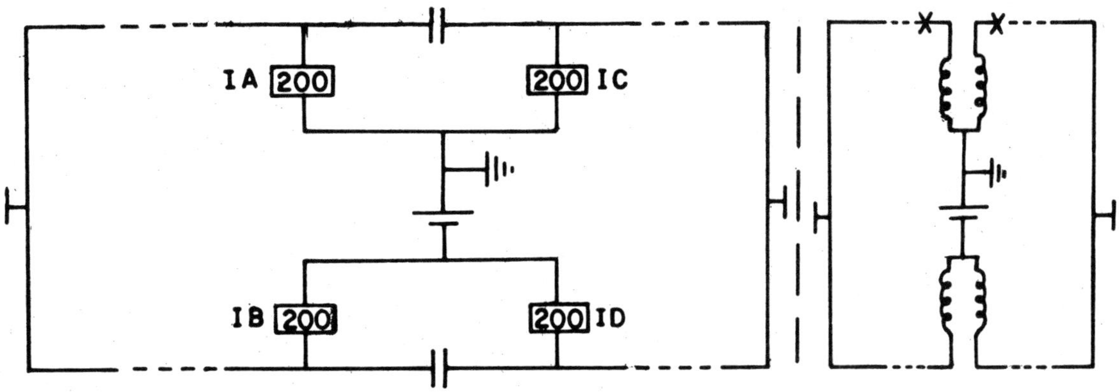

Fig. 4.12 (a) Stone Transmission Bridge **(b)** Hayes Transmission Bridge

In Stone transmission bridge, the transmitter feeds the current for one telephone via the impedance coils *IA, IB* whilst the supply for the other telephone is taken via the impedance coils *IC, ID*. The low resistance and high impedance of these coils are utilised respectively for DC signalling relays and prevention of speech current through battery. The two sides of the transmission bridge are completely isolated for direct current by the insertion of two coupling condensers of 2 μF or 4 μF, allowing an easy passage for the transmission of speech current, with a transmission loss of 1.2 db for 800 Hz for 2 μF condenser and 0.5 db for 4 μF condenser.

In the Hayes transmission bridge, separate DC is supplied to each telephone. The four windings of a transformer are wound on a common core for audio frequency transmission, in such a way that when the transmitter on the left is excited, the voice frequency circulates through the windings *A* and *B* only via battery inducing corresponding emfs in windings *C* and *D*, thus circulating the speech current through the right hand loop. Again if the transmitter on the right side is used for speaking, the voice frequency current in *C* and *D* produce similar current in the left hand loop.

Thus in Stone transmission circuit, the speech currents are prevented from passing through the battery, whilst in Hayes transmission bridge, the speech current is passed only through the battery.

Signalling: Singalling is done between the subscriber and the telephone exchange for the following purposes. (a) a calling signal from the subscriber to the exchange for outgoing calls, (b) a clearing signal from the subscriber to the exchange to indicate the end of conversation, (c) a ringing signal from the exchange to called subscriber for incoming calls, and (d) an answering signal by the called subscriber to indicate that he has attended the call. As the process of signalling depends on the type of telephone exchange, these will be described with the respective exchanges.

Sub-station apparatus and switch board apparatus. Telephonic communication is required between subscribers at different places joined by lines and cables from subscriber's station to the switch board apparatus situated in the telephone exchange. The apparatus are installed at the sub-station and the telephone exchange with some standard practice. As the switch boards with magneto, CBs and semi-CB types are nowadays more or less obsolete, we shall go through these types only with a passing mention and describe in detail the more used switch boards for modern practice like the CB and automatic types.

Sub-station apparatus: There are two items in a sub-station apparatus — the first is the telephone instrument, and the second is the line termination in a sub-station. There are two methods of joining various part of a magneto or other types of bell, viz. series method and shunt method as shown below in Fig. 4.13(a) and (b).

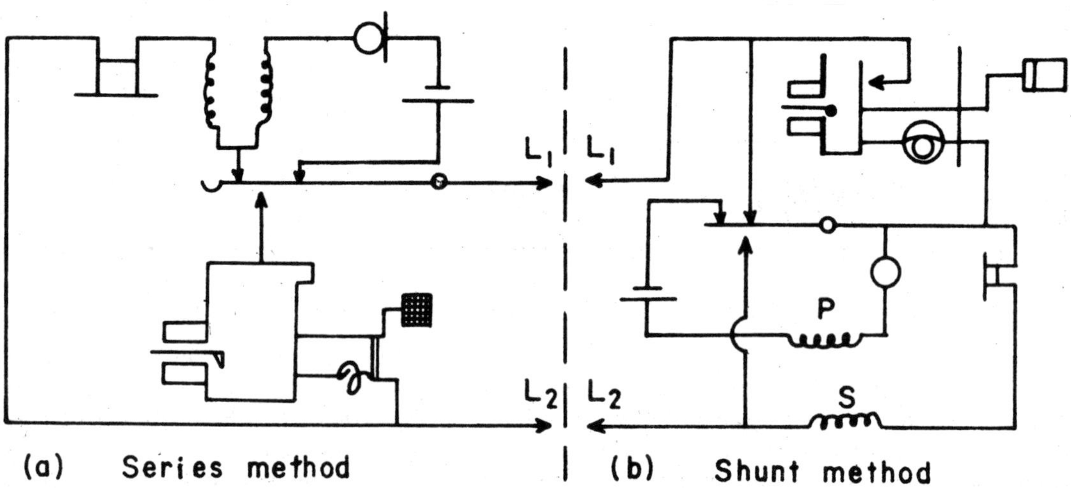

Fig. 4.13: Magneto Instrument Connections at Sub-station

In Fig. (a) the movement of the switch hook after taking the receiver off as shown, connects the speaking circuit and breaks off the ringing circuit, while in (b) the same movement does not break either circuit but simply short-circuits the ringing part of the instrument and opens a short-circuiting contact in the speaking circuit as shown.

Of the various types of C.B. sub-station circuits, a few are shown below in Fig. 4.14.

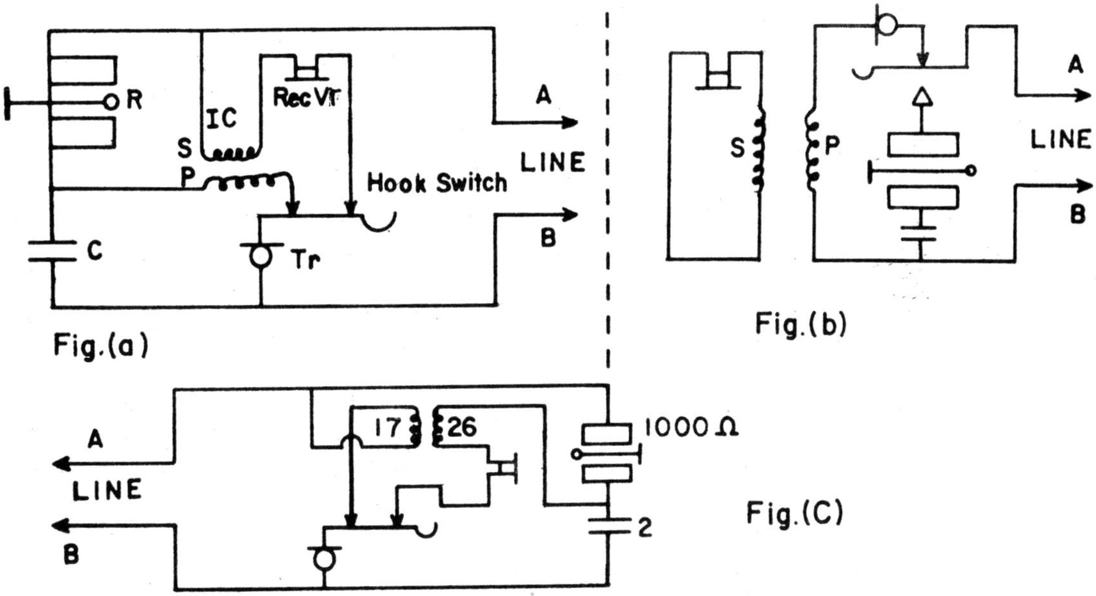

Fig. 4.14: C.B. Sub-station Instrument Circuits

In Fig. (a) it will be seen that the transmitter, receiver, and the secondary coil are in the direct line circuit and the transmitter primary winding of the induction coil and the condenser are in the local circuit. In Fig. (b) the primary coil and the transmitter are in the main loop while the receiver and the secondary winding of the induction coil form the local circuit. In Fig. (c) the primary (17 Ω) coil of I.C. and the transmitter are in the main loop circuit when the switch hook is up, and the receiver with 26 ohm winding of secondary of the I.C. transmitter and the condenser are in the local circuit. It may be seen that the transmitter is included in both the circuits. Although there is another direct path for the current through the secondary, the receiver and the ringer coils, it is almost equivalent to a break due to its very high impedance for quickly varying speech current. The receiver is wound to 60 ohm winding and the normal resistance of a transmitter is about 55 ohm.

4.4 Subscribers Line Termination

The telephone lines are terminated at the sub-station on the window terminal, from where they are taken inside the room via a protective device unit, which contains special sockets for fuses, lighting arrestors or two carbon strips separated by mica foil, with three holes having an earthen centre plate as shown below in Fig. 4.15.

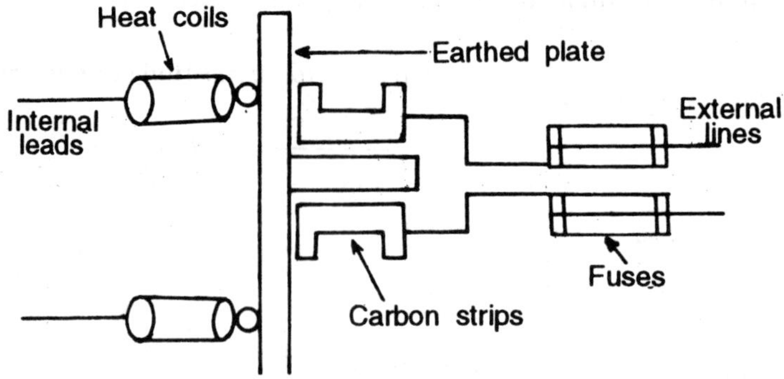

Fig. 4.15: Protective Device Unit at Sub's Place

There are also two heat coils of disconnection type fitted as shown. The whole unit is fitted with a cover in a single hole fixed screw. The base of the protector unit is of porcelain and accommodates two cylindrical line fuses rated 1.5 A. The two heat coils are of resistances 4.5 to 5.5 ohm operating with 500 mA within 60 seconds. An earth connection is made at sub-station through a water supply pipe or tubewell pipe, etc.

Exchange Apparatus: This consists of the cable termination chamber, main distribution frame via protective devices, intermediate distribution frame, switch board apparatus, power plants including secondary batteries with charge-discharge arrangements and ringing current supply from mains and battery, standby engine generator for meeting prolonged power failure. Then there are relay racks, meter racks and special apparatus bay, etc.

Cable Termination Chamber: The subscribers' lines from different areas are brought into the telephone exchange through a cable chamber to the main distribution frame via the protective devices fitted with protectors. The cables enter through one end of the telephone building plinth adjacent to the main cable route and the base of the chamber. Its high walls are built with water proof material and with high basement so that flood water cannot enter the chamber. The underground cables from the cable chamber are taken directing into the M.D.F. room near the the outside wall, just above the cable chamber. The cables are carried on bearers through short length of iron pipes to the foot of the verticals in the MDF room, so that the pipes project 2 to 3 inches above the floor to prevent damage to cable as shown below and then the underground cables are terminated to the protective devices. The cables are supported on bearers racks or ducts if necessary from the cable chamber to the M.D.F. room as shown in Fig. 4.16.

If the telephone exchange switch board is with multiple answering facilities, the subs' lines are considered during the basy hour according to traffic originated from each and they are distributed over the main jack field in front of each position by jumping from MDF to IDF calling equipment in front of operator's position, in such a way that the work load an each operator is equally divided. For this an intermediate distribution frame is required to be in stalled after the protective devices of the M.D.F., so that by jumping the lines from calling equipment to multiple equipment, the purpose of equitable distribution of work may be served. Thus the geographical area order is conveniently converted into numerical order within the exchange, by jumpering both the sides of a terminating tag block, one side from the MDF and the other side to the IDF as shown below in Fig. 4.17, where the number in the brackets indicates the number of calls originated by the sub during the busy hour.

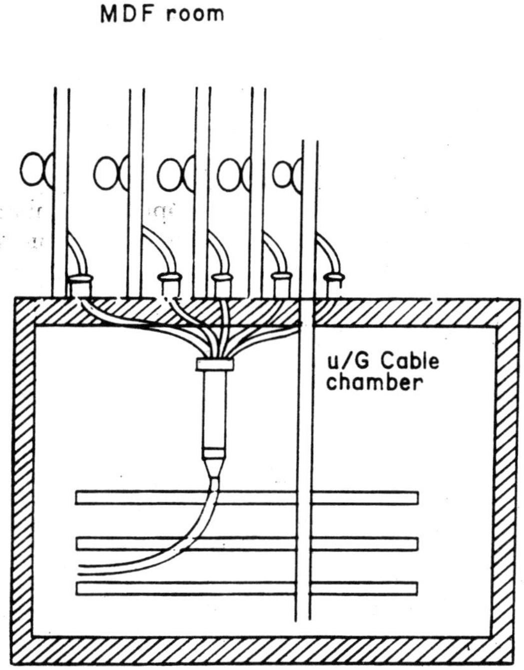

Fig. 4.16: Underground Cable Chamber to M.D.F. Room

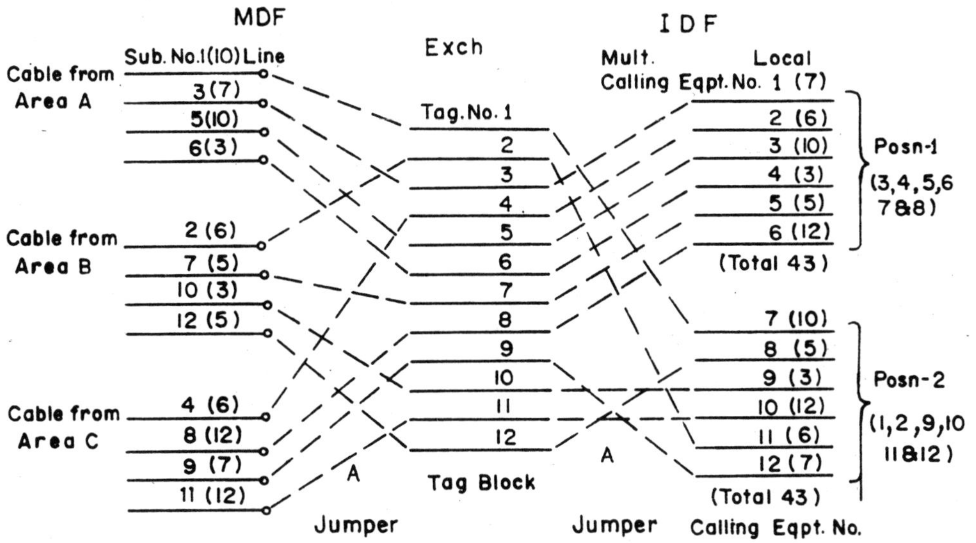

Fig. 4.17: Arrangement of Subs' Numbers from Geographical Order to Numerical Order by Jumpering from MDF to IDF

The jumpering of lines from fuses to heat coils and carbon strips in the cable termination in MDF and jumpering of calling equipment to multiple equipment in IDF are shown in Fig. 4.18.

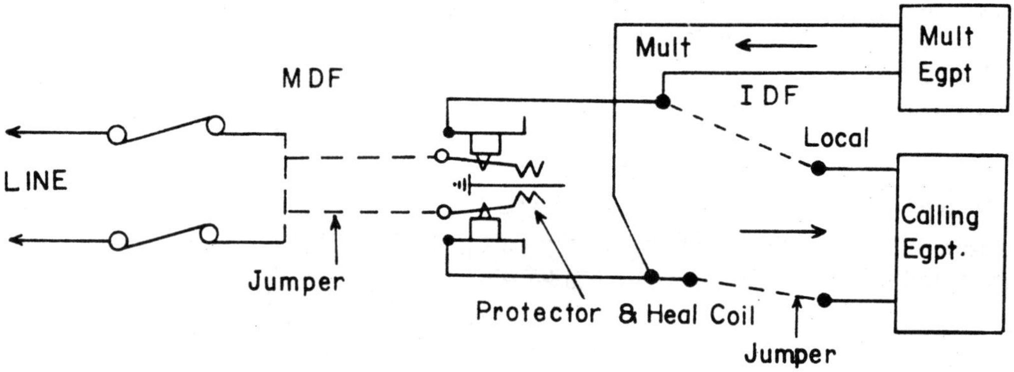

Fig. 4.18: Protective Devices in MDF to IDF

Just like local lines, the junction lines are also terminated on the line side of the MDF and are jumpered to protectors or connection strips on the exchange side, from where a bank of tie cables extend them to the multiple side of IDF. For purpose of giving the required signalling facilities, a group of relays are jumpered and a second set of jumpering connect the relay group to the outgoing junction multiple or answering jack via IDF local as shown below in Fig. 4.19.

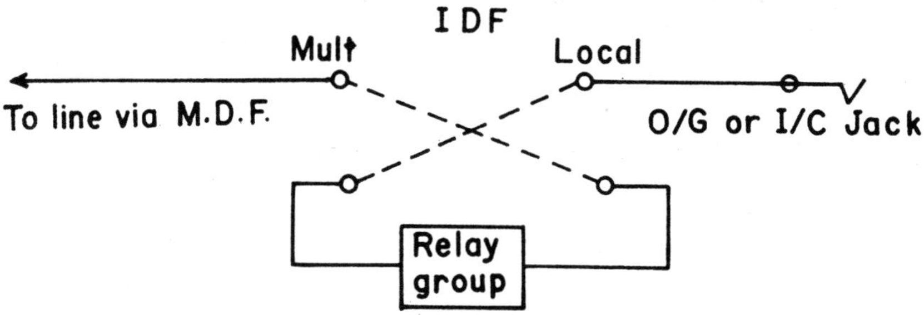

Fig. 4.19: Cabling of Junctions

Thus the junction answering jacks and the outgoing junction multiple are both cabled to the local side of the I.D.F., whereas the subs' outgoing multiple is cabled to the multiple side of the IDF. Both the M.D.F. and I.D.F. are installed in such a way that all the jumpers run through jumper rings horizontally on the local side of the frame and vertically on the multiple side. In exchanges less that 2000 lines capacity, combined distribution frame (CDF) are installed where the lower part is equipped with fuse and protective devices, whilst the upper part serves as IDF.

A separate earth connection is provided in the telephone exchange by driving vertically downwards a number of six feet long, one inch diameter galvanised steel pipes on the base of the underground cable chamber and connecting the spills together by means of 19/16 strand copper cable, which is enclosed in a lead pipe to avoid corrosion and taken in to the exchange via underground cable chamber and is terminated on the MDF.

4.5 Design of Manual C.B. Switch Board

All switch boards for manual operation have the same type of equipments, like key and plug shelves, jack fields, relay shelves, switch board cords, pulley weights and cord fasteners, etc. In Fig. 4.20(a) below is shown the cross-sectional view and in Fig. 4.20(b) the front view of a manual switch board. The keys and plug shelves are placed at a standard height from the floor in a horizontal position accommodating the speaking and the ringing keys in the key shelf which is hinged along its back edge to allow its raising from the front for maintenance work. After the key shelf in the horizontal position is placed the plug shelf, where in two rows the answering cord plugs with corresponding calling cord plugs fitted at one end with three conductor switch board cords (white for tip, blue for ring and tinsel for sleeve conductor) with brass pulley weights (either single or double pulley system as shown) through which the cord is threaded and kept at tension with other ends fastened to the cord fastener, either directly or through fixed pulleys. Each exchange line is connected to a drop indicator or call lamp provided on the lamp-jack strip on the vertical field of the switch board and arrangements are made for connecting one pair of the lines together by plugging into the jack the answering cords on seeing the glow of the calling lamp and on putting the speak key to ascertain from

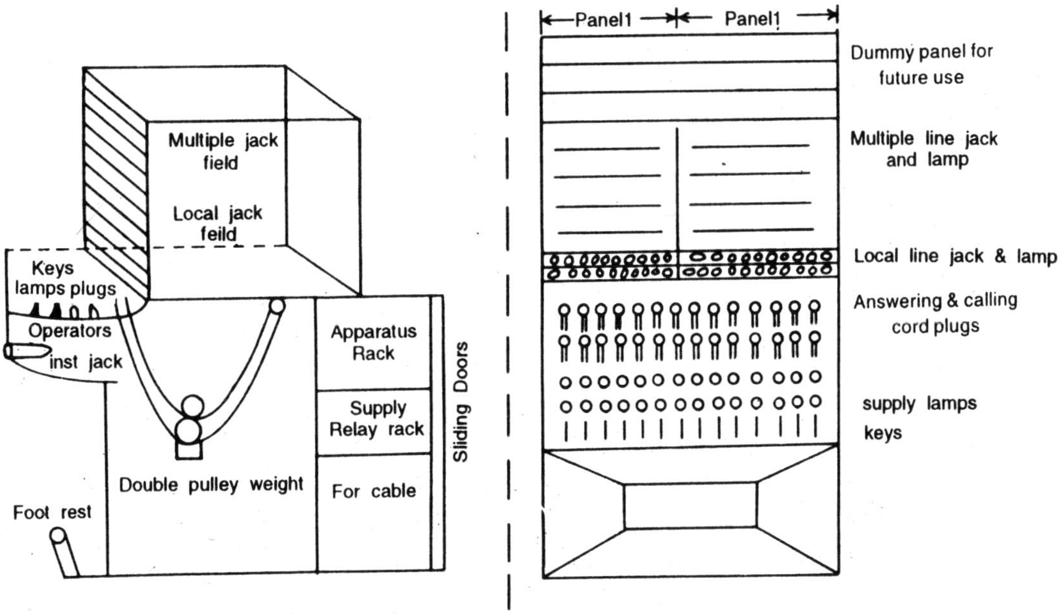

Fig. 4.20(a): Cross-section View of Manual Switch Board

(b): Front View of Manual Switch Board

the calling sub the number he wants to be connected with and then join the same with the corresponding calling cord just on the back of the answering cord to the jack of the called number who gets a ring either automatically or by operation of manual key as per arrangement made.

Above the horizontal key and plug shelf, stands on a vertical position the jack field containing the line jacks and subs calling lamps or indicators either above or below the corresponding jacks.

There are mainly two types of jack strips, one in the 3-point branching jack in which insertion of a plug makes a contact with the springs of jack and the other is the five or eight-point break jack, in which the insertion of a plug not only makes contact with the line springs but also breaks the normal circuit as in the case of five point break jack or changes over with auxiliary contact springs as in the case of eight point break jack. Similarly the keys may be divided into two types, the lever keys where contact member is the one moved by means of a lever, and the plunger key where the contact members are moved by means of a plunger of V-shape. Both types may be either locking, i.e. remain operated until restored or they may be non-locking, i.e. they are to be held in the operated position or they will automatically restore to their normal position by spring action. Transfer circuits are provided by interconnecting various operators positions, so that each operator can operate all the calls of other switch boards adjacent to him. But in most of the modern C.B. switchboards, the top portion of the answering jack field is generally provided for multiple jack field multiplied to the answering jacks of the other switch boards in the same suite, as the cords which are of 72 inches length cannot reach further. This is called a six panel multiple system. But in stressing, a lot of time and distance is required in six panel arrangement, which tell upon the rate of call in busy hour for which a four panel multiple is arranged for each 1600 lines.

We have seen how the cables enter in the underground cable chamber and then are terminated in the MDF room on the ground floor. Below is shown the layout of a C.B. exchange for MDF, IDF, relay rack, special apparatus rack and meter rack, etc. with test desk arrangements all located in one spacious hall called the apparatus room on the ground floor just above the underground cable chamber. The arrangement of plinths for battery, motor generator, ringing machines and other distribution bus bars are made in the power room, generally adjacent to the apparatus room as shown in Fig. 4.21.

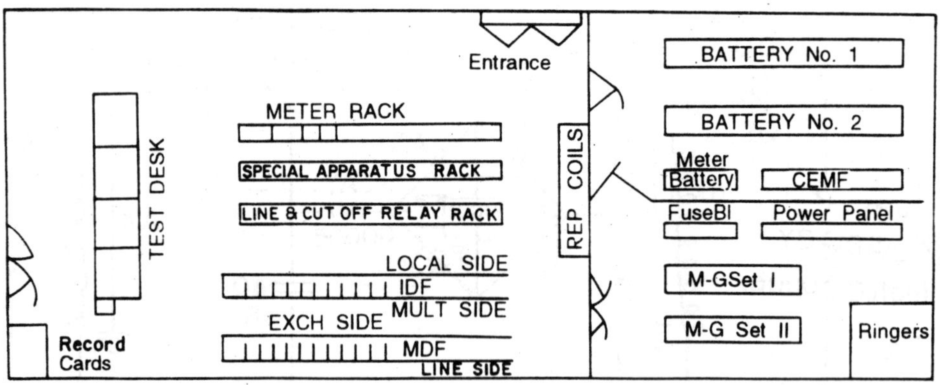

Fig. 4.21: Ground Floor Apparatus Room Layout for C.B. Exchange

Generally the manual switch boards are installed in the first floor. But if it is a small exchange with only one storey building, the switch boards are installed in an adjacent hall of an *L* shaped building for convenience as far as it is feasible. But the arrangements of the manual C.B. switch boards remain the same as shown below in Fig. 4.22.

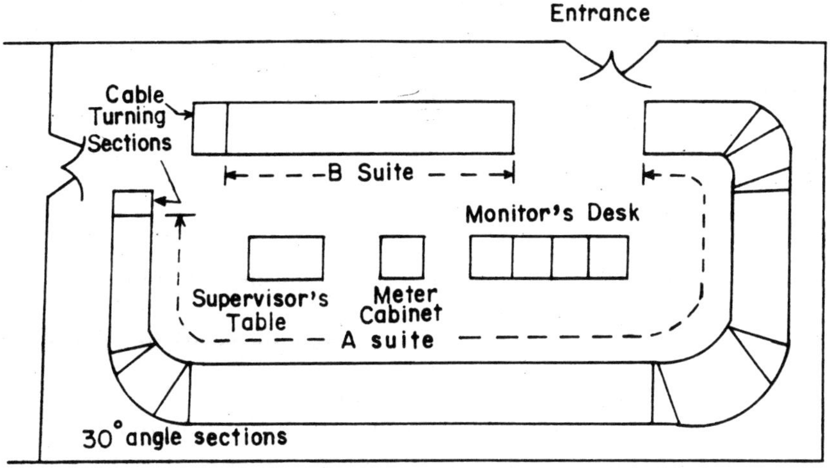

Fig. 4.22: Manual Switch Room of C.B. Exchange

A diagram for the cabling arrangement in the intermediate distribution and main distribution ranges for the subscribers' line and a cross-sectional diagram for C.B. exchange apparatus room showing the main cable rooms is given in Figs. 4.23 and 4.24, respectively.

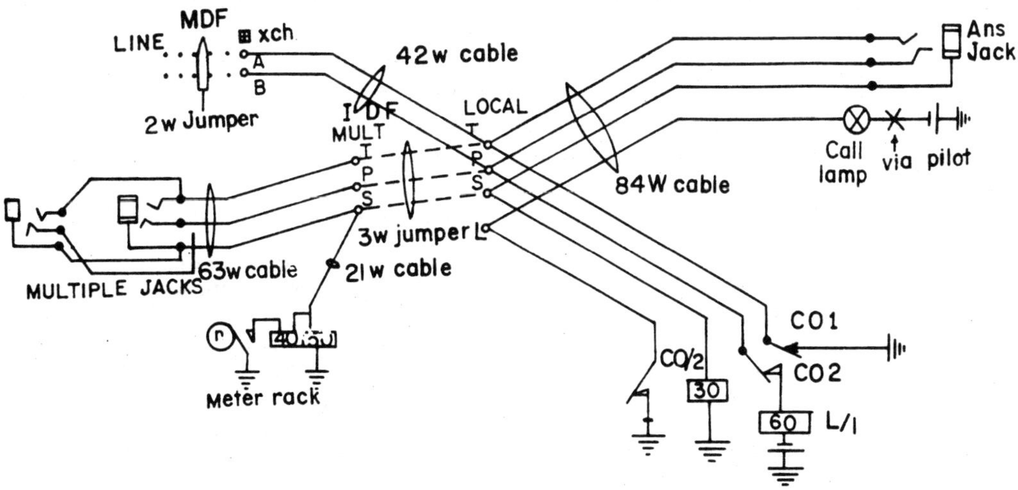

Fig. 4.23: Cabling Arrangements for MDF

As a matter of fact, the necessary exchange apparatus connected with the switch boards are generally accommodated in five separate rooms if possible for convenience. (i) All the switch boards are generally kept in two separate suites, one with the subscribers lines and the other with special services such as 'local assistance', complaints, directory enquiry, time, special information, trunk enquiry, morning alarm, etc. These two suites with monitors' desk and supervisors' desk are accommodated in the same spacious hall called switch room. (ii) The batteries are accommodated in a separate room called battery room. (iii) The power plants such as motor generator sets, ringer machines, both mains operated and battery operated are accommodated in the power room, where the battery charge discharge boards along with other accessories are also fitted. (iv) The meter racks, special apparatus racks, line and cut-off relay racks, junction relay racks, I.D.F., M.D.F. along with test desks, record cards, etc. are accommodated in one spacious hall called the Apparatus room just above the cable chamber. (v) Lastly, the engine alternator of diesel oil or petrol engine set with its associated accessories and power boards, etc. are accommodated generally in a separate small building or hut called the engine room to minimise the vibration in the main building and engine noise to the operators working on the switch boards with their head sets on.

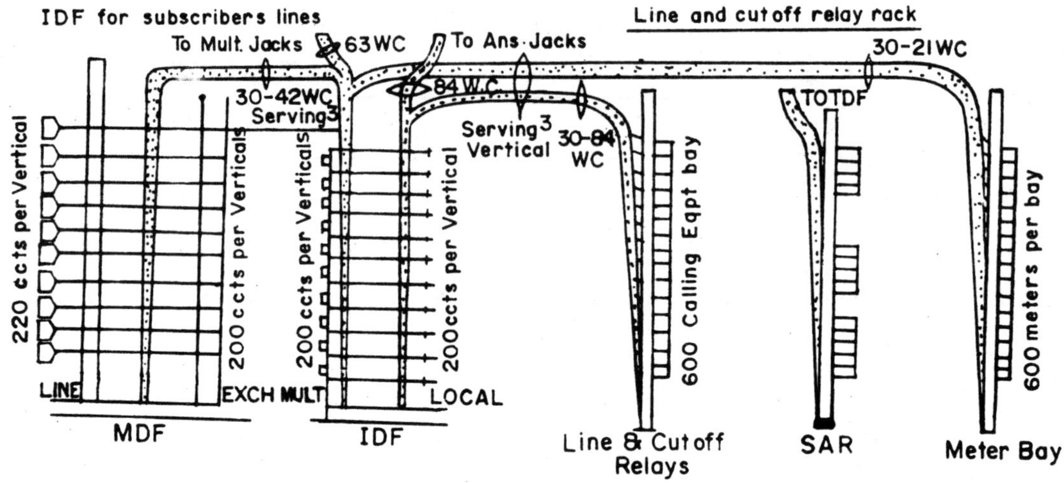

Fig. 4.24: Exchange Apparatus Room

4.6 Local Line Signalling in Manual Switch Boards

Signalling in local telephone system is done either by DC signalling or by AC signalling. The DC signalling is shown in Fig. 4.25.

When press button P is pressed, the current from battery B flowing through coils C_1, C_2 joined by a soft iron yoke Y produces inductive current, which attracts the armature, producing a sound when it strikes the gong. The spring action of the armature at once pulls it back, but the inductive effect of the coils again attracts the armature to produce another striking on the gong of the bell and thus the bell goes on ringing.

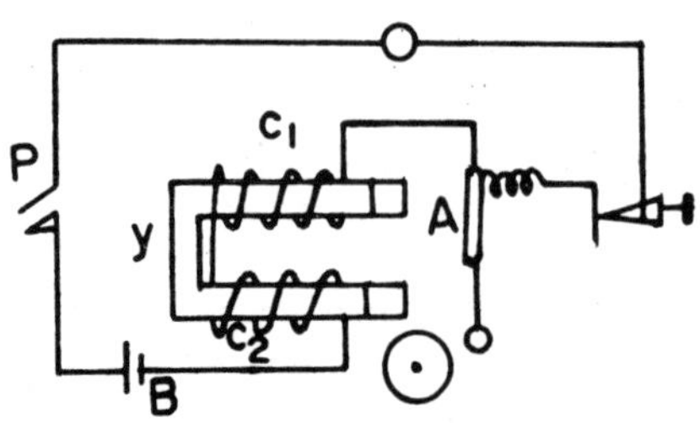

Fig. 4.25: Trembler Bell (DC)

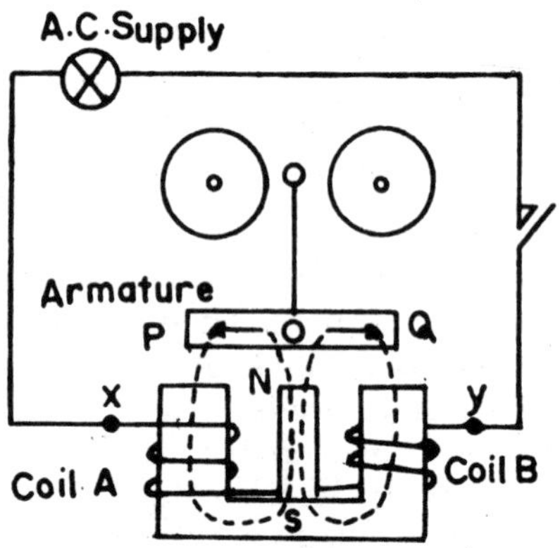

Fig. 4.26: A.C. or Magnetic Coil

In AC signalling, as shown in Fig. 4.26, the two coils *A* and *B* coupled at one end by a soft iron yoke, at the centre of which is attached a permanent magnet NS is shown. Near the centre of the free ends of the coils, the armature *PQ* is pivoted at its centre. If now an AC is passed through the coils *A* and *B* from terminals *x* to *y*, the magnetomotive force due to the current opposes the flux due to permanent magnet at the air gap *P* and assists that at *Q*, due to which the flux density at *P* is reduced as that at *Q* is increased. For this armature moves over to right and the hammer attached to the armature strikes the right gong producing a sound. The next moment when the current reverses due to its alternating nature, the tractive force is increased at *P* and reduced at *Q* and the hammer attached to the armature strikes the left gong producing another sound. These sounds go on as long as the press button remains pressed. Such bells are also called magneto bells. In the diagram in Fig. 4.27 is shown how the two local battery telephones with DC signalling are used for speech as also for signalling with a trembler bell.

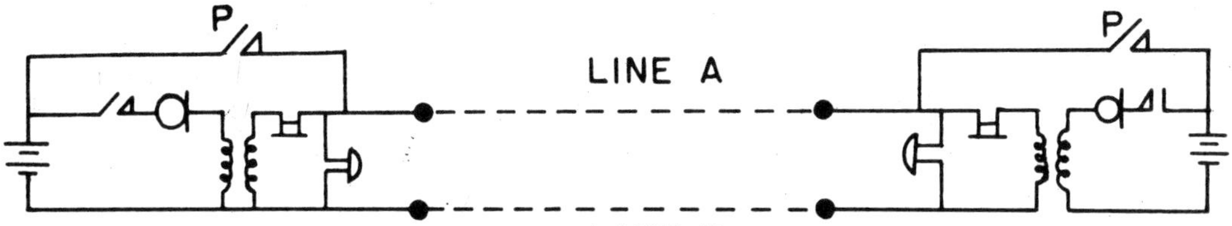

Fig. 4.27: Two Local Battery Telephones with Direct Current Signalling

Here the circuit in addition to providing normal speech between the two telephones connected by the lines *A* and *B* for transmission of speech, uses a trembler bell for signalling from the speaking battery to ring one telephone from the other telephone by operation of a press-button key *P*. Below is given the diagram as shown in Fig. 4.28 showing the circuit for the C.B. telephones with AC signalling.

Here the calling indicator *L* is connected to the *B* line and the loop is completed by an earth placed on *A* line in the exchange. On answering a call, the indicator and the earth are removed from the speaking pair and a C.B. transmission bridge is provided in the cord circuit. which may be either of impedance coil type or of repeating coil type. In the former type, the impedance element connected to A-wire serves as a supervisory indicator, while in the latter type, a separate supervisory indicator is provided in series with the repeating coil and the *A* line. In both the types, the supervisory indicator works so long as the call is in progress and is released when the subscriber puts off his phone. Two types of cord circuits are used by the telephone operators in joining a calling sub and a called sub. They are the bridge control cord circuits and the sleeve control cord circuits. Again the bridge control cord circuits may be either with negative supervisory signal or with positive clearing signal.

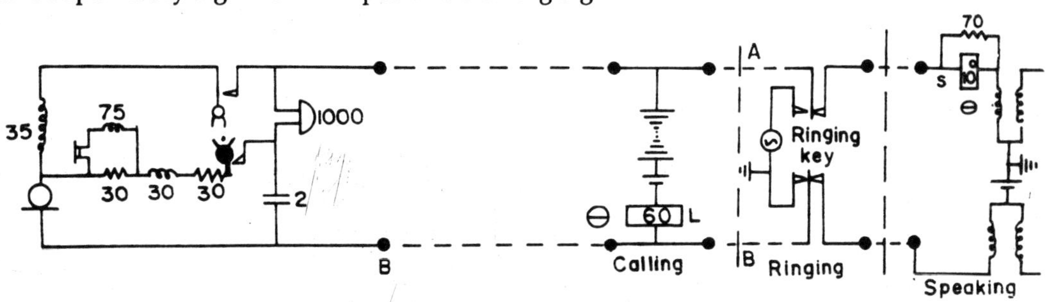

Fig. 4.28: Calling, Ringing and Speaking Conditions in CB Exchange

The bridge control cord circuit with negative supervisory signal is used to inform the operators when the joined parties have left their phones. In such manual switching systems, where the transmission bridge and the signalling relays are included in the cord circuit are known as the bridge control systems. In the diagram shown below in Fig. 4.29, a condenser bridge is depicted between the answering and calling cords.

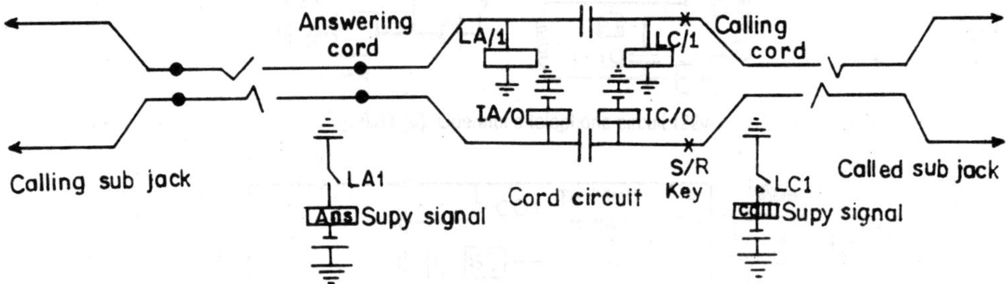

Fig. 4.29: Bridge Control Cord Circuit with Negative Supervisory Signal

The impedance coils *LA* and *LC* are connected to the tip of the cord circuits and are in the form of relays, the contacts of which control the answering and calling cord supervisory signals, respectively. Since the clearing signal is a disconnection in the subs' loop, the cord circuit supervisory signals are energised for as long as the subscribers' receivers are off the hook. This form of supervisory signal is called negative clearing, which can be made positive clearing by introduction of a third connection on the plugs and jacks as shown in Fig. 4.30.

In this case, the contacts of relays *LA* and *LC* are arranged either to disconnect or to shunt the supervisory signals during the conversation period and at the end of the call the contact LA_1 and LC_1 restore and complete the circuit via the third conductor of the plug and jack through the earth. Thus as soon as the telephone subscriber hangs up his receiver, the supervisory signal remains until the operator withdraws his plugs from the jacks. This type of positive clearing is nowadays the standard procedure for most of the public exchanges provided with supervisory relays.

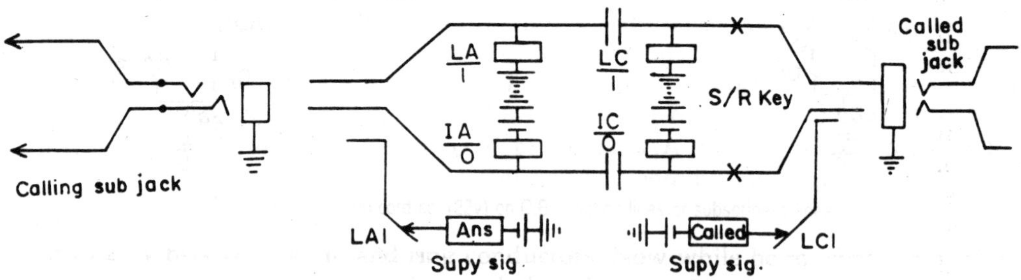

Fig. 4.30: Bridge Control Cord Circuit with Positive Clearing Signal

But in all the automatic and trunk exchanges, the signalling relays and the transmission bridges are placed in the line termination circuit, the cord circuit being only a pair of conductors with speak and ring keys. Since it is necessary in this system to pass the clearing signal from the line terminations to the cord circuit via the sleeve, the method of this type of cord circuit is called the sleeve control system as shown below in Fig. 4.31.

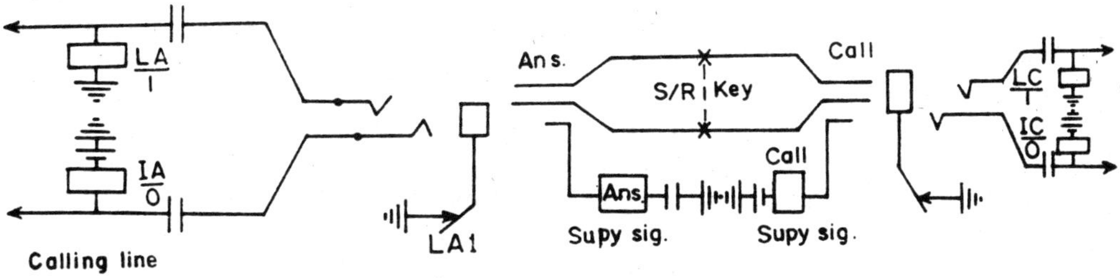

Fig. 4.31: Sleeve Control System

At the answering side of the cord circuit, the relay *LA* remains operated so long as the calling loop is there and releases when the conversation is over and the subscriber leaves his receiver. LA_1 completes the circuit via the brush of the jack and the sleeve of the plug to operate the answering supervisory signal.

At the calling side of the cord circuit, similar action takes place. Thus two transmission bridges are required here per connection, as compared with the single bridge in the cord circuit of bridge control system. The sole difference between the bridge control and sleeve control system is, that in the former, the supervisory relays are in the cord circuit, whereas in the latter the supervisory relays are in the line termination and the cord circuit supervisory signal is controlled from the line termination via the sleeve lead of the plug.

For an operator to put through a call in a switch board fitted with multiple jack field, it is essential to provide with an engaged test field facility to avoid double connection by the operator. The universal method is to provide a potential at the bush of the engaged line, so that an operator when taps on the bush of that multiple jack by the tip of a calling plug hears a click. To explain this we require to go through the following circuits as shown in Fig. 4.32 (a), (b), (c) and (d) respectively for (a) operator's telephone circuit, (b) subscriber's line circuit, (c) cord circuit for sub's line, and (d) the circuit condition during the testing of engaged line as given in Fig. 4.32.

The operator's head set current is regulated by a retard coil of 165 Ω in the primary circuit and an antiside tone arrangement in the secondary circuit. Now in case of non-multiple switch boards for 22 V, the 2 uF condenser in operator's head set is charged to 22 V p.d. and on inserting the answering plug into the calling jack of the answering field, the operator ascertains the number of the wanted subscriber from the calling subscriber and puts through the call to the required jack by inserting his calling plug into that jack. But if the switch board is of multiple jacks, the operator has to test for engaged condition of the wanted number. The bushes of the multiple jacks of a subscriber's line are connected generally to earth at the cut-off relay of the line circuit. If the line is engaged as shown in Fig. (d), the battery is supplied from the sleeve circuit of the engaging cord. Thus the resultant sleeve circuit current produces a potential drop of 3.5 V by the cut-off relay and thus the bushes of all the multiple jacks have a negative potential of 3.5 V from earth in accordance with the potential drop in the cut-off relay. An equivalent circuit for the relevant parts of engaged test conditions is shown in Fig. 4.32(e) below, where point B represents the bush of the engaged line with a potential of about 3.5 V below earth potential.

When the operator puts his speaking key on in the cord circuit position, a momentary current flows in the secondary circuit of his head set by which the 2 μf condenser in his head-set is

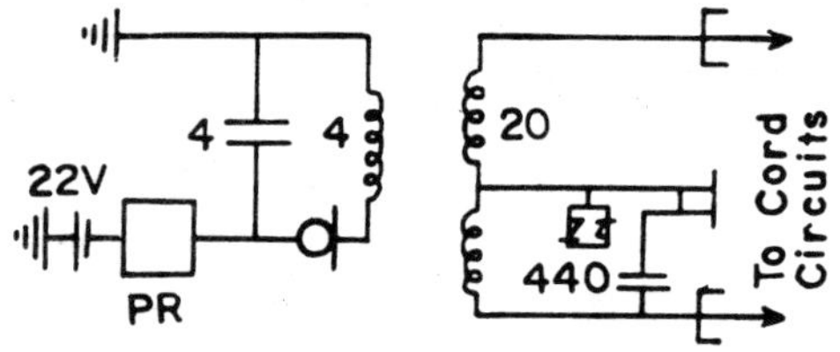

Fig. 4.32 (a): Operator's telephone circuit (22v)

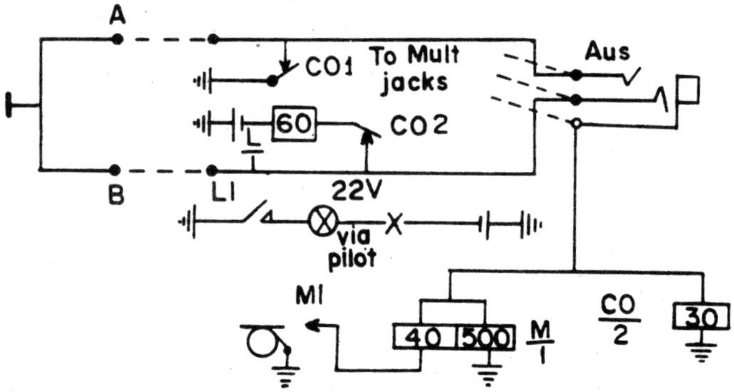

Fig. 4.32 (b): Subscriber's line circuit (22v)

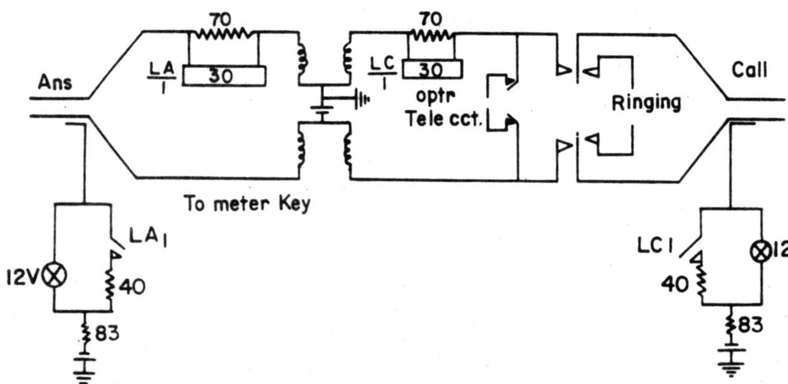

Fig. 4.32 (c): A position cord cct (22v) on C.B. junction lines or subscriber's lines

charged to 22 V between the tip and ring conductors. Now while he touches the bush of an engaged line with the tip of his answering cord plug, this negative potential on the bush causes a current to flow from earth at the repeating coil via 23 ohm winding and the supervisory relay in series and finally to battery through the sleeve circuit of the engaging cord. The current at the tip of the testing cord is found about 66 mA and the potential at the bush of the multiple jack is now found to be 2.9 V negative instead of 3.5 V before the application of the testing plug. Thus the p.d. across tip and ring of cord circuit and hence across the operator's telephone set is now 22 – 2.9 = 19.1 V. Thus the condenser in the operator's head set discharges until its p.d. falls to 19.1 V. and the discharge current flows throughthe secondary windings of the induction coil and the opertor's receiver and produces a loud click in the ear of the operator signifying thereby that the wanted line is already engaged and he intimates the same to the calling subscriber.

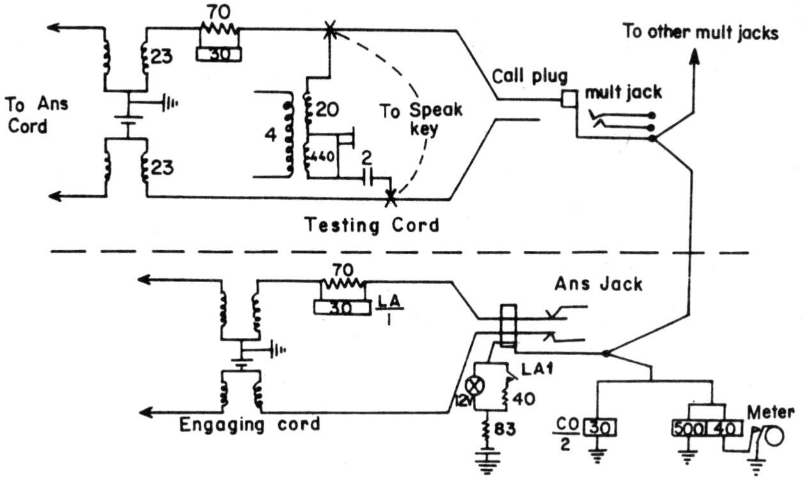

Fig. 4.32 (d): Circuit conditions during the testing of an engaged line.

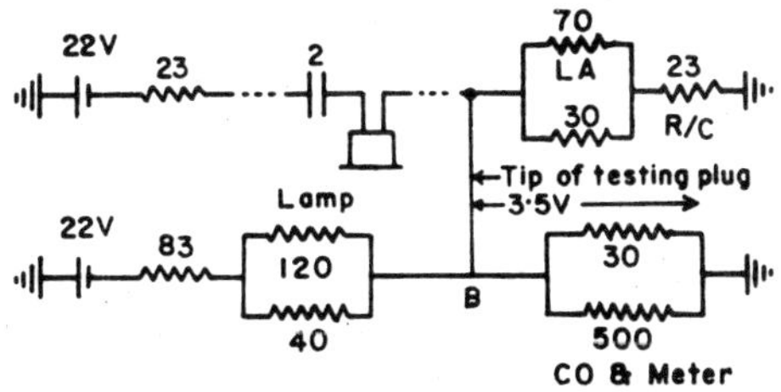

Fig. 4.32 (e): Equivalent CCT of engaged Test condition.

Metering: In C.B. exchanges, separate meters are provided for each subscriber for recording the local and junction-fee calls. As soon as a call is finished, the operator depresses the plunger type key relating to the answering cord where the calling sub's meter makes one step, debiting one unit fee against the calling subscriber's number.

Pilot and Night Alarm: During night hours, say from 10 PM to 5 AM, the traffic remains light, for which it is convenient to man the switch board with skeleton staff by providing pilot and night alarm circuits as shown below in Fig. 4.33.

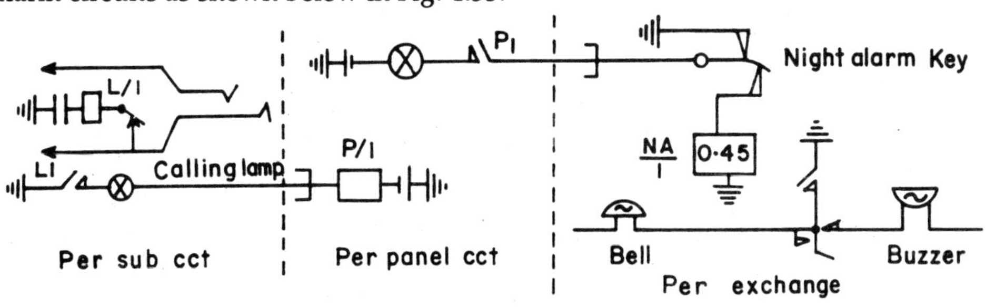

Fig. 4.33: Pilot and Night Alarm Circuits

All the calling lamp battery is supplied through a common 1.25 ohm pilot relay P. When any subscriber lifts his phone, his line relay L completes a circuit for the corresponding calling lamp through P relay. On operation of P relay, P_1 completes the circuits for a visual indication of pilot lamp. On putting down the night alarm key, the NA relay is operated at that time and an audible alarm is at once heard by the operator which takes quick action.

Local telephone with intercommunication switch: For the purpose of having the benefit of discussion with other members of an office, extension telephones are used with the main telephone connected directly to the exchange, so that proper reply can be given to the calling subscriber after ascertaining the exact position. For this an intercommunication switch is required with buzzer press button signal arrangement from main to extension and vice versa, so that (a) apart from the main telephone connected to the exchange, (b) the main telephone may be connected to extension, (c) the main telephone may be connected to extension and exchange held, or, (d) exchange may be connected to extension as shown in the diagrams below in Fig. 4.34.

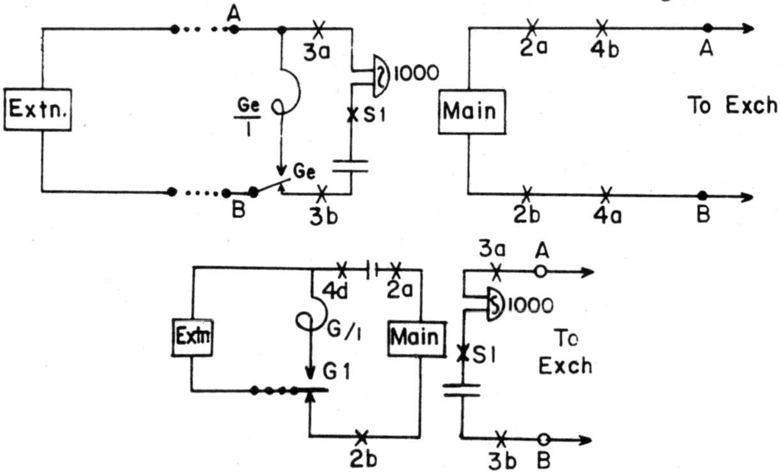

Fig. 4.34 (a): Main to Exchange. (b): Main to Extension

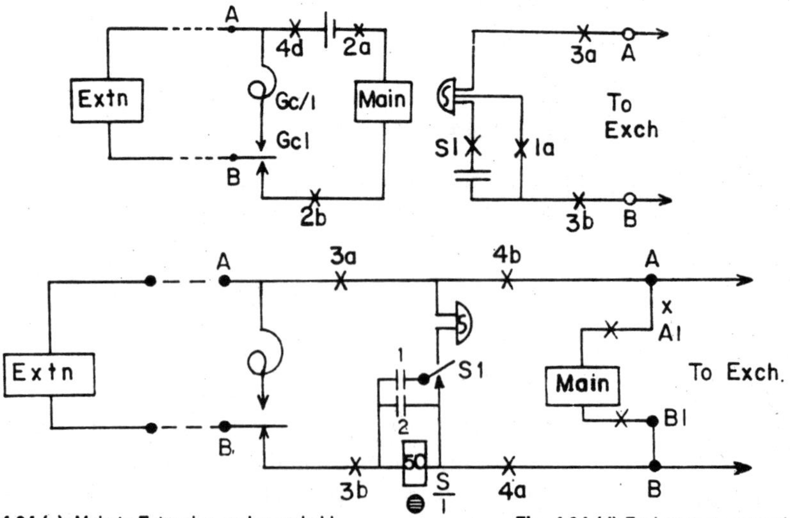

Fig. 4.34 (c): Main to Extension-exchange held

Fig. 4.34 (d) Exchange to extension

Fig. 4.34: Different Positions of Main and Extension Telephones

Simplified connections of Bell Set No 20: To facilitate this a complex switch called BPO bell set No. 20 is used with a four position key as shown below in Fig. 4.35.

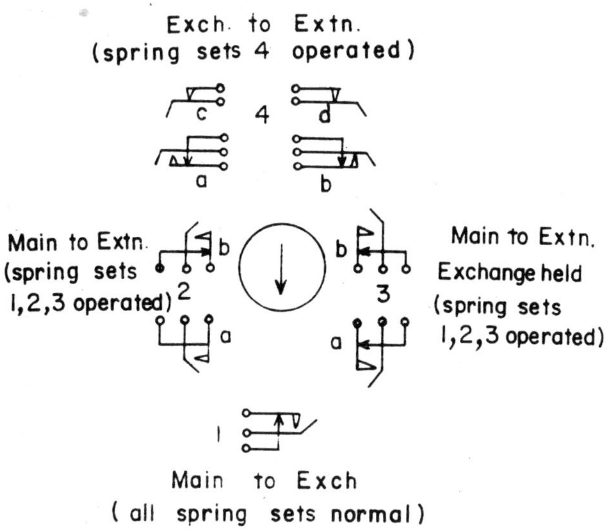

Fig. 4.35: 4-position Key of Bell Set No. 20

(a) On position 1, the main telephone is connected to the exchange when all spring sets are normal and when the extension telephone is terminated on a magneto bell, so that the extension may be called by a generator. (b) On position 2, the main telephone is connected to the extension when spring set 2 and 3 are operated and when a magneto bell is connected to the exchange line to receive a call. (c) On position 3, the main telephone is connected to the extension and the exchange is held when spring sets 1, 2, and 3 are operated. (d) On position 4, the exchange is connected to the extension telephone when spring set 4 is operated and when a supervisory in the bell set shows that the extension to exchange call is going on.

The rotary switch contacts are shown above with its four sets of springs for the four-position key as seen from the rear. For this bell set no. 20, Plan 7 with a single extension instrument is generally used. For Plan 7a which is a variation of Plan 7, straps of the bell are arranged to give secrecy on extension to exchange calls.

Similarly, Plan 5 and Plan 5a (secret) are utilised to avoid provision of private branch exchange, when a second extension is wanted. But one defect of Bell Set No. 20 is that the subscriber can irregularly hold the exchange equipment by leaving the switch on the exchange hold position. This difficulty is overcome by using bell set No. 39, where it is not possible to leave a holding condition on the exchange after the call is finished, where the holding condition is removed as soon as the receiver is replaced by introducing additional cradle switch contacts on the main telephone connected in the 100 ohm hold circuit.

Thus plans are devised to meet the requirement of the subscriber. There are other types of plans like Plan 8, where a single extension can be connected to any of the two exchange lines to avoid provision of a switch board for three telephones with two exchange lines. Plan 9 gives the facility for using one telephone with two exchange lines. But when more than two extensions are required, we use a small switch board called private manual branch exchange.

The original design of such switch boards were cordless type only. But they could be used for a few number of telephones only. Nowadays only double cord switch boards are used for private manual branch exchanges, where the one cord of a pair is used for answering a calling subscriber and the other cord is used for calling the wanted subscriber, just as is done in case of exchange switch boards and keys are provided in the cord circuit for speak and ring by the operators.

Private Manual Branch Exchanges: When a subscriber requires facilities for intercommunication which cannot be met by extension arrangements amongst his own office in the same premises in various floors and rooms, and at the same time when the subscriber wants to make public exchange calls and trunk calls, etc. from all the telephones from his office, he uses a private manual branch exchange (PMBX) in which case with a limited number of exchange lines, he can have as many extensions as required to meet his needs. Such PMBXs may be of three types, namely, the cordless switch boards, cord type switch boards, and multiple switch boards.

All modern PMBXs are designed to have through clearing from extension to the public exchange, i.e. when an extension subscriber replaces his phone after finishing the call, a clearing signal is given to PMBX operator as well as the public exchange switch board operator.

Cordless switch boards: There are four types of standard cordless switch boards for terminating exchange lines on drop indicators and extension lines on eyeball indicators. They are

$$\frac{1+3}{4}, \frac{2+4}{6}, \frac{3+7}{12} \text{ and } \frac{3+9}{12}.$$

The designation

$$\frac{3+7}{12} \text{ means } \frac{\text{Eqpt for 3 exch line + 7 extension}}{\text{Total capacity of except extension}}$$

These cordless type PMBXs are used in C.B. and automatic exchange areas with a local power supply of 12 V. The smallest cord type of switch boards, for use in C.B. and automatic areas is $\frac{5+20}{25}$ PMBX. The larger standard cord type PMBXs are $\frac{10+30}{65}$, $\frac{10+50}{65}$ and $\frac{10+60}{18}$. The maximum capacity in No. A type of PMBX is 160 exchange lines and 800 extensions, where inter-switch board lines and private wires are fitted in the exchange line multiple with the following facilities:

(1) Through clearing from extension to public exchange is the standard and both safety lamps on PMBX light when the extension clears.

(2) Individual positive supervision on extension-to-extension call with separate battery feed to the calling and called lines.

(3) Ring and dial facility on calling cord and ring back facility on answering cord.

(4) Standard click on engaged test provided.

(5) Direct dialing from extension to public exchanges possible, etc.

The smaller PMBXs are to work on a local battery of 12 V by means of primary cells or power leads from main exchange power unit.

In case of larger PMBXs they are served by 24 secondary cells installed in the premises or from power leads from the main exchange power unit.

Cordless PMBXs: Cordless types of PMBXs are also in use in local battery areas, where no transmission bridge is required but a supervisory relay R is used in each connecting link and it is isolated from the exchange by a 2 μF condenser as shown in Fig. 4.36 below for cordless switch board for local battery areas for one connecting link (key contacts omitted).

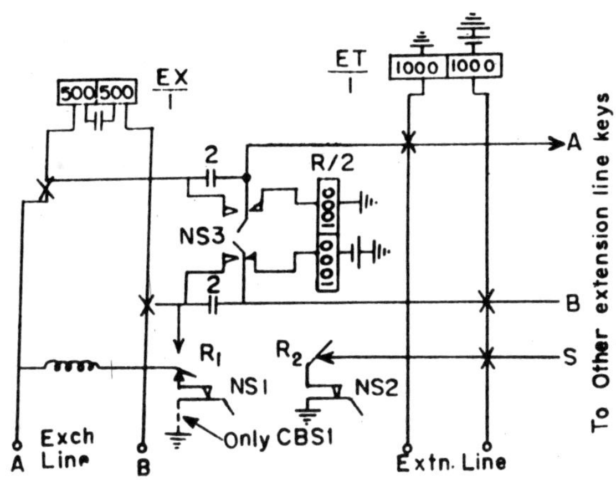

Fig. 4.36: Cordless Switch Board for Local Battery Areas·

Contact R_2 provides a reoperative circuit for extension indicators via switch board key to give positive clearing signal, and contact R_1 puts a 250 ohm hold coil across exchange line when extension receiver is off, thus providing a holding loop to the exchange. The night switch (NS) key disconnects the relay R and short-circuits the condensers, thereby an extension can call the exchange direct. In CBS1 areas, the back contact of R is wired to earth for giving normal clearing signal at the end of a call (A line earthed).

Cord circuit for 25 line PMBX boards: The cord circuit for a 25 line PMBX switch board is shown below in Fig. 4.37.

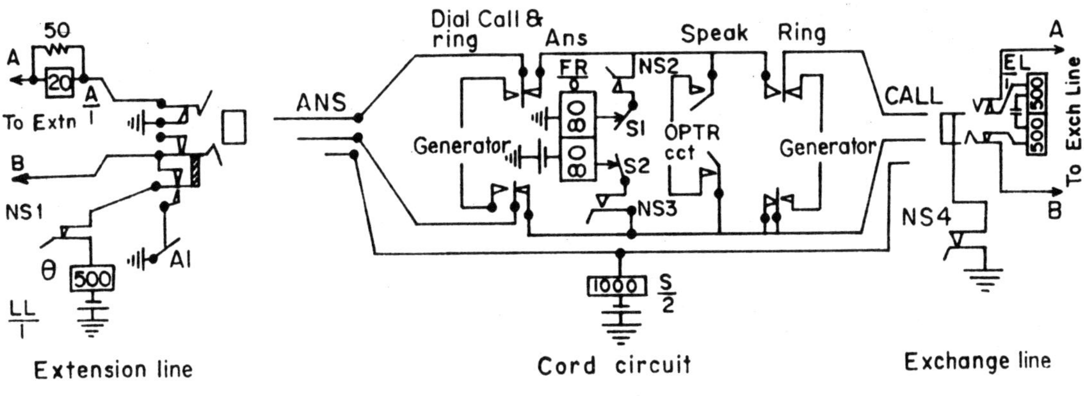

Fig. 4.37: Cord Circuit for 25-line PMBX Switch Board

It has a capacity for five exchange lines and 20 extensions with a maximum of eight cord circuits provided with supervisory indicators. An individual transmission bridge is arranged for the cord circuit and disconnected at exchange calls at S_1 and S_2. While the bush of each exchange line jack is earthed, no earth is given to the bush of extension line jacks. Thus when a plug is inserted in an exchange line jack, relay *S* operates and removes the transmission bridge. The supervisory relay *A* here is associated with extension line circuit and at the end of a call, A is released due to break in extension loop and contact A_1 extends the extension indicator's circuit to give positive clearing signal, until the answer plug is removed. NS key disconnects extension indicator, transmission bridge and alarm bell.

Cord circuit for 65 and 180 line PMBX boards: These are shown in Fig. 4.38 below. Of the two models, the first gives an ultimate capacity of 15 exchange lines and 50 extensions and the second gives a maximum capacity of 20 exchange lines and 160 extensions. Both the models are similar, with the difference that the 180 line switch board has a capacity of 18 chord circuits, while the other model has 15 cord circuits, and that the jack field of the larger board is arranged in two panels with some difference in positions of the exchanges line boards. The main difference between this switch board and 25 line switch board is the provision of supervisory indicators in the cord circuits. Hence separate indicators are provided for answering and calling cords. These indicators are negative in operation, i.e. the indicator discs are visible while the call is going on and disappear when the call is finished. This is the reverse of positive singalling indicators in cordless and 25 line PMBXs. The extension indicators are doll's eye type whilst the exchange indicators are drop type. The bushes of all exchange line jacks are earthed as shown in the diagram, but the extension jacks are not earthed. On extension to extension calls, the relay S is not operated due to absence of earth on sleeve and (80 + 80) ohm retard coil *FR* is connected to speaking conductor. The supervisory indicators A and B operate to the calling and called extension loops and give a visual indication that the call is in progress.

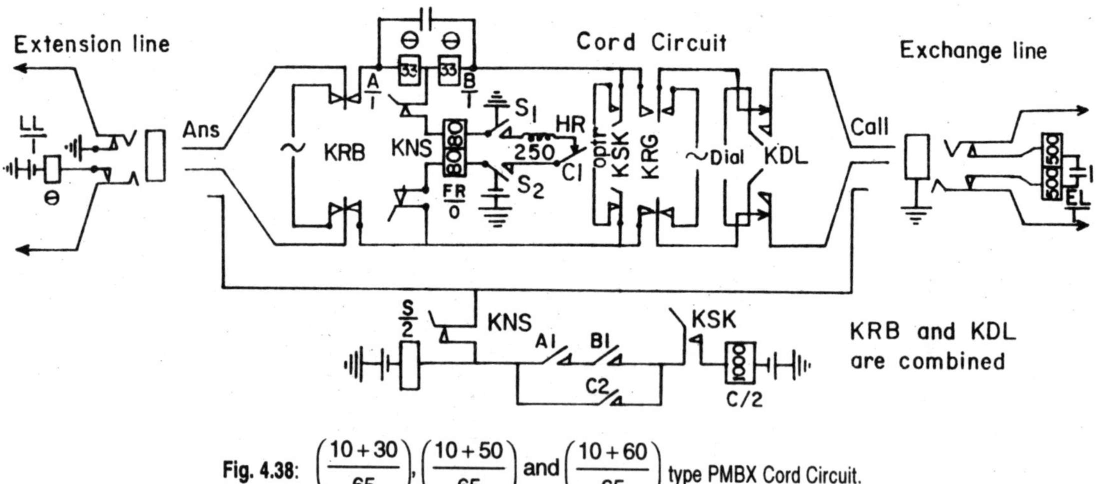

Fig. 4.38: $\left(\frac{10+30}{65}\right), \left(\frac{10+50}{65}\right)$ and $\left(\frac{10+60}{65}\right)$ type PMBX Cord Circuit.

On exchange line calls relay *S* operates through the earth on the bush of the exchange line jack and the contacts S_1 and S_2 remove the transmission bridge from the tip and ring wire and instead put a holding loop *FR* and *HR* in series. When the extension receiver is off the hook, relays *A* and *B* operate in series from the current in the loop to the exchange, A_1 and B_1 complete a circuit for relay C, which locks via the speak key contacts and C_2 to earth in the sleeve circuit. C_1 removes the holding loop from cord circuit. The night service key KNS on operation removes the transmission bridge and breaks the operate circuit of relays S and C.

When the subscriber requires bigger number of exchange lines, multiple type PMBXs called No. 9 type are given. In it each position with two panels has a capacity of 20 exchange lines and 200 extensions with 17 cord circuits. There are no separate answering jackets, but each line is multiplied on a four panel repetition basis. The cord circuit of No. 9 type PMBX is practically the same as that of 65 and 180 line switch board.

The basic unit of No. 1A type PMBX is a two panel single position section, the key shelf of each being fitted with 12 cord circuits with wiring for a maximum of 16 cord circuits per position. Each pair of cord circuits is provided with a separate speak and ring key and a ring back/dial key, with separate supply lamps for answering and calling cords. Each jack field panel has a capacity for 40 exchange lines in strips of 10 jacks and 200 extension lines in strips of 20 jacks. Each position can therefore operate for 80 exchange lines and 400 extensions. All lines are multiplied on a 4-panel repetition basis. The apparatus racks are arranged in two units, the first containing the connection strips for termination of cables from the MDF multiple, together with long extension line relays, fuse panels, and exchange line relay sets, and the second is a supplementary rack to contain relay sets, when there is a very high number of exchange lines, private wires, etc.

4.7 Principle of Signalling on Junction Lines and Trunk Lines

Various types of DC and AC signalling are used in junction and trunk lines. On DC signalling, they may be (a) C.B. signalling, (b) differential signalling, and (c) loop signalling; on AC signalling, they are (a) generator signalling, and (b) V.F. signalling.

C.B. signalling: We know that in a C.B. system the transmission bridge provides battery potential to one conductor and the earth to the other conductor in a local exchange, when one subscriber is joined to the other through cord circuit. These conditions of DC signalling over the junction lines is called the C.B. signalling as shown in Fig. 4.39.

The maximum junction loop resistance is dependent on the sensitivity of calling and supervisory relays and also on the voltage of the signalling battery. The loop resistance of the junction workable with primary cells exchange is only 800 ohm. But if secondary cells are used, the loop resistance of the junction between those exchanges are given below:

(1) 22 V and 24 V exchange — 1300 ohm junction loop.
(2) 40 V " — 3600 " " "
(3) 50 V " — 4800 " " "
(4) 60 V " — 6500 " " "

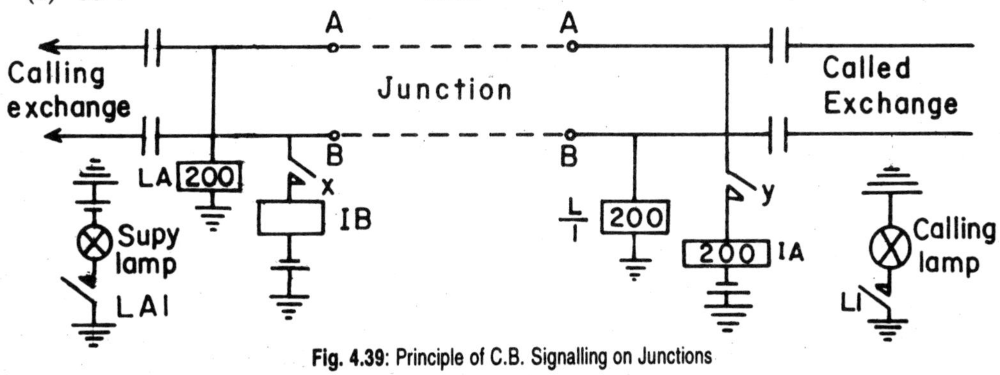

Fig. 4.39: Principle of C.B. Signalling on Junctions

When the originating exchange operator inserts the plug into an outgoing junction jack, connection X is closed and negative battery is sent out over B wire to the distant exchange where relay L is operated and L_1 lights the calling lamps, on seeing which the called exchange answers. Then the connection Y is closed and battery is returned to the originating exchange over A wire to operate the ordinary cord circuit supervisory relay LA, whose armature LA_1 disconnects the supervisory lamp at the calling exchange. Then the operators at both ends exchange informations and put through the called sub to the calling sub. Thus in C. B. signalling, the two conditions are: (a) battery is supplied over the B wire to call the distant end, and (b) the battery is returned on the A wire for supervision at the originating end.

Differential signalling: This is used on long junctions where line loop resistances are too great for C.B. signalling as shown below.

Signal Voltage, V	Loop resistance, ohm
20	3,200
30	7,200
40	11,200
50	15,200

Here the DC signalling lines are virtually connected in parallel, though for speech retard coils are used to separate the two conductors. The line current in the calling condition is reduced to zero by an opposing emf at the distant end, when the called exchange answers. The principle of differential signalling is shown in Fig. 4.40.

When the calling exchange operator inserts a plug into the outgoing junction jack, the contact X changes over and battery is applied to the A and B wires in parallel via one winding of DR and both the reatard coils of I. Also the battery current flows through the other windings of DR through resistance YA to earth, thus equalising the line current. But as the two currents are equal and opposite, due to the magnetically opposing windings of DR, the relay remains inoperative.

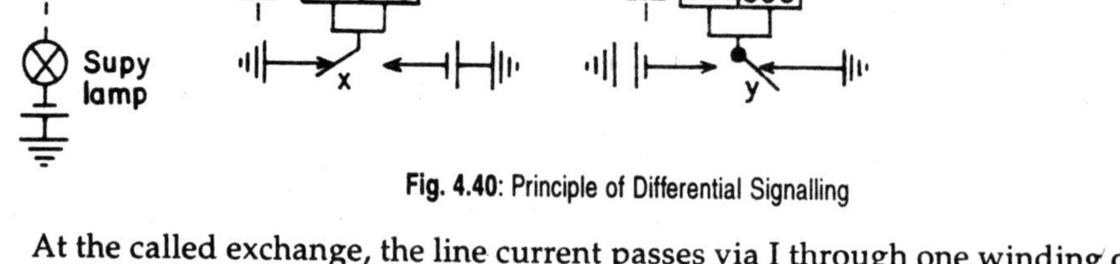

Fig. 4.40: Principle of Differential Signalling

At the called exchange, the line current passes via I through one winding of DR to earth at contact Y and as there is no current in the second winding for neutralising the same, DR operates and its contact DR_1 lights the calling lamp when the called exchange operator answers. Then the contact Y is made to change over and the battery is now connected to DR and opposes the line current. But as the batteries of both the stations are of the same emf, the line current is reduced to zero, thus disconnecting the supervisory lamp at DR1 at the calling exchange. However, the relay DR remains operated by the current flowing through its second winding through YA to earth.

When the called sub finishes, Y releases and the resultant line current neutralises the effect of the current in the local circuit of DR relay at the calling exchange, thereby relay DR releases and DR_1 gives a clearing signal when the calling operator withdraws his plug. The release of X restores the circuit to normal at the calling end, whilst the stoppage of line current releases the DR relay at the called exchange and gives a clearing signal.

Loop signalling: Loop signalling is used between two exchanges when the calling exchange is automatic and the distant called exchange is manual and these are joined by junction lines. The loop on the *A* and *B* lines at the automatic exchange with battery and earth supply via a calling relay at the manual exchange gives the calling conditions and a reversal of the battery and earth returned from the manual exchange when the called subscriber replies, gives the supervisory signal at the calling end as shown below in Fig. 4.41.

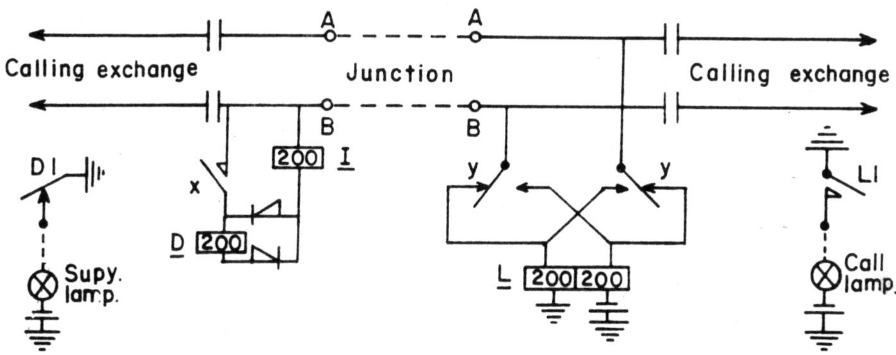

Fig. 4.41: Principle of Loop Signalling on Junction Lines

As soon as the originating exchange operator inserts his calling plug into an outgoing junction, the *X* contact is closed and the *A* and *B* wires of the junction get looped via the impedance coil I and D relay, energising the L relay at the called exchange and thereby lighting the calling lamp through the made contact L_1. The D relay does not operate due to the rectifier during the calling condition. When the called exchange operator answers on seeing the call lamp glowing, the change over Y contacts reverse the line current and L relay coil connections and D relay now operates and puts out the supervisory lamp by disconnecting the contact D_1.

When DC ringing procedures fail on long physical lines, magneto ringing or AC ringing is used by means of a generator, the principle of which is shown in Fig. 4.42.

On insertion of calling plug into the outgoing junction of the calling exchange, the contacts X apply 17 c/s generator ringing AC current to the junction, which operate the line relay *L* shielded by a rectifier at the called exchange. The contact L_1 of L relay operates a relief relay LL, which locks itself by its own contact LL_1 and lights the calling lamp by its contact LL_2. When called exchange wants to ring off the calling exchange at the end of the call, contact Y changes over and ringing current from the called exchange operate the AC relay L in the calling exchange and the operations are repeated and LL_2 contact lights the clearing lamp at the calling end. Thus the magneto or generator signalling at the junction line is bothways and can be used for long distance trunk lines also, up to a resistance limit of about 5000 ohm.

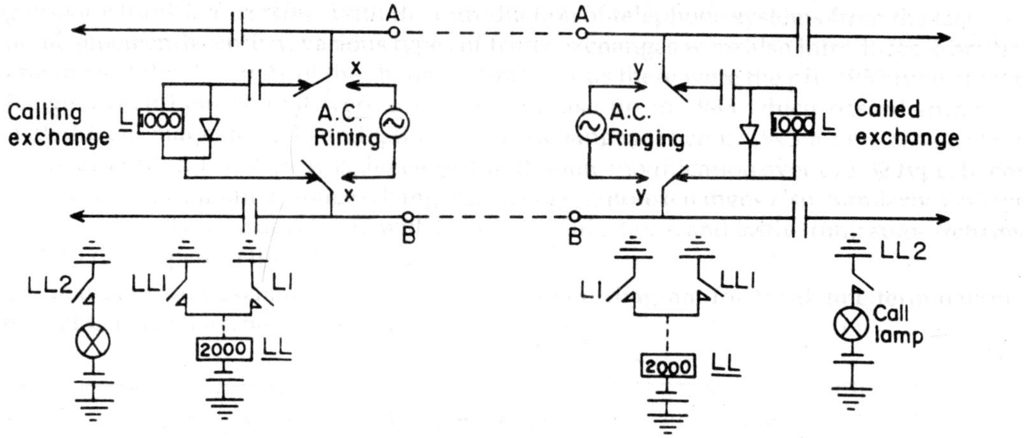

Fig. 4.42: Principle of Generator Signalling or Junctions

V.F. Signalling: Voice frequency signalling is mainly used in underground long-distance circuits with loaded cables and specially in carrier channels, as the magneto ringing which uses continuous high voltage and current of the maneto generator may damage the delicate elements of the loaded cable and carrier channels. Therefore a 17 c/s ringing from a magneto generator is first converted into a continuous 500 c/s tone, which is subsequently converted into 500/17 c/s by an interruptor panel and sent to the loaded underground cable junction or carrier channel. At the distant end it is again reconverted into 17 c/s ringing current to operate the exchange relays. A delay device is used to reduce the false operation on frequencies other then 500 c/s for a considerable length of time. The principle of a V.F. ringing converter is shown in Fig. 4.43.

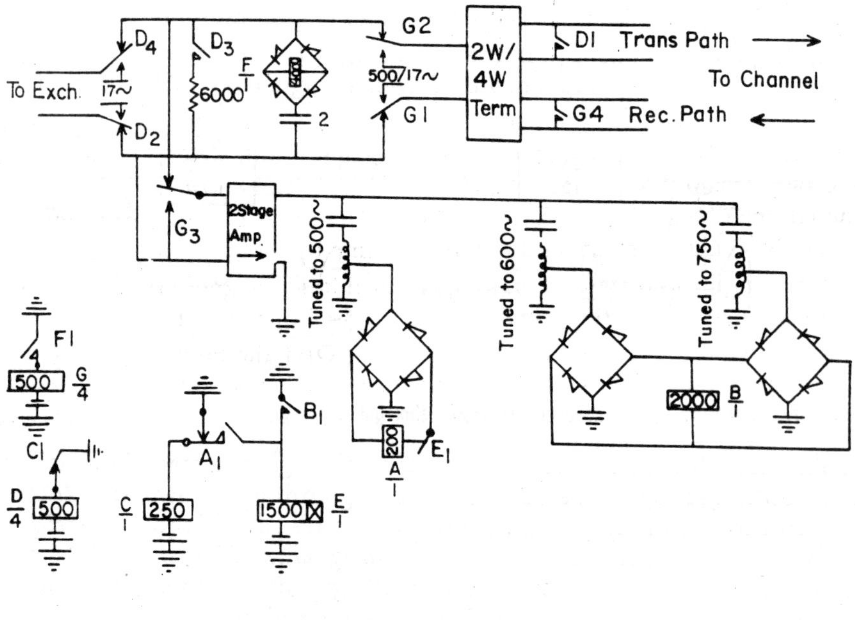

Fig. 4.43: 500/17 c/s Conversion Circuit

The exchanges at both ends of junction, trunk lines or carrier channels use 17 cycles ringer generators for sending 17 c/s ringing, which is converted to 500/17 c/s tone to be transmitted in the junction or trunk or channel as follows. 17 c/p ringing current from the exchange operator operates F relay and contact F1 operates G relay whose contacts Gl and G2 supply 500/17 cycle tone to be transmitted via 2W/4W panel in the transmit path. This 500/17 cycles tone when received at the other end enters a two staged amplifier. The output of this amplifier has three tuned circuits, the first being tuned to 500 c/s, the second to 600 c/s and the third to 750 c/s. The relay A is connected through a rectifier bridge to 500 c/s tone circuit, whilst relay B is connected to both 600 c/s and 750 c/s circuits to separate the true 500 c/s tone from the unwanted tones. *A* relay operating releases C and contact C1 then operates D relay and contact D2 and D4 supply free 17 c/s ringing current to operate the exchange relay. Similar operations take place when the distant end operator send 17 c/s ringing.

4.8 Unidirectional and Both way Jack-ended Trunks and Junctions

The junctions are terminated in jack strips with calling lamps below the answering jacks in the switch board panel in separate rows for incoming and outgoing jacks. In big C.B. exchanges, separate positions are kept for dealing with incoming junction calls and these positions are called B-positions, to differentiate from A positions, which are provided primarily to deal with outgoing calls from local subscribers. A diagram for unidirectional jack-ended junctions is given in Fig. 4.44.

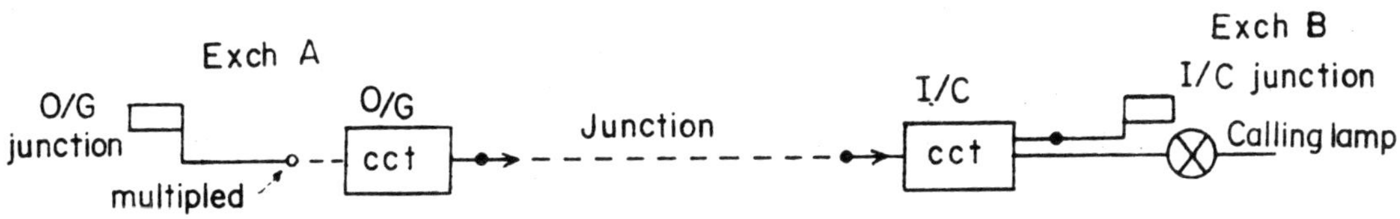

Fig. 4.44: Jack-ended Unidirectional Junction

As the unidirectional jack ended junctions are used in small groups of five or so, it becomes uneconomical at times to spare a separate operator only for the purpose and as such recourse is taken to both way jack-ended junctions as shown below in Fig. 4.45.

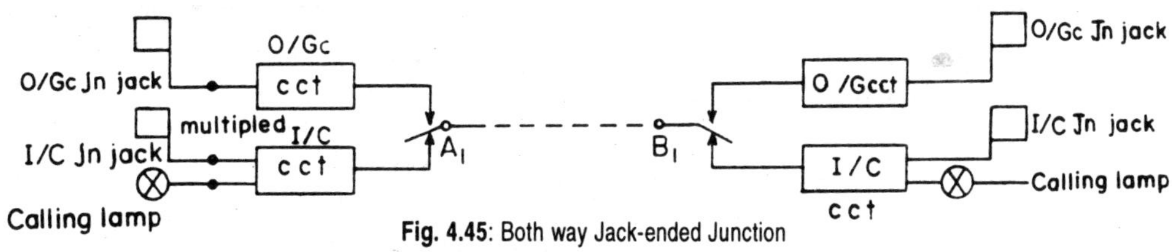

Fig. 4.45: Both way Jack-ended Junction

Ordinarily the junction is switched to incoming equipment at both ends. When a plug is inserted into an O/G jack at end A, the changeover contact *A*1 connects the O/G jack to the junction to clear the traffic to the *B* end of the junction terminated to the I/C jack via I/C circuit, thus lighting the calling lamp to draw the attention of operator there. Similar sequence of operation takes place when a plug is inserted into the O/G jack at end *B*.

Long distance trunk line working: With the introduction of telephone systems from the last quarter of the nineteenth century, various types of trunk exchanges were also introduced from time to time to meet the demands of the changing times. Thus there were the pre-1932 type of trunk exchange switch boards, that were exclusively in use up to 1943 which passed trunk calls orginated from magneto, CBS and CB types of exchanges. Then in 1943 a new type of trunk switch board called T-43 type was developed with some modification over pre-32 type, to cope with trunk calls originating from exchanges including auto-exchanges also. Similarly T-53 type trunk switch boards were developed with further modifications, and lastly trunk auto-exchanges have been developed in the late sixties.

For manual trunk exchange operation, a simplified diagram for trunk line termination in respect of C.B. local exchanges is shown in Fig. 4.46.

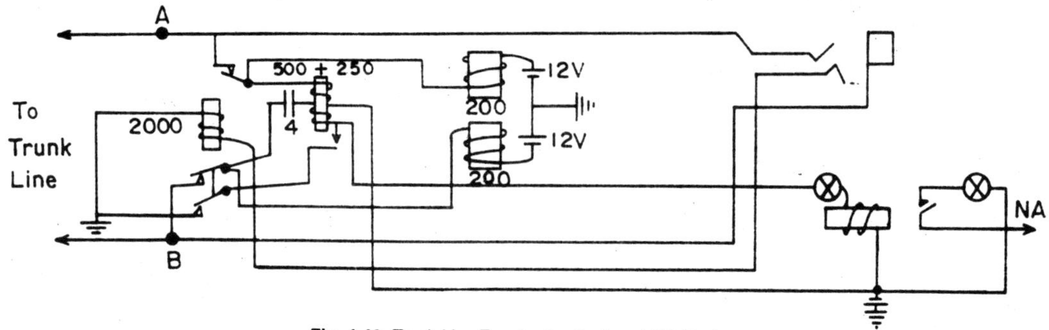

Fig. 4.46: Trunk Line Termination for Local CB Exchange

A locking relay is used to receive calling current from the distant end and is cut out when a calling plug is inserted into the jack by means of a 2000 ohm relay connected to the bush of the jack. Calling is done by generator alternating current and clearing by means of loop direct current. Calls from a trunk exchange to other CB exchanges are put through order wire junctions outgoing from the trunk exchange. An order wire junction outgoing from a trunk exchange to a local 40 V C.B. exchange is shown below in Fig. 4.47.

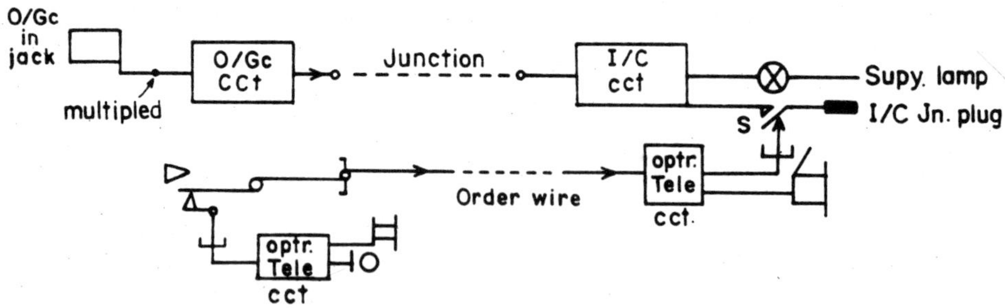

Fig. 4.47: Plug-ended Order Wire Junction

A special circuit known as plug-ended order wire junction is provided, in which the operator in the trunk exchange to put through a call to the *CB* exchange, depresses a plunger type key, which places the operator's headset in direct communication with the incoming junction line operator at the distant *CB* exchange over the order wire. The trunk operator then requests the *CB* exchange operator to put through the required number; when the *CB* exchange operator calls back to the trunk operator the number of the free junction to be used for the purpose. Both the operators then complete the connections simultaneously at both ends by plugging into the junction assigned. Trunk cord circuit for T-32 type trunk exchanges is shown in Fig. 4.48.

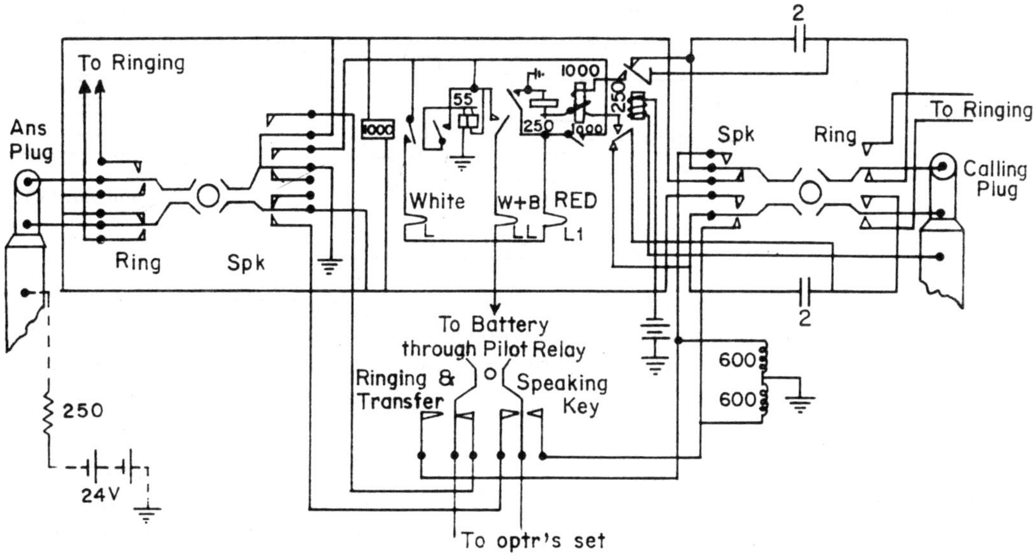

Fig. 4.48: Trunk Cord Circuit for T-32 Type Trunk Exchange

For identification, generally the answering plug is fitted with black cover, while a calling plug is fitted with a red cover. The answering and the calling plugs have their own speak and ring keys. When the transfer key is in its normal position, the answering side speak key on being thrown to speak position, bridges the telephone set across the cord, enabling the operator to speak on a trunk line or tap a connection already set up. But the calling speak key, on being thrown to speak position, disconnects the calling plug from the answering plug and by throwing on this key, the operator can speak on a junction to a subscriber without interfering the subscriber connected to the other end of the trunk cord circuit. The 600 + 600 ohm coil connected to the outer contacts of the calling speak key is to provide for taking a busy test click on junction lines, etc. multiplied round the switch board. Lamp L gives the clearing signal from the distant end, L_1 gives the signal controlled by the subscriber connected by the calling plug via a trunk junction to a local exchange, whilst L_2 lights when an alternating current ring-off from either the trunk or trunk junction is received and remains alight by means of locking relay until released by throwing off the speaking key of answering plug.

When both the answering and calling plugs of the trunk cord circuit are inserted into the other trunk line jacks, the 250 ohm relay connected to the sleeve of the calling plug is inoperative due to no connection on bush of trunk line jack. The 2 μF condensers are short-circuited and the only signalling apparatus consists of the 1000 ohm relay bridging the line and responds to either ring off alternating or clearing battery current and controls lamp L. But when the calling plug is inserted to a trunk junction, the 250 ohm relay operates through the 250 ohm resistance connected to the trunk junction jack. Then the short-circuits are removed from the two condensers and across the junction side of them are connected a differentially wound relay (1000+1000 ohm) with its center connected to a 250 ohm relay to earth for responding to ring off from a subscriber in magneto-exchange controlling lamp L_2. But when a connection is made to a C.B. exchange, the 250 ohm relay connected to the sleeve of the calling plug controls lamp L_1.

4.9 Time Checking of Trunk Call Periods

A calculograph is used to punch the trunk call tickets with dial prints, once when the call is commenced and again when the call is finished, to indicate the period for which the call is to be charged for. A simplified circuit diagram explaining the principle of working of a calculograph is shown below in Fig. 4.49.

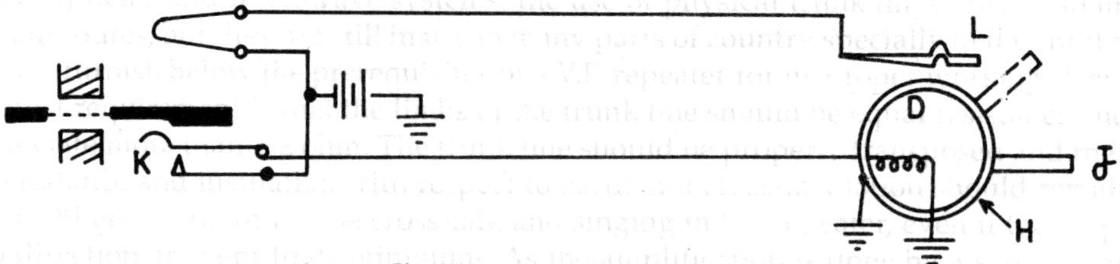

Fig. 4.49: Automatic Time Clock Circuit

In this case, *H* is a soft iron drum driven by means of an electromagnet and power gear actuated in response to the seconds impulses of an electric clock. The drum revolves once in every 24 minutes. A disc with a projection J at one end is mounted on the same pinion as the drum and can turn within an angle of 50 degree about the pinion. If the key *K* is pulled out, a circuit is completed through coil *D* which on magnetising the drum causes the disc to revolve when the projection J travels towards the contact *L*. As the end of three minutes the projection meets *L* and completes a circuit for the lamp to glow. The drum continues to rotate but the disc slips and is held against the contact *L* until the key is pushed in. Thus the circuit of coil D breaks and the soft iron drum is demagnetised when the disc with its projection *J* comes back to its normal position.

This mechanism is fitted on the trunk board at the centre of the key shelf of each section, with its clock face upwards fitted with two handles at the right and left. When a call begins, the record ticket is put in slot under the clock face and the right hand handle is pushed back when a dial diagram is printed with a triangular dot showing the hour and an arrow indicating the minutes, as shown when the call began.

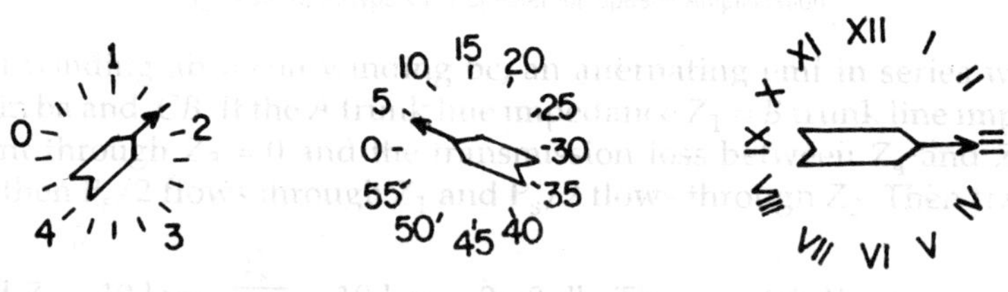

Fig. 4.50: Calculographer Dial Prints

In this case, the left hand dial revolves once in five minutes and the middle dial once per hour, with their arrows normally pointing to zero and are controlled by the left hand handle. When the conversation is over, the ticket is again put in the slot and the left hand handle is pushed back when impressions of minutes elapsed since the first standing are printed as shown in Fig. 4.50. The above figure shows that the connection was made at 8.15 am and last for 5 + 1½ = 6½ minutes.

Trunk line operating methods: There are three methods of operating trunk lines. These are (a) No delay working, (b) Delay working, and (c) Demand working.

(a) In No Delay working sufficient circuits are provided for all the calls to be completed in order of their booking times and priority, such as ordinary, urgent, important, most urgent and immediate, etc. as soon as they originate.

(b) In Delay working, the trunk calls are at first booked and later on the trunk operator rings the subs to put through the calls in turn, according to priority on the route.

(c) In Demand working, bulk of the traffic is cleared within a predetermined period, say one minute, the sub remaining on the line when an effort is made to put through the call. Thus 85 per cent of busy hour trunk traffic is cleared within a minute of demand and the remaining 15 per cent is cleared on delay basis.

Trunk exchange and toll exchange: In the early thirties, all trunk traffic on open wire trunk lines were cleared on delay basis. But now due to the availability of sufficient trunk outlets due to the installation of carrier systems, coaxial system during the forties and fifties, and during the sixties the metro wave system, etc., no delay and trunk demand working procedures are in general use since then. Now exchanges within 50 miles radius of a zonal switching centre are worked on no delay basis through the zone centre exchanges, which are called exchanges and the circuits are called toll circuits. Such toll areas where the local exchanges are mainly auto exchanges are generally called auto-manual exchanges. But the trunk lines of length greater than 50 miles are taken as long-distance circuits and are generally terminated in a separate exchange called trunk exchange in big cities. Both the toll and trunk calls are cleared on tickets, which are separately charged against the calling sub.

Trunk line commutator: Just as the local lines enter a telephone exchange through MDF via the protective devices, so also the physical trunk lines enter a trunk exchange commutator through the local Repeater Station Distribution Frame (RSDF) via the projective devices and jumperings, etc.

A diagram for Trunk line Commutator is given below in Fig. 4.51.

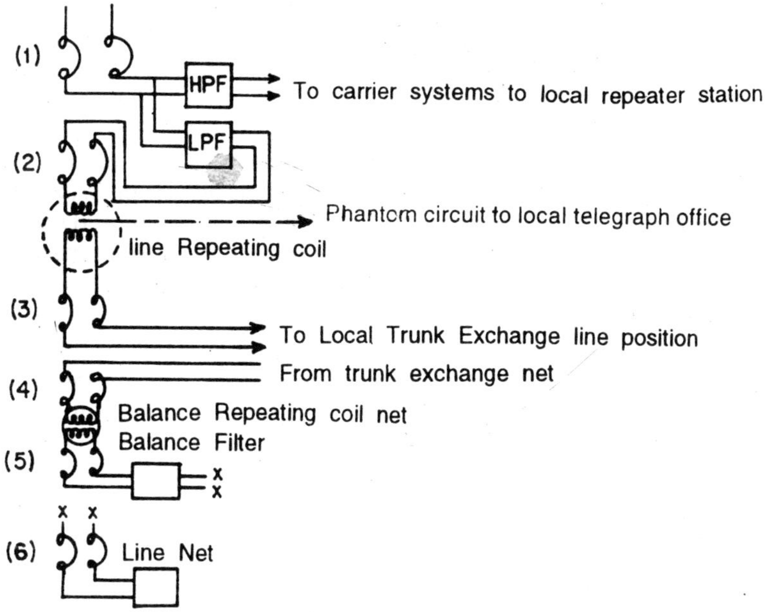

Fig. 4.51: Trunk Line Commutator

4.10 Use of V.F. Repeater on Physical Trunk Lines

Up to late thirties, when bulk of the trunk traffic was cleared on physical trunk lines, voice frequency repeater played an important part for amplifying the attenuated speech current and regenerating the ringing current, which also considerably decreased to operate the ringing relay at the other end of the long trunk lines. Though due to the installation of carrier systems, coaxial systems and microwave systems, the use of physical trunk lines are not so important on main routes, but they are still in use in many parts of country specially in the rural areas. As such we furnish below the prerequisites of a V.F. repeater for its proper working. The first and foremost requirement is that the limbs of the trunk line should be equal resistance and should not deviate more than ± 2 ohm. The trunk line should be properly transposed and the conductor resistance and insulation with respect to earth and cross-insulation should remain within limits. Otherwise there will be cross-talk and singing in the repeater, even if the amplifiers in each direction are kept to its minimum. As the amplification is done by a separate amplifier, the speech current in each direction is separated before entering the repeater, so that they can be separately amplified undirectionally in the amplifier and then sent to the same trunk line in opposite directions. For this a two-way two-element repeater using a differential transformer consisting of a four-wire termination set, usually known as Hybrid coil, is used to convert a two-wire circuit into a four-wire circuit and vice versa at each end of the V.F. repeater. Its wiring is shown below in Fig. 4.52.

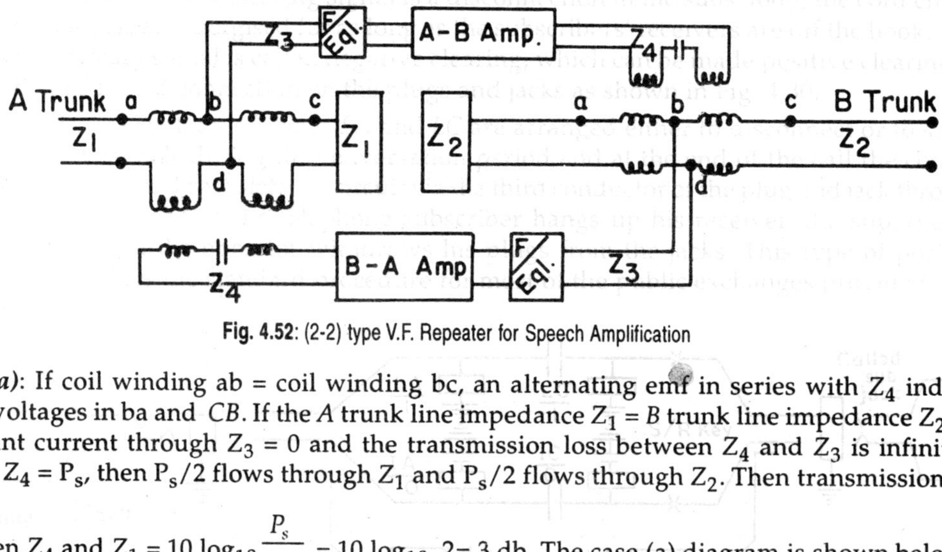

Fig. 4.52: (2-2) type V.F. Repeater for Speech Amplification

Case (a): If coil winding ab = coil winding bc, an alternating emf in series with Z_4 induces equal voltages in ba and *CB*. If the *A* trunk line impedance Z_1 = *B* trunk line impedance Z_2, the resultant current through $Z_3 = 0$ and the transmission loss between Z_4 and Z_3 is infinity. If power $Z_4 = P_s$, then $P_s/2$ flows through Z_1 and $P_s/2$ flows through Z_2. Then transmission loss

between Z_4 and $Z_1 = 10 \log_{10} \frac{P_s}{Ps_2} = 10 \log_{10} 2 = 3$ db. The case (a) diagram is shown below in Fig. 4.53 (a).

Case (b) : If emf is in series with Z_3 as shown below in Fig. 4.53(b), then with same argument, we can show that the voltage induced by ab in Z_4 = voltage induced by bc in Z_4 and as such their effects are cancelled and no current flow through Z_4. Therefore the transmission loss between Z_3 and $Z_4 = \infty$ and the same between Z_3 and $Z_2 = 3$ db.

Case (c): If emf is in series with Z_1 as shown in Fig. 4.53 (c), then ab and ba are on the same core having equal number of turns n_1 in each coil and if n_2 is the number turns in ef coil in such a way that n_1: n_1: n_2 = 1:1:$\sqrt{2}$, then if impedance across ab = impedance across bd (Z_3), then we find that p.d. across ab = p.d. across bd.

But p.d. across ab = p.d. across bc

p.d. across bc = p.d. across bd.

i.e. d and c are at the same potential and therefore no current flows through Z_2.

Again since equal powers flow through ab and Z_3 and the same power is transformed through Z_4, therefore same power flows through Z_3 and Z_4.

Therefore the transmission loss between Z_1 and Z_3 = 3db and the transmission loss between Z_1 and Z_4 = 3 db.

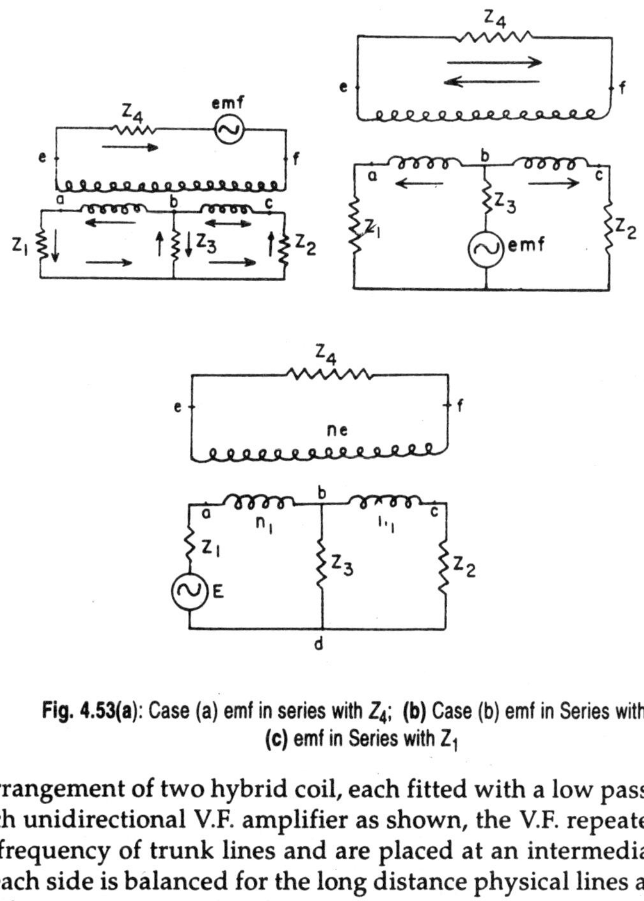

Fig. 4.53(a): Case (a) emf in series with Z_4; **(b)** Case (b) emf in Series with Z_3; **(c)** emf in Series with Z_1

With such arrangement of two hybrid coil, each fitted with a low passfilter and equaliser at the input of each unidirectional V.F. amplifier as shown, the V.F. repeater is used for amplifying the speech frequency of trunk lines and are placed at an intermediate station, so that the impedance on each side is balanced for the long distance physical lines and the net impedance on each side is also properly simulated.

The ringing voltage which drops down due to the long line used on each side is reenergised by regeneration separately for each direction at the V.F. repeater station at the middle, when the terminal stations ring by using CB signalling as shown below in Fig. 4.54.

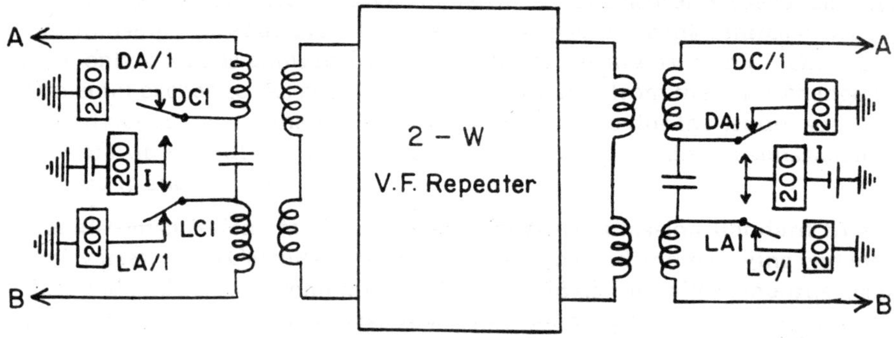

Fig. 4.54: Repetition of CB signalling at an intermediate V.F. Repeater

Now in the intermediate V.F. repeater station, the LA relay receives the calling condition on B-wire and the contact LA1 retransmits the battery via retard I over the B wire to the other distant trunk exchange. Similarly a sepervisory signal received on the A wire operates the relay DC, which through its contact DC1 returns the supy signal over *A* wire to the orignating exchange. In the reverse direction, the calling and supy signals are similarly repeated by relays LC and DA.

The cord circuit plugs used for putting through long distance trunk call on physical lines via V.F. repeater consist of a special double pin plug, which introduce a loss of .5 db in each direction and brings the line and net in circuit for proper balancing. A filter with a pass band of 300-2400 c/s with an equaliser followed by it is inserted at the input of each directional amplifier, so that the filter prevents the entrance of frequencies beyond the V.F. pass band and the equaliser equalises the unequal attenuation of the physical line and filter to different frequencies in the V.F. pass band. Thus we find that with the above arrangement, ring from the east will operate a relay which will send 17~ ringing from the V.F. repeater station towards west and similarly ring from the west will be sent after regeneration to the east, thus protecting the ringing voltages to pass through the amplifiers. Also of the V.F. amplifiers, one will amplify the speech in the east-west on A B direction while the other will amplify the speech in the west-east or B - A direction only.

Singing point: Before using the V.F. amplifier, the following tests are required to be taken: (a) internal balance test, (b) station balance test, and (c) the singing point test, so that the unidirectional amplifiers in each direction may be placed at the minimum step to avoid having noise called singing due to overloading of the amplifiers, i.e. the margin of stability must be found in each direction. If due to any reason the net impedance is not properly simulated with the line impedance, then some energy will be reflected back to the sending station. This phenomenon is called reflection.

If P_s = power sent from one end and P_r = power reflected back at the sending end due to unbalance of line with net, then 10 log $\frac{P_s}{P_r}$ is return loss or singing point of the repeater. If the line is very nearly balanced by the net, $\frac{1}{2}$ P_s goes to net and is absorbed there, $\frac{1}{2}$ P_s goes to the line. Since line and net are not perfectly balanced, some power, say, x returns to the sending end. Then S.P = 10 $\log\frac{\frac{1}{2}P_s}{x} = 10\log\frac{P_s}{2x}$. If x = 0, then S.P. = ∝, i.e. the repeater does not sing even at infinite amplification. If $x = \frac{1}{2}P_s$, i.e. if all power is reflected back, then S.P. = 0, i.e. the repeater sings even at very low amplification. So we see that the singing point of a V.F. repeater may vary from O to ∝. Thus singing point may be considered as a degree of balance between the line and the net.

Now the transmission equivalent of a circuit can be improved by increasing the gain of the two amplifiers, but a stage will come when further attempt to improve to increase the same will make the circuit unstable, i.e. make it sing at this marginal stage. The difference between the transmission equivalent of the circuit at its normal operating condition and this maximum stage when singing begins is called the margin of stability of the circuit.

We know, S.P. = 10 log $\frac{P_s}{2x}$ where P_s is the power sent and x is the power reflected back at the sending end. Now of this power, x, $\frac{x}{2}$ is dissipated at the output circuit and $\frac{x}{2}$ appears at the input. Then if the transmission loss between the output and input circuit is denoted by T, then

$$T = 10\log\frac{P_s}{x/2} = \log\frac{2P_s}{x}$$

$$= 10\log\frac{4P_s}{2x} = 10\left[\log 4 + \log\frac{P_s}{2x}\right]$$

$$= 10x.6 + \log\frac{P_s}{2x} = 6 + S.P.$$

To be more precise, considering the repeater cord circuit loss etc., the transmission loss = SP + 7 db.

If one net is imbalanced and one is balanced, there will not be any singing, but if both are unbalanced, there will be singing. But if the total gain = total loss, there will be no singing. But if the gain exceeds loss, singing will occur and if loss exceeds gain, singing will occur at first and will gradually die out. Thus if the total transmission loss $T_E + T_W$ and total gain = $G_E + G_W$, then the V.F. repeater sings when $G_E + G_W \geq T_E + T_W$, or the limiting condition is that the repeater should start singing although it has not yet started singing when

$G_E + G_W = T_E + T_W = S_E + 7 + S_W + 7$

$= S_E + S_W + 14$ db.

where S_E is the singing point east and S_W is the singing point west.

Transmission standards: Two transmission standards, namely, the national and international standards are followed in putting through a trunk call from a local exchange to a distant exchange on trunk lines:

1) ***National Standard :*** The overall transmission equivalent betwen two primary trunk centre

= 6db and the average transmission equivalent between a primary and a secondary trunk centres = average transmission equivalent between two secondary trunk centres = 12 db. A transmission loss of only 4.5 db is allowed from a primary or secondary trunk exchange, to a very small departmental exchange. As such the transmission loss from the 1st exchange to the last exchange should not be greater than 12 + 4.5 + 4.5 or, 21 db. To this the sender sub and wanted sub's loop loss of 7 db + 4.5 db has to be added. That is, in all, 32.5 db transmission loss may be allowed from sub to sub.

2) *International transmission standard* : International transmission standard from different cities to Poona varies. From Calcutta it is 13.5 db + sending end and receiving end losses i.e. 20 db for sending end and 18 db as compared to 20 db and 16 db for CCIF standards, respectively.

Junction line transmission standards from trunk sw Board to auto exchange should not exceed 4.5 db and for local and toll traffic loss up to 6.5 db per junction is allowed.

5

Automatic Telephone Systems

In automatic telephone systems, the subscriber's telephone has a dial controlled by a rotatable disc called finger plate with holes corresponding to the numbers 1, 2, 3, 4, 5, 6, 7, 8, 9 and 0 fitted on the front side, below a finger plate with each digit below each hole, as shown in Fig. 5.1(a).

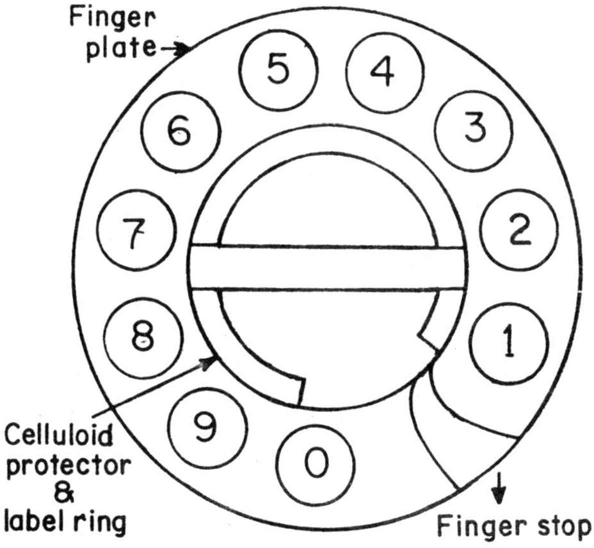

Fig. 5.1(a): Front View of Dial

The finger plate is rigidly fixed to a central spindle, and below this finger plate is the main spring box having a flat clock spring, one end of which is fixed to a central spindle. The stop screw keeps the spring in a partially wound condition to stop the rotation of the finger plate, which returns to its normal position by the spring action automatically. The main gear wheel drives the governor gear and a helical spring in the governor gear forms a coupling between the small gear and the worm wheel as shown in Fig. 5.1 (b).

There are two wings of the governor, each having a small weight at its free end, to rub over the inner surface of governor cup when moved outwards by the centrifugal force of rotation. The friction limits the speed of the governor and thus the rotating speed of the dial. The rear view of the dial is shown in Fig. 5.1(c).

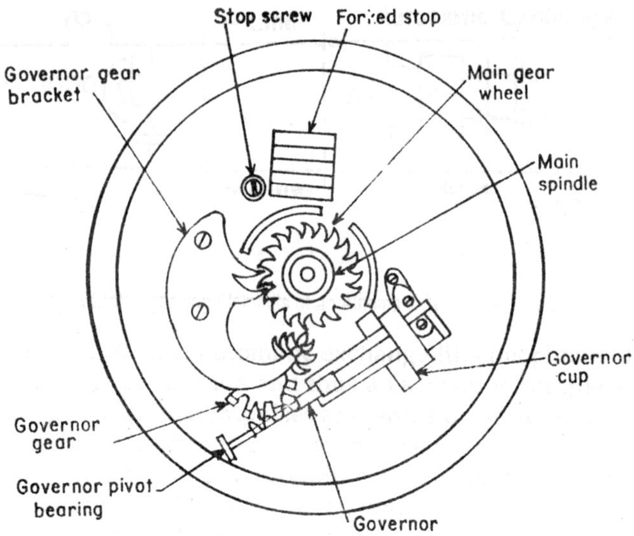

Fig. 5.1(b): Dial with Finger Plate and Numbering Removed

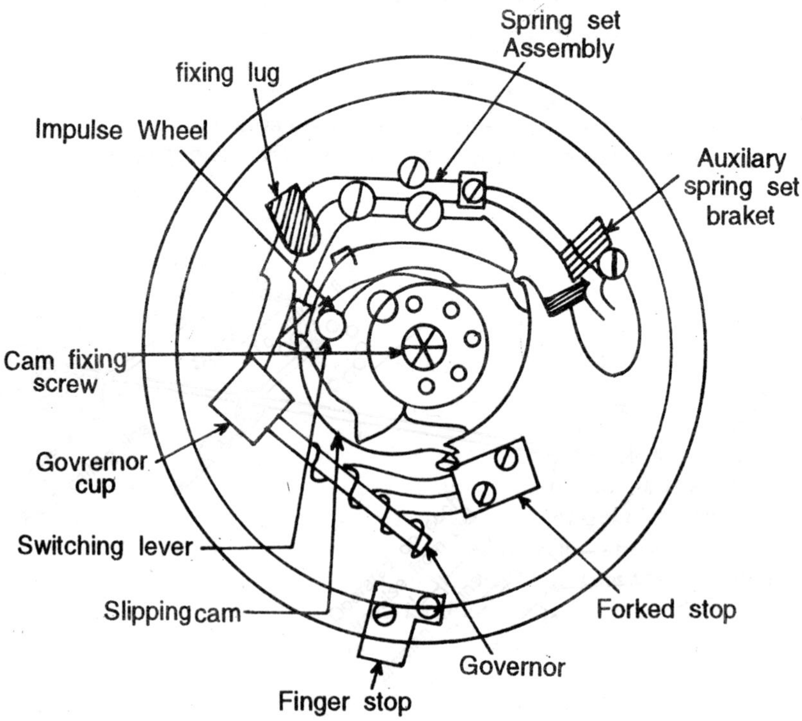

Fig. 5.1(c): Rear View of Dial

The impulse wheel rigidly fixed to the central spindle has ten slots at same angular distance on the finger plate. The switching lever is fixed to the central spindle and between it and impulse wheel a slipping cam and assembly of washers are loosely fixed on the spindle. The off-normal springs give switching facilities when the dial is moved from its rest position. Normally the switching springs are held apart by the switching lever and on moving the finger plate, the switching lever also moves away and contact is made by the switching springs. As the finger plate rotates, the helical spring in the governor gear unwinds and prevents high speed for the governor till the projection of the slipping cam strikes the forked stop.

Thus, by dialing the coded number of a subscriber of a particular exchange, a caller can raise up a particular subscriber he wants to contact. In an automatic telephone system, the calling subscriber controls the electromagnetic switches on a + wire and a - wire of the speaking pair and a third conductor is required for holding, releasing and other purposes. Correspondingly, in the manual exchange, a calling subscriber is put through to a called subscriber by an operator on A and B wires and a sleeve wire is required for supervisory purposes.

For this the following basic functions are required to be automatically completed by an automatic exchange for satisfactory performance:

1. The calling subscriber must wait for getting a dial tone after lifting his telephone from the hook switch and then only he should dial the desired number.
2. The calling subscriber must understand after dialing the number whether ringing tone is going on to ring the bell, or the number is engaged. In the first case he should get a ring back tone and in the second case he should get an engaged tone. And if the number is unobtainable, then he should get a number unobtainable tone.
3. The ringing current must be cut off the moment the called subscriber lifts his phone and the call must be metred.
4. A guard circuit must be provided to protect intrusion by other subscribers from the start to the finish of the call.

Of the four tones that are required to be supplied from an automatic exchange, namely the dial tone, ringing tone, busy tone and number unobtainable tone, the dial tone consists of a 33 c/s continuous tone that indicates to the calling subscriber to begin dialing. The ringing tone is a 133 c/s tone which energises the level of a called subscriber 0.4 sec on, 0.2 sec. off and 0.4 sec on and 2 sec off. A busy tone is a 400 c/s interrupted for 0.75 sec every 1.5 sec, i.e. .75 sec on and .75 sec off and a number unobtainable tone is a continuous tone of 400 c/s.

5.1 Types of Telephone Systems

Of the various types of automatic telephone systems, there are mainly three types in use in India and these are: (i) the Strowger type of step-by-step selection, (ii) Cross-bar type of selection used by B.P.O. Practice, and mainly (iii) the Cross-bar type used in Belgium and usually known as the Penta conta type of cross-bar system. More recently, the digital electronic telephone exchanges are also being installed, which will be dealt in a separate chapter.

Principle of step-by-step Strowger's system: The signalling process in Strowger's system is shown in Fig. 5.2(a) below.

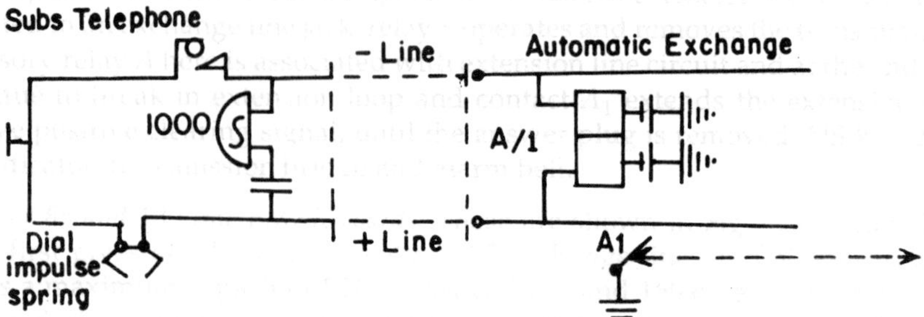

Fig. 5.2(a): Signalling Elements of an Automatic Telephone Circuit

Suppose a subscriber in a 100 line automatic exchange lifts his phone off the hook switch, then the cradle switch contact CS is made and thus a DC loop is completed to seize the auto equipment in the exchange. When dial impulses are sent, a uniselector provided for each subscriber with 10 sets of contacts is seized and moved forward under the action of the electromagnet step-by-step by a ratchet and pawl coupling arrangement. The loop circuit is broken by the dial the same number of times as per dialled digit and the break impulses are received by the relay A as shown in Fig. 5.2(a) above and its contact A1 repeats the impulses to the selector control circuit. The mechanism of the control electromagnet via a ratchet and pawl arrangement is shown in Fig. 5.2(b).

At the end of dialling, the calling subscriber's lines are connected to the wipers moving over the bank contacts the same number of times, energizing the stepping magnet, so that the associated pawl and the ratchet wheel step the wiper to the proper position, where it is held tight by a release detent. At the end of the call when the calling subscriber puts back his phone on the cradle switch, the CS contact is broken and the release magnet operates, withdrawing the release detent from engagement with ratchet and thereby the wiper assembly returns to the normal position by spring action.

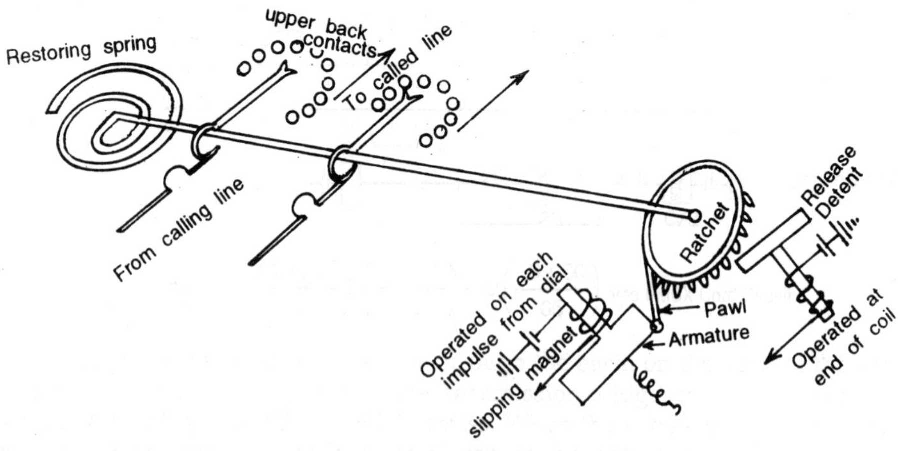

Fig. 5.2(b): Principle of Step-by-Step Selection

Now the coupling arrangement of ratchet and pawl may either be forward-acting selectors in which case the wipers are moved during the operation of the electromagnet armature, or reverse-acting switches where the attraction of the electromagnet armature gives tension to a restoring spring that releases the armature when the restoring spring moves the wipers. A forward-acting ratchet and pawl drive and a reverse-acting ratchet and pawl drive arrangments are shown as follows in Figs. 5.3(a) and 5.3(b), respectively.

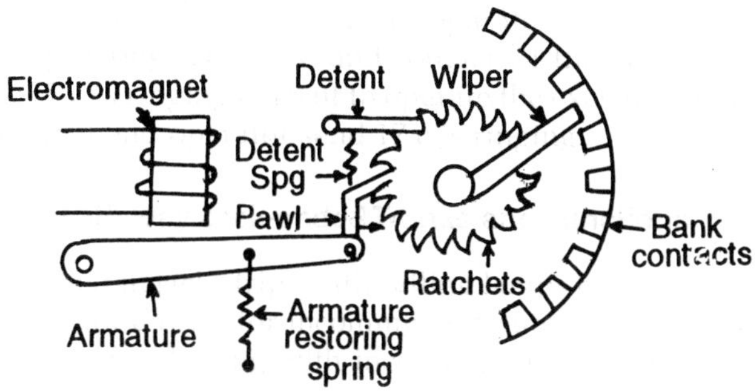

Fig. 5.3 (a): Forward-acting Ratchet and Pawl.

In Fig. (a) for forward-acting ratchet and pawl, the armature is pivoted at one end and carries a pawl at other end engaging a ratchet wheel rigidly fixed to the wiper assembly, so that each energisation of the electromagnet makes the pawl fitted with a spring to move the ratchet wheel one step forward. A detent holds the ratchet wheel and the wiper assembly in position after each step and the armature spring pulls back the armature to its normal position at the end of each impulse. In Fig. (b) for the reverse-acting ratchet and pawl, the pawl is made to slip over the teeth of the ratchet when the armature operates and to engage with the ratchet on the release stroke of the armature.

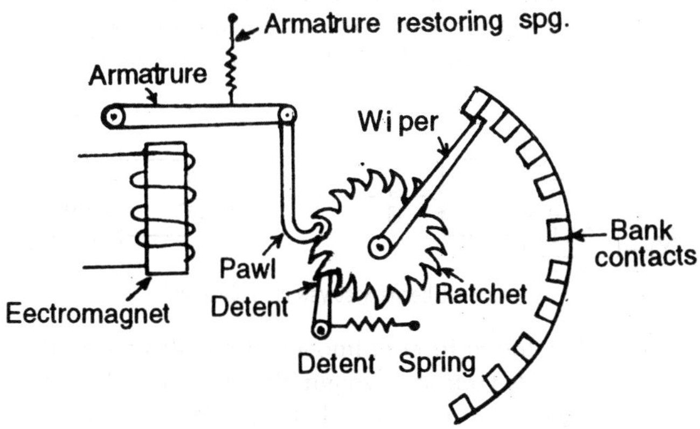

Fig. 5.3 (b): Reverse-acting Ratchet and Pawl.

Uniselector : The arrangement of parts of an uniselector is shown below in Fig. 5.4 below.

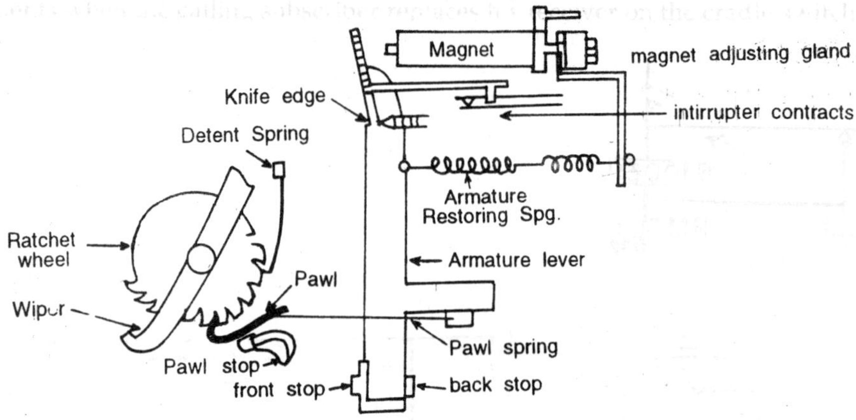

Fig. 5.4: Arrangement of Essential Parts of an Uniselector

Uni-directional selecting mechanism (uniselector) is of a step-by-step type with ratchet and pawl arrangement to move the wipers round the bank used for connecting the subscribers' line to the first selector and for other purposes. The uniselector bank consists of a number of semi-circular bank assemblies. Thus a 24-point uniselector provides 25 working outlets or 24 outlets and a home position. Under self-driving conditions, the reverse-acting ratchet and pawl arrangement is found better for uniselector working. The wipers are assembled on a tubular extension of ratchet wheel and the wiper assembly rotates when the tip of one wiper engages with one side of the bank contact and the tip of the other wiper engages with the other side of the same contact. The magnet is of two coils with glands for adjustment of armature air gap. The armature is pivoted on a knife edge support and in normal position, the armature lever rests on its back stop. Its restoring force is given from the helical springs with tension adjusting screws. The pawl engages with the underside of the ratchet wheel and is adjusted so that the pawl is made to slip over the ratchet when the armature is operated and to engage with the ratchet on the release stroke of the armature.

The uniselectors are nowadays used mainly on individual subscriber's line and thereafter all the group selectors from first group to the final group are of two-motion type.

Two motion type selectors: Strowger type two-motion selector is shown in Fig. 5.5.

Two racks *VR* and *RR* for the vertical and rotary motions respectively are carried by the vertical shaft *S*. These are engaged by two pawls *VP* and *RP* which are operated by the two electromagnets *VM* and *RM* for vertical motion and rotary motion. The vertical motion is against gravity and the rotary motion is against the restoring spring *RS*. The two detents D_1 and D_2 are pressed against the racks *RR* and *VR* by springs S_1 and S_2, respectively to keep the shaft in position. But when at the end of the call the release Rel *M* is operated, the detents D_1 and D_2 are withdrawn and the switch rotates by the spring *RS* and then drops to it normal position under the action of gravity, when at the first vertical movement the 'off-normal' springs 'ONS' are operated. The wiper *W* consists of two springs P_1T_1 and P_2T_2 insulated from each other and make contact on double projections insulated from each other numbering 1 to 0, both on vertical and horizontal rows. These double contacts are called line bank, while there is a single contact wiper (not shown in figure) mounted on the same shaft *S* above line wiper *W*, which make contacts on a second set of single projections placed similarly above the line bank. These are known as private wiper and private bank. In this Strowger system, the selector mechanism in which the wipers move consecutively in two planes, first vertical and then horizontal, is called a two-motion selector.

The two-motion selectors are of three-magnet type, i.e. with vertical, rotary and release magnets. Now when a subscriber lifts his phone to make a call, his uniselector searches over the available trunk outlet for the first free selector and as soon as a free selector is found, a group selector is stopped to the desired level and hunts over the contact of that level to search for the first available trunk to the next switching stage. The time available to hunt for, find and seize a free trunk is limited to the period between two successive impulse trains, i.e. interdigit pause of some 400 m sec.

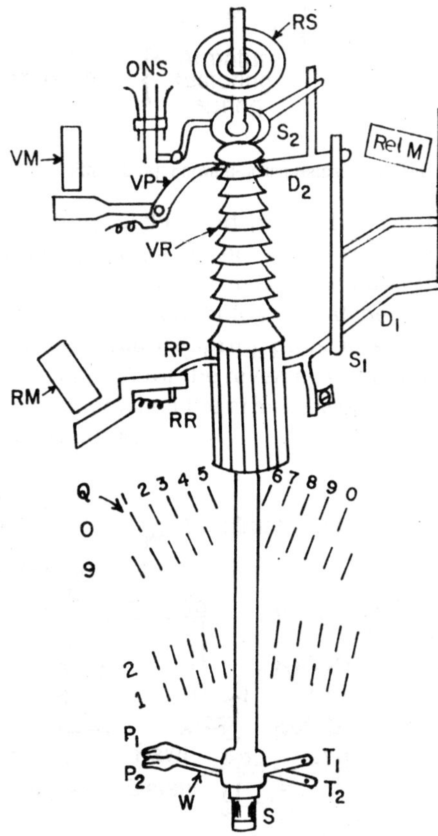

Fig. 5.5: Strowger Two-motion Selector

An impulsing circuit is required for a selector for controlling the movement of its wipers, a hunting circuit for searching and testing over the outlets and a switching circuit for switching the caller to a selector of the next stage on a free outlet. This is shown below in Figs. 5.6 (a) and (b). The other required items are the return of tones, the guarding of the calls during setting up and conversational period against intrusion and the holding of connection till the release of all equipments when the calling subscriber replaces his receiver on the cradle switch.

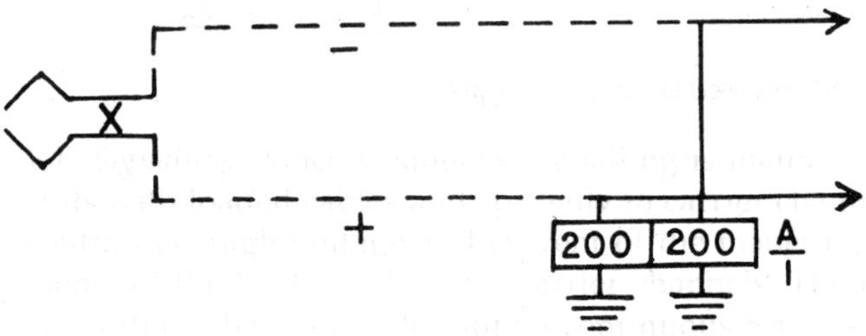

Fig. 5.6 (a): Impulsing Circuit

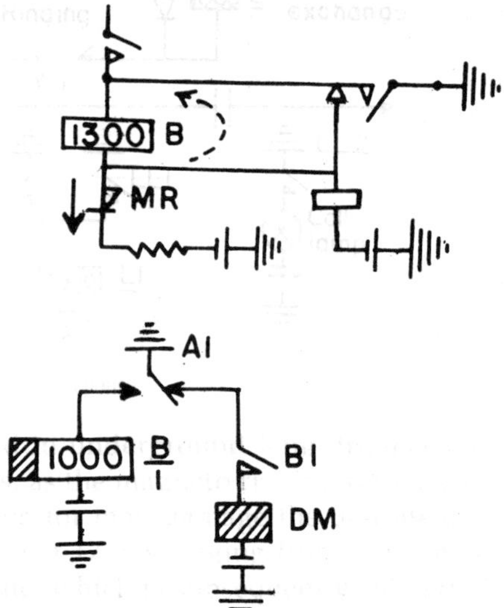

Fig. 5.6 (b): Separate Battery fed for Relay B

In Fig. (a), a single changeover contact is used to control relay *B* and to repeat the impulses to the selector driving magnets. The relay *A* operates from the subscriber's loop and *A*1 energises *B*. Then *B*1 makes the circuit for the selector driving magnet to step at each release of *A*1 by the received impulses. The other contacts of relay *B* provide the holding and guarding conditions for the period till the holding circuit is disconnected at *A*1 contact. A metal rectifier may be fitted across relay *B* in addition to a slug to decrease decay of flux on release. Thus, a modified circuit with a short circuited relay *B* without a slug may be used with one side of relay *B* connected to a separate battery via protective resistance of 200 ohm and a metal rectifier *MR*.

The impulses from a subscriber's dial are utilised to control the various movements of a step-by-step selector, but on certain calls these impulses are distorted and require correction or regeneration. In director system, these impulses from subscriber's dial do not control the switch mechanism but are used as signals to special apparatus, which on receiving the dialling code, sends out corresponding trains of impulses to route the call to the proper place. Thus, when a call is set up from one exchange over junctions to another exchange, a third wire is not usually used for private wire for holding and control purposes. For this a separate holding circuit for switching train in each exchange is provided by a relay bridged across the line to be controlled by the dial. This transmission bridge in turn requires repetition of the impulses from one side of the bridge to the other, causing an operate lag of each relay A slightly less than its release log and thus the impulses are subjected to cumulative distortion due to unequal lags causing large break periods and shorter make periods for the distortion. In order to minimise this effect of cumulative distortion, a common form of impulse repetition bridge is used by maintaining the transmission path by provision of two MF condensers, as shown belew in Fig. 5.7.

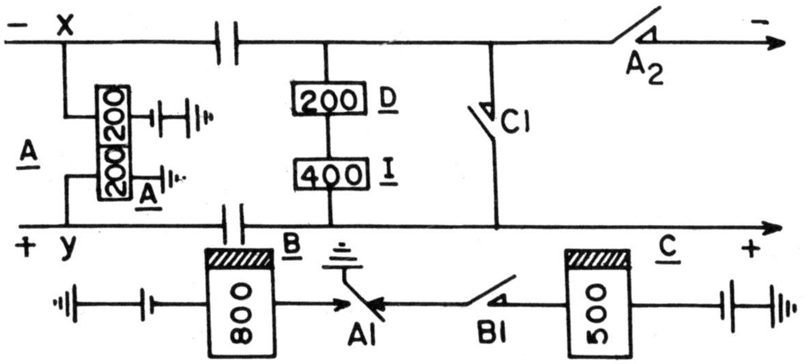

Fig. 5.7: Capacitor Type of Impulse Repetition Bridge

On seizing the circuit, we find the relay *A* operates relay *B* through its contact A_1, and B_1 operates relay *C* and at the same time A_2 gives loop through D_1 and *I* relays to the outgoing junction seizing the selectoı at the distant exchange. The relays *D* and *I* are of very high impedance and have no function during the impulsing as they are cut off during dialling by providing an impulse train cover signal from relay *C* contact. Thus when relay *A* first releases, the contact A_2 breaks the loop to the distant exchange, and simultaneously relay *C* operates through A_1 and B_1 closing contact C_1 during the first break impulse thus short-circuiting *D* and *I* relays to provide for the subsequent impulses a low impedance loop. When the impulse train is over, the relay *C* releases and C_1 contact opens, so that relays *D* and *I* can respond to signals sent on the junction from the distant exchange. Now if we consider the effect of the bridge capacitors, we find on operation of C_1 contact the 2 μF line condensers are connected in series the impulsing relay of the bridge and their resultant is reduced to a capacity of 1 μF across relay *A* when the dial contacts are open, the potential difference between *X* and *Y* wires is 50 V, and the same is reduced when the impulse springs close and the reduced value depends on the junction line resistance. Thus it is found that the bridge condensers begin to discharge at the start of the make period and thus delay the change of potential at *X* and *Y*, thereby increasing the operate lag of relay *A*. When the contacts reopen after the make period, the bridge condensers charge to the battery potential, delaying the release of the relay and tending to decrease the break period. Thus neutralisation of the two effects takes place and thus the degree of distortion is also decreased depending on the value of line resistance in relation to that of impulsing relay coils.

5.2 Testing, Hunting and Switching Circuits

Various methods are applied in different systems for automatic search over available outlets. For example, in the Western Electric Rotary system, the wipers are rotated by motor driven shafts and the selector is coupled to the drive system by electromagnetic clutches. In some other systems, the automatic search is obtained by a ratchet and pawl drive of an electromagnet with machine generated impulses. In B.P.O. systems the automatic search is obtained by interruption of driving magnet current from contacts associated with driving magnet armature, which is known as Self Drive Circuit as shown below in Fig. 5.8.

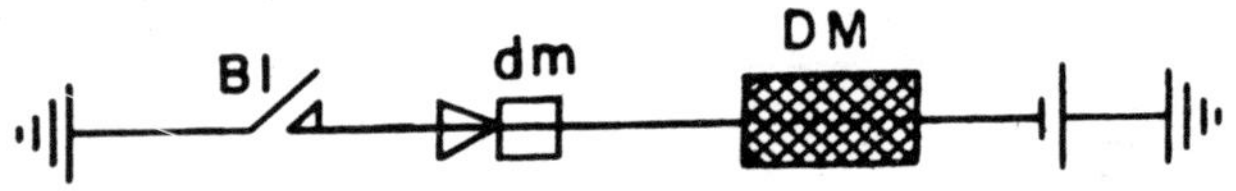

Fig. 5.8: Self Drive Circuit

For hunting purpose, contact B_1 closes the circuit to operate the driving magnet *DM* via the interrupter contact *dm* mechanically coupled to the driving magnet armature in such a way that the contacts open when the armature is attracted and at a late point in the stroke, contacts *dm* break, disconnecting the magnet circuit. The armature is released and *dm* springs remake and the magnet is re-energised and the process is continued until a free outlet is found.

Testing: Now we know that in automatic exchange, in addition to a speaking pair of wires, a third wire is required for the setting up and holding purposes and for testing and metering also. Two types of testing ciruits are used, namely, (a) battery testing, and (b) earth testing. Battery testing circuit switches only when a battery potential is found on P-wire and other conditions of earth or disconnection is regarded as engaged. An earth testing circuit switches only to a battery or disconnection on P-wire and considers an outlet as engaged, if there is earth on P-wire.

A simple form of earth testing circuit is shown below with bridging private wiper in Fig. 5.9.

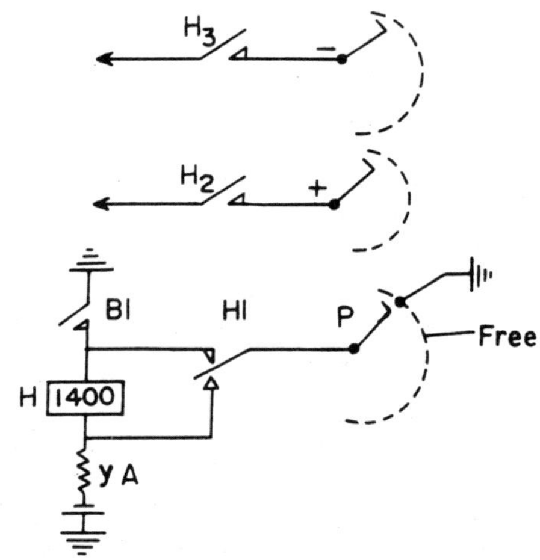

Fig. 5.9: Earth Testing Circuit with Bridging P-wire

Here, the engaged condition is a full earth on the *P* bank contact whilst the disengaged outlets are either disconnected or may have battery potential. At the start of the testing, the contact *B*1 closes and as long as *P* wire faces an earth, *H* relay cannot operate due to earth on both sides of its coil. When *P* wire faces a free outlet, the absence of earth on *P* bank operates *H* relay. Thus *H*1 after disconnecting the testing circuit, extends earth from *B*1 for guarding and engaging the outlet just seized and the contacts *H*2 and *H*3 put the speaking pair through while resistance *YA* prevents blowing off fuses during search on engaged outlets. To provide for automatic hunting with earth testing over engaged outlets to search for a free trunk, a selector driving magnet *DM* with its interruptor springs (dm) in series is placed in position of protective resistance *YA* as shown in Fig. 5.10.

The driving magnet *DM* is energised via *H*1 and *dm* on P-wire bank and towards the end of the forward stroke of the armature, the interruptor spring *dm* breaks the magnet circuit allowing the restoring springs to move the wipers to the next contact and so on till a free bank contact is found, where the absence of earth makes the *H* relay to operate, whereby contact *H*1 extends the guarding earth to the seized outlet and *H*2 and *H*3 contacts put through the call to the next stage.

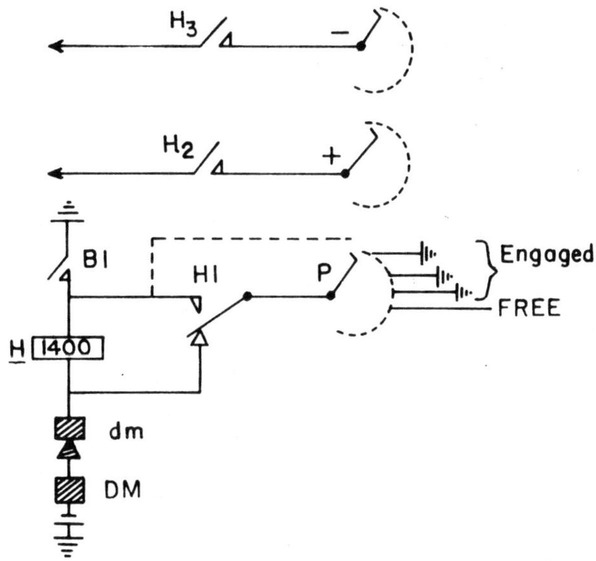

Fig. 5.10: Combined Self-drive and Earth-testing Circuit

After finishing the testing of the outlets, let us go through the diagrams of hunting and switching circuits and through the diagrams of uniselector, group selector and final selector circuits as also the various types of transmission bridges to follow the further sequences of operations for holding speaking, releasing and metering circuits.

Hunting and switching: For hunting and switching, a circuit operation is to be done in the following sequences: (a) relay *H* is preoperated before hunting begins, so that it can release on finding the first free outlet, (b) relay *C* is released when rotary hunting begins, (c) *H* releases on finding a free outlet allowing *C* to reoperate and complete switching of the call, (d) on reoperation of *C* relay, *H* reoperates and holds on earth on *P*-wire when the call is finished and the subscribers leave their phone. The removal of this earth allows *H* to release, which in turn releases *C* and allows the selector to restore to its normal position. The rotary hunting with pre-operated testing relay is shown in the diagram in Fig. 5.11.

On seizure of the circuit, *B* operates and closes contacts *B*1, *B*2, *B*3 and *B*4, operating relay *C* which is held until the end of vertical stepping. At the first vertical step, *N*1 contacts close operating *H* relay on 500 ohm coil to earth via *C2*. *H4* offers an independent holding circuit without C2. Due to the stoppage of current pulse in its low resistance coil at the end of vertical impulse train, C releases and at C2 completes the rotary magnet circuit from earth at *H4* via *R1*, *N1*, and *C2*, so that wipers rotate under self-interrupted drive till the circuit breaks by the release of H4. Now at each rotary step, the holding circuit for relay *H* is broken when *R1* contacts open. When the outlet is found engaged, an alternative holding circuit is given through 2000 ohm coil from earth on *P*-wire via *H1*. When a free outlet is found, H releases during the open period of *R*, and *H4* restores disconnecting the rotary drive circuit whilst *H3* supplies a temporary earth to the P-wire. *H5* relasing reoperates C and locks it through C1 contact. Contacts C3, *C4* and *C5* complete the switching operation and *C2* reoperates *H* which holds the earth on *P*-wire till the call is finished and by the release of relay *B*, the holding circuit of relay *C* is now transferred to the control of *H5* contact.

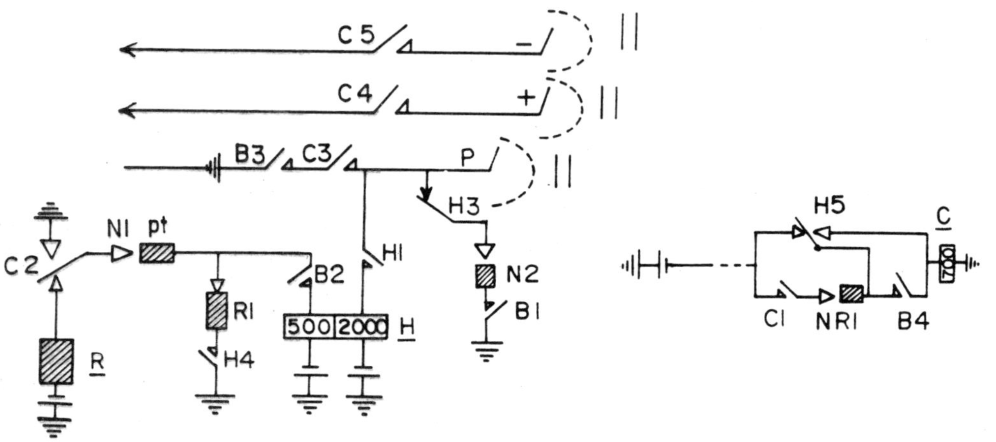

Fig. 5.11: Rotary Hunting with Pre-operated Testing Relay

5.3 Holding, Speaking and Release Circuits

There are two methods of applying the holding condition, i.e. forward holding and backward holding, depending on the position of the bridging relay of the transmission bridge. If the bridging relay is placed in the first link of switches, then the remaining switches in the same exchange are held by the earth applied to the first switching point extending forward over the switching point to the *P*-wire. But if the switching relay is located in the final selector, then the earth is returned backward on the *P*-wire to hold all preceding switches of the train. But in the junction calls it is uneconomical to provide three wire junction for speaking pair and *P*-wire. For this the bridging relay across the speaking pair at each exchange maintans earth on *P*-wire. This method is based on the location of transmission bridge in the most economical position as required by the impulse repetition. We have already seen the function of capacitor type of transmission bridge, but for impulse repetition purposes, the transformer type of bridge is more efficient and its circuit is given below in Fig. 5.12.

When the called subscriber replies, the ring trip relay (F) operate and F_1 contact disconnects the ringing tone and both the subscribers can talk now.

Application of number unobtainable tone to spare lines: All the spare or unallocated final selector multiple numbers are usually connected to number unobtainable tone (NUT) as shown in Fig. 5.27.

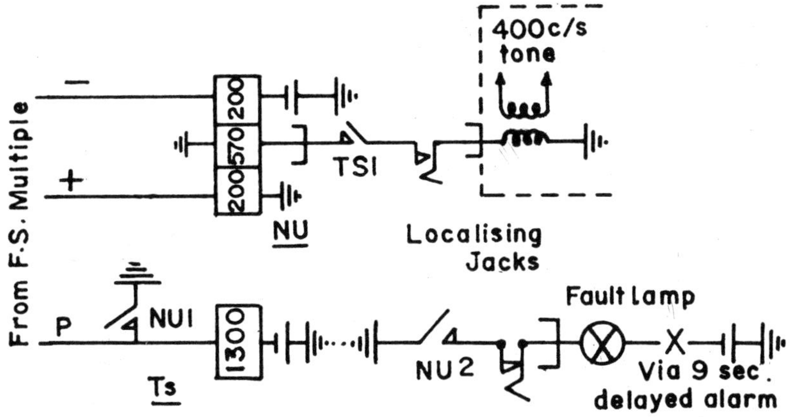

Fig. 5.27: Method of Applying Number Unobtainable Tone to Spare Line

Here one 1300 ohm winding of relay *TS* (having four windings each serving one hundred lines) together with one *NU* relay is provided per hundred lines of the exchange multiple. The -, + and *P*-wires of *NU* tone equipment are multipled as required to all spare lines in groups of 100 and if any of these lines are dialled, the final selector switches to the battery at *TS* relay, operating the ring trip relay on the - wire of *NU* relay when *TS*1 contact applies *NU* tone to 570 ohm winding of *NU* relay and thence through the transformer action of the winding to the calling party.

5.6 Traffic and Trunking

After going through the preliminary working principle of Strowger's auto telephone exchange, we require to know a bit of the traffic to be handled by the exchange to provide a reasonable standard of service to the subscribers by an equitable distribution of traffic to all selectors at every stage. As it is not possible to foretell the number of calls that will originate at any time, we make use of the mathematical theory of probability to determine the probable traffic in different situations.

Busy hour calling rate: To design an auto-exchange, the first thing required is to find the busy hour calling rate, i.e. the average number of calls originated per subscriber during the busy hour. It varies from 0.3 for a rural exchange to 1.5 for a very busy exchange. For other exchanges of average busy hour, it varies from 0.5 to 1 normally. The next important thing to know is the duration of calls. The total time for which a circuit is held up for a call is known as the holding time, which is the sum of the operational time and the conversational time. These two items, namely the busy hour calling rate and the duration of calls may be obtained from traffic meters and traffic recorders.

It is observed that the busy hour varies from day to day and also to some extent, with the time of the year. In our country, the busy hour for local traffic is taken from 10 to 11 in the morning, when all the offices open and shops in the market are in full swing. But the busy hour for local traffic is not always the same for that of trunk traffic, which generally occurs once at about 12 noon to 1 P.M. and again at about 4 to 5 P.M. to ascertain the market rates at the climax period and also at the close of the day.

For a manual exchange, the volume of traffic determines the number of cord circuits required per position and the number of circuits in every junction route. The quantity of such equipment depends not only on the number of calls but also on their duration. Generally to clear a total of 140 calls of three minutes duration in busy hour, seven cord circuits are sufficient but if the duration of call is six minutes, a total of 14 cord circuits will be required to clear the calls. Thus for manual exchanges, for the purpose of staffing the postions, it is the number of calls that plays the main part, but for auto-exchanges the functions of both the number of calls and the duration of calls are necessary to determine the quantity of switching equipment.

The calls of all types are compared with a standard unit call based on a local call between two subs on a *CB* switch board and is supposed to require 18 sec for operating time including the time taken by the operator in handling the call and the time taken to perform some accessory functions. The equated call values for various types of calls are given below.

Local call at manual exchange of 5000 lines multiple

= 1 unit call

Local call at manual exchange of between 2000 lines multiple

= 0.9 unit call

Local call at automanual exchange directly connected to a line in service multiple = 1 unit call

Junction or toll call connected over order wire circuit = 1.35 unit call

Junction or toll over signal circuit direct to a manual exchange = 1.90 unit call

Demand Trunk call through an intermediate exchange = 12 unit call, and so on.

Traffic unit (T.U.) means the rate of flow of traffic during the busy hour. It should not be confused with the 'unit call', which is a measure of the total traffic and not the rate of flow of traffic. Thus if

A = total traffic flow (in T.U.s) during busy hour

N = number of calls in busy hour

T = average duration of call in hour, then

1 call of 1 hour duration = 1 traffic unit

N calls of 1 hour duration = N traffic units

N calls of T hour duration = NT traffic units.

Now if the total number of calls originating in an exchange during busy hours = 6000, and if the average holding time of call = 3 minutes then the holding time $T = 3/60 = 1/20 = 0.05$ hour and No. of calls (N) = 6000.

$\therefore$ Traffic flow = 6000 x 0.05 = 300 T.U.

5.7 Trunking Arrangement in a 4-digit Exchange

In a 4-digit exchange using u/s, first Grp., second Grp. and final selectors as shown in Fig. 5.28..

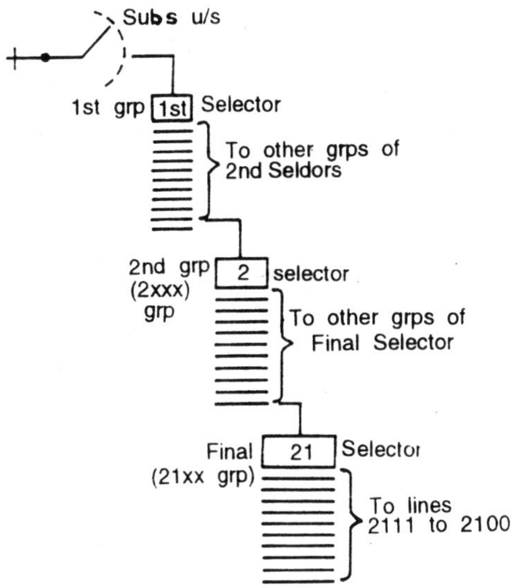

Fig. 5.28: Trunking Arrangement of 10,000 Line Exchange with Subs. u/s

A call may fail due to insufficient equipment at any stage, like (a) u/s failing to find a free first selector, (b) a first selector may fail to find a free second selector, (c) a second selector may fail to find a free final selector. Thus there are three places in a 4-digit auto exchange, where a call may be lost due to insufficient equipment. If the first or second group selector fails to get a free outlet, the wipers step to the 11th rotary position and a busy tone is returned to the calling sub.

The proportion of calls that are lost during the busy hour due to insufficient switching equipment is known as the 'grade of service'. The general practice is to specify a grade of service of 1 lost call in 500 per switching stage, so that the standard grade of service = 0.002. Now if three calls are lost for each 500 calls originated, then the overall grade of service = 0.006, and so on.

If the total number of selectors at any stage is not greater than the number of bank contacts in the preceding stage, then it is observed that the caller has access to all switches of the next stage. This condition is called 'full availability', and the method of connecting outlets, one selector to the selectors of the next stage is called 'interconnecting'.

Traffic capacity of a full availability group: For the design of an auto exchange we are to know the number of trunks required to clear a known traffic with a known grade of service. The number of trunks gives an idea of the number of selectors for the next stage and so on. Although various formulae have been devised by Erlang, Lely, Grinsted, Molina and others to determine the number of switches required to carry a given amount of traffic, their differences are small as compared to actual work and as such we describe the method adopted by Erlang to find out a formula for a full availability group, giving relation between the proportionality of lost calls, number of outlets tested and the average volume of traffic.

Erlang found out that

$$B = \frac{\dfrac{A^N}{L^N}}{1 + A + \dfrac{A^2}{L^2} + \dfrac{A^3}{L^3} + \ldots + \dfrac{A^N}{L^N}}$$

where B = proportion of lost calls, i.e. grade of service

A = average traffic in T.U., and

N = Number of trunks.

If in a system of interconnecting, it is not possible for a selector to have access to all the other selectors of the next rank, then in this case it is called 'limited availability' condition. It is not practicable and economic to provide full availability condition, when traffic and hence the number of first selectors still further increases, where it becomes necessary to divide the traffic into two or more groups, each having an access to individual series of first selectors or to divide the traffic into a convenient number of groups, each with some individual trunk and the rest are common to a number of groups. But it has been observed that less traffic is carried by later choices of individual groups. In such cases, it is observed that while interconnecting individual trunks are used for the early choices of each group followed by a system of progressive commoning, so that traffic from more and more groups is offered to the later choices. Such system of interconnecting is called 'grading', where the traffic offered to the later choices is increased. Since grading assumes that the traffic offered by each group to later choices is small, it follows that the trunks of a graded group should always be tested in the same order. For this the switches should be of homing type, i.e. they should always search over the trunk in the same order. Thus a case of simple grading is shown below, where a total traffic of 6 T.U. emanating from a total of 10 outlet 200 uniselectors and proceeding to two motion first selectors. The selectors are arranged in two groups, each with five shelves of 20 uniselectors. The first six choices of each group are trunked direct to individual first selectors, whilst the remaining four outlets of the two groups are commoned and trunked to a further four first selectors as shown in Fig. 5.29.

It can be seen that the traffic of 6 T.U. can easily be cleared by segregation of the traffic into two limited availability groups, each offering 3 T.U. with first six choices of each group being joined direct to individual first selectors and the remaining four outlets of the two groups being commoned together and joined to four first selectors.

Similarly, for 360 subs, u/s can be arranged in six groups each serving three shelves of 20 u/s and 1.8 T.U., so that the whole traffic of 10.8 T.U. can be cleared with 27 trunks by grading as follows, with an availability of 10, as shown in Fig. 5.30.

In the same way, a total traffic of 10 T.U. can be cleared by a simple grading of four groups, with an availability of 10 with 23 trunks as shown in Fig. 5.31.

In the above cases, we have restricted to availability of 10 for trunking arrangement from selectors. For a larger volume of traffic, however, dual testing facilities over two banks, and thereby increasing the availability to 20, are provided. Subs u/s are of 25 contacts providing 24 outlets and a home position. It will be seen that different alternative gradings can be obtained with different number of individuals, pairs threes and commons as shown below in Fig. 5.32 for a six group grading with 31 trunks for a total traffic of 13.2 T.U. and an availability of 10 for each group.

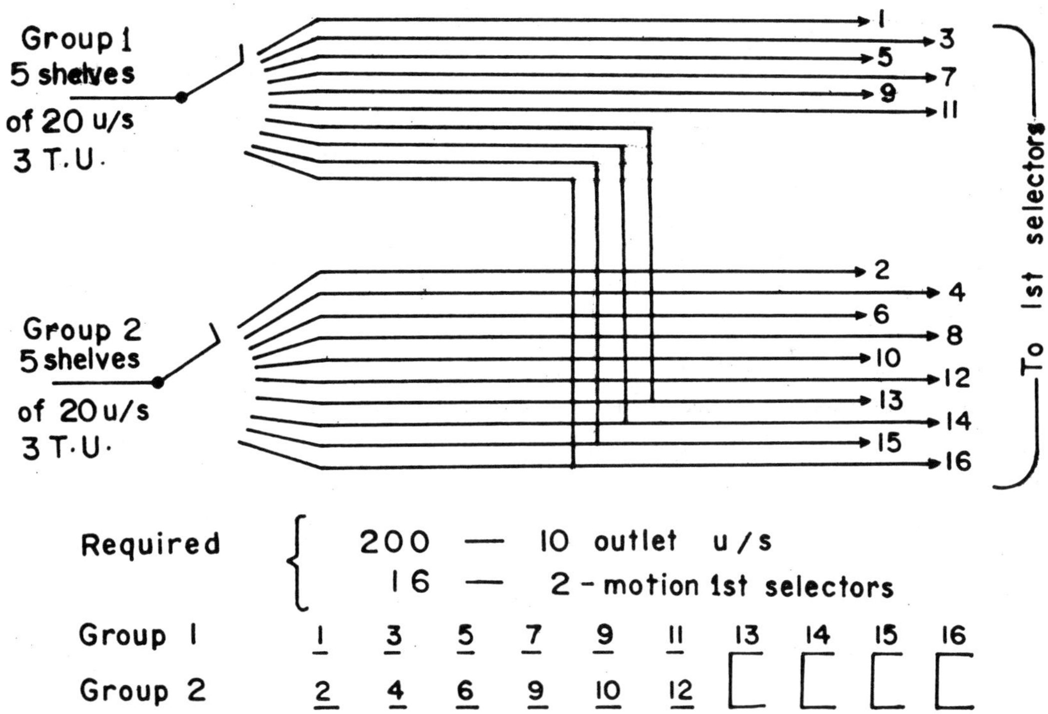

Fig. 5.29: Simple Example of the Principle of Grading

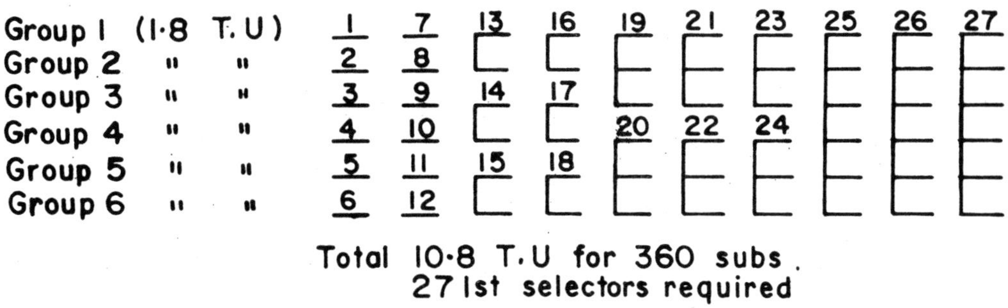

Fig. 5.30: A Practical Example of Grading

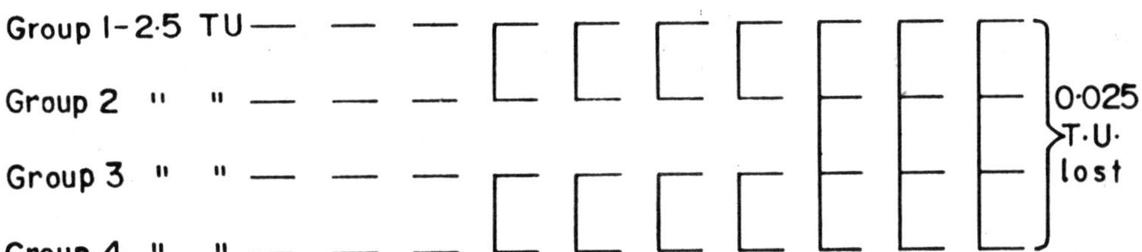

Fig. 5.31: An Alternate Example of Grading

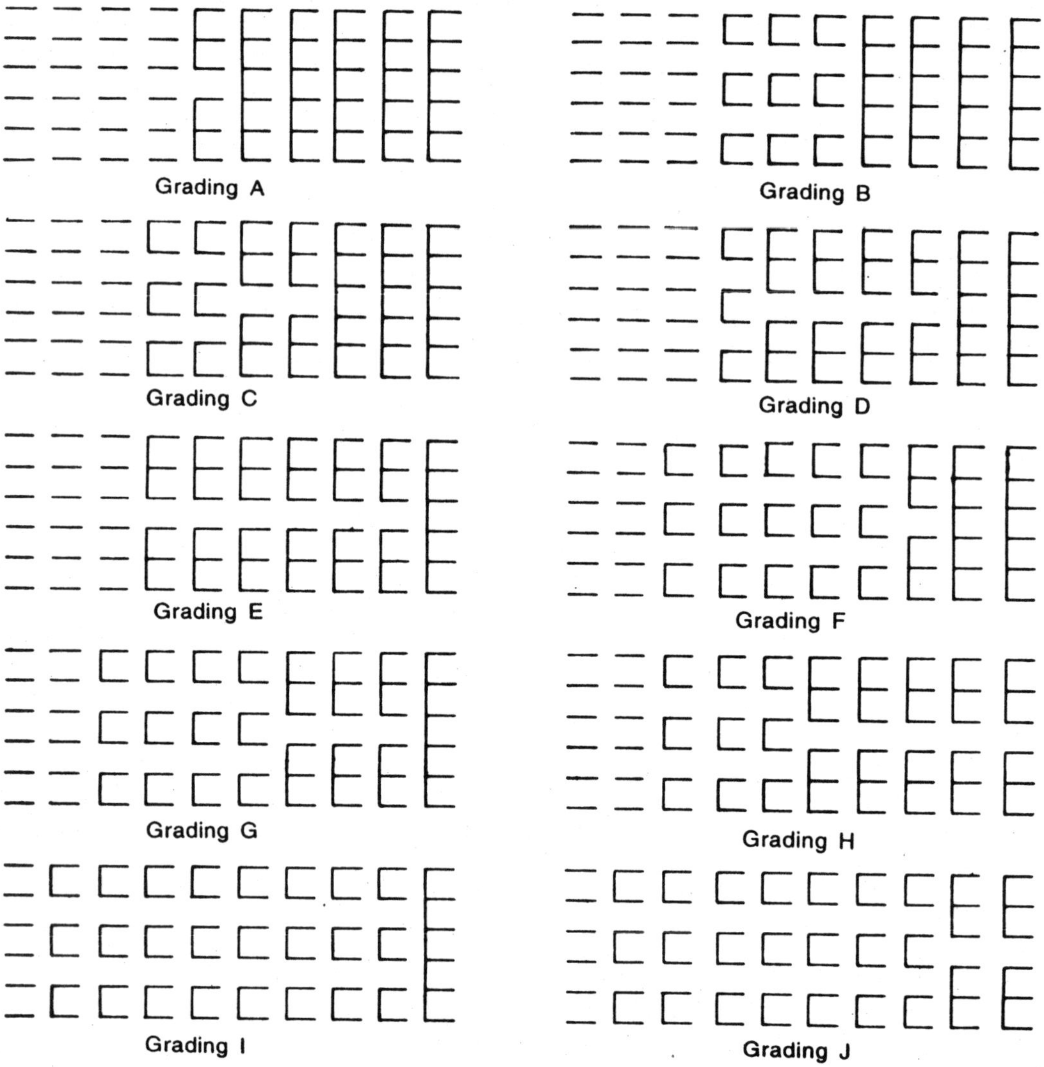

Fig. 5.32: Relative Efficiency of Alternative Grading with Same Number of Trunks

Now arranging the alternatives with their individuals, pairs, threes and commons in a table, the sum of their successive differences is given in Fig. 5.33.

It can be observed that the alternative C gives the minimum sum of the successive differences and is the least grading.

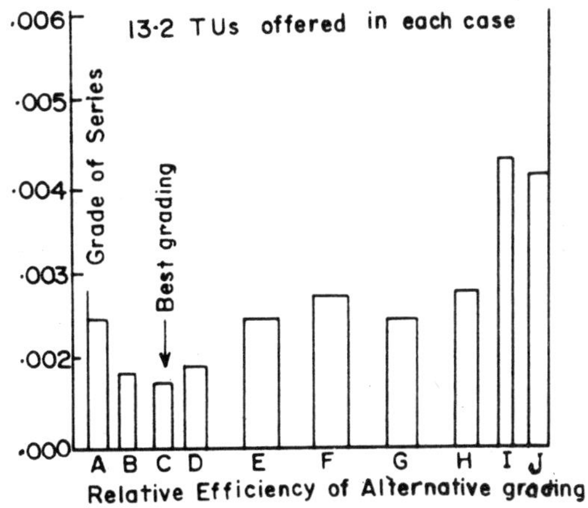

Fig. 5.33: Relative Efficiency of Alternative Grading

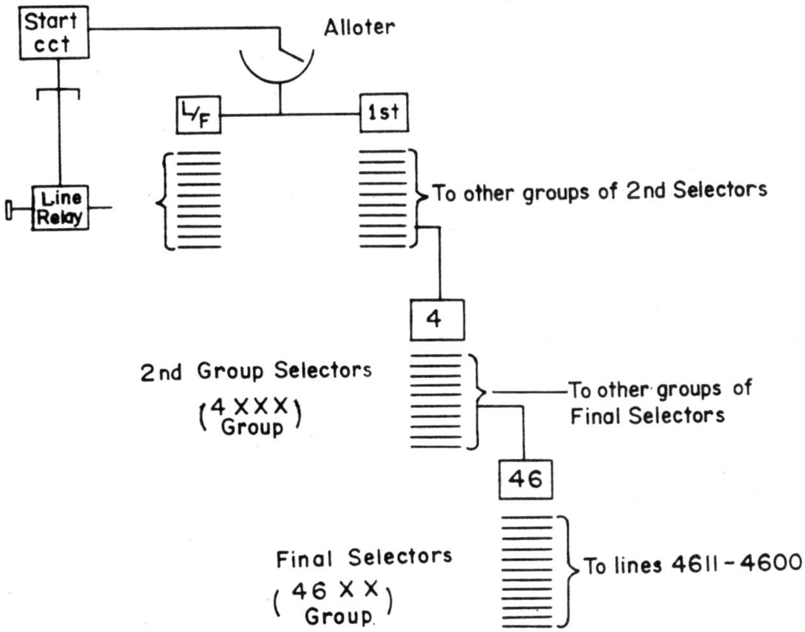

Fig. 5.34: 10,000 line exchange with line finders

5.8 Non-director and Director Strowger Exchange System

Non-director exchange using 2-motion line finders for 10,000 lines is shown in Fig. 5.34.

Here, the lifting of the calling subscriber's receiver energises the start circuit. A free first selector and its associated line finder are usually pre-selected by the allotter and the start condition causes the line finder to search for a marking condition on its bank due to the calling line. After seizure of the line, the call is switched to the first selector and proceeds exactly the same way as in an exchange with subscribers uniselectors.

In a 4-digit scheme for 10,000 lines, level 1 is avoided in dialing the initial number for misrouting, resulting from false initial impulses, 0 level is generally allotted for obtaining access to operators on auto-manual exchanges, while level 9 is connected to special services position for miscellaneous services such as phonogram, enquiries, speaking clock service PBX, test desk, etc. Thus as a result of these restrictions, a 4-digit exchange can have a maximum capacity of 7000 lines theoretically, but practically maximum capacity of 6500 lines, leaving some positions for outgoing junction routes from selector levels to some spare postions of lines for emergency use, can be utilised. But the above limits do not apply in case of Director system, where special 'code selectors' are introduced in front of numerical selectors, to cater for junction dialing codes. In such circumstances, the full 10 levels of the first numerical selectors are available for subscribers' numbers.

If the non-director exchange is situated within 15 miles of the centre of a director exchange area, in addition to levels 0, 1 and 9 the level 7 is reserved for giving access to the director exchanges. Also level 8 is reserved for calls to exchanges in multi-fee metering area within 5-15 mile belt. So the practical capacity of a 4-digit non-director scheme in such situation comes down to 5000 (or 6000) lines, whilst the capacity of a 5-digit scheme is ten times this number.

A typical trunking diagram of a 4-digit non-director exchange is shown in Fig. 5.35.

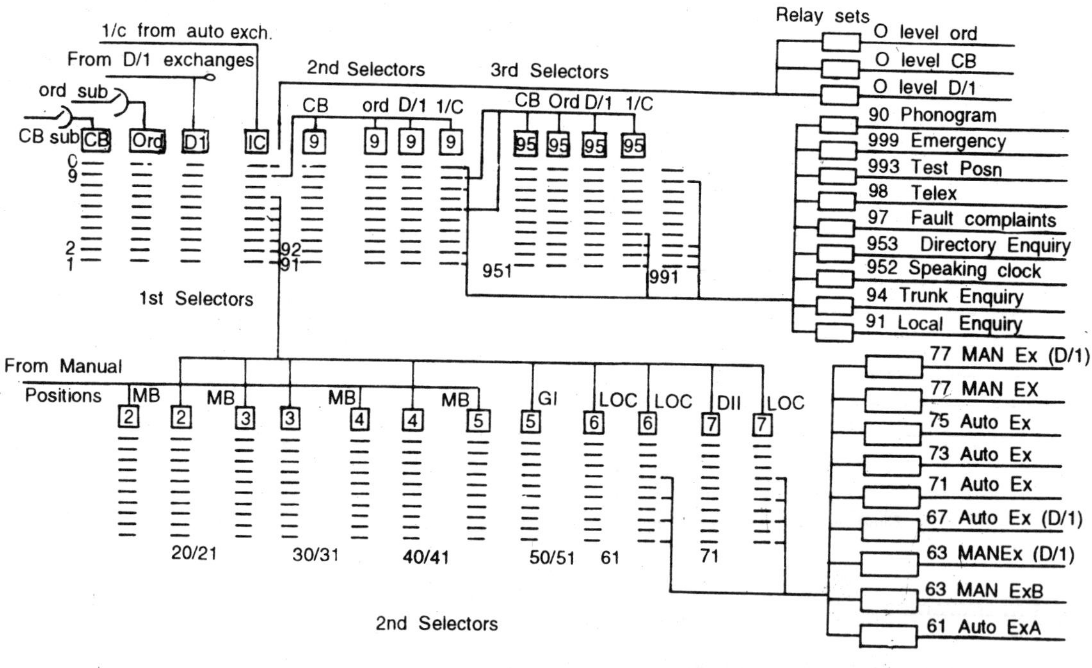

Fig. 5.35: Typical trunking diagram of 4-digit non-director exchange

The first selectors are divided into four groups. The first group (CB), serves the coinbox subscribers. The second group (Ord), serves ordinary noncoinbox subscribers. The third group (D/I) serves incoming junctions from distant exchanges.

The trunking scheme provides for seventeen 200-line final selector group or a total of 3400 lines. Spare levels are kept for future growth. There are four separate groups of circuits from level 9 of the first selector trunker to four distinct groups of second selectors, providing splitting up of levels, both on second selector banks and in some cases on the third selector for all the special services position. In all outgoing junctions to manual and auto exchanges, suitable relay sets are provided at appropriate selector levels, but the trunking arrangements depends on the junction and local traffic, type of automanual sw. Bd. etc. Direct circuit from manual board to second selector groups are provided to auto-manual operator, who dials only three digits to get the required subscriber. More troubles are experienced in providing non-director trunking scheme from main exchange to satellite exchanges for necessary holding condition for the main exchange train of selectors and to discriminate between local calls and trunk calls. Moreover, extension of dialing and introduction of multimetering in non-director areas require additional equipment in satellite exchanges and large number of dialing codes. But though a bulk of outgoing traffic from a non-director exchange is to directly connected exchanges, a smaller volume of traffic requiring tandem dialing through a distant centre can be handled satisfactorily at auto-manual board. For such cases, a tandem dialing scheme for trunking is provided, as shown in Fig. 5.36.

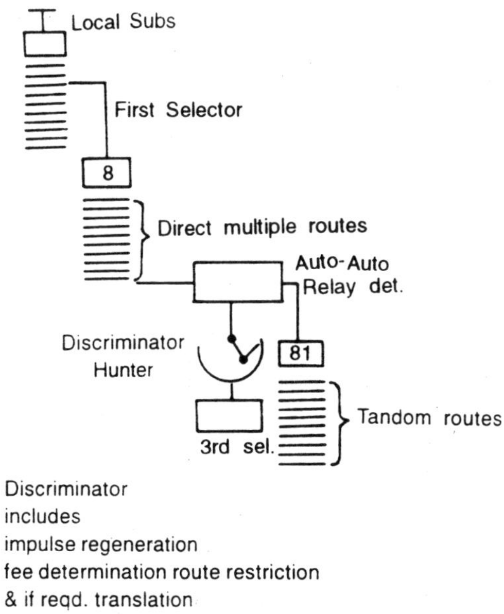

Fig. 5.36: Method of obtaining tandem dialing in a non-director exchange

A special auto-to-auto relay set is inserted between level 1 of the 2nd selectors and third selectors. This relay set is provided with a uniselector, which serves as a hunter to seize a discriminator circuit, whenever a call is originated and includes impulse counting units to provide fee determination and route restriction facilities. It also includes impulse regeneration to increase the number of links over which tandem dialing is possible. With this trunking scheme, three digits are required to route the call to the selected tandem switching centre. A translation equipment is put in the discriminator circuit, so that for a three-digit code, the first two digits will position the first and second selectors and the third digit is passed into the discriminator to give a maximum of 100 routings, more than which is not required for a non-director exchange.

But the additional digits to route a call via tandem will be required on calls from some exchanges and not on calls from other exchanges.

Thus, the scheme of trunking for non-director exchange defeats the primary condition to provide a uniform dialing code from all exchanges in the area and a flexible exchange trunking scheme.

Director Exchange System: These difficulties are, however, overcome in a director system, where the piece of equipment which provides the routing code translation is called a director and hence the system has become known as the director system of switching to distinguish it from non-director system of switching tandem and sub-tandem exchanges. In large cities a main tandem centre is provided with a number of sub-tandem centres just to decrease the length of junction links by linking the sub-tandem centres, just to decrease the length of junction links by linking the sub-tandem exchanges as required with one another and to the main tandem central exchange.

Translation: The process by which the equipment of an exchange will, on receipt of a dialing code from the subscriber, transmit the necessary impulse code to route the call to the desired exchange is known as translation. It has got the following advantages. (a) The routing code dialled by a subscriber can be uniform throughout the area. (b) the number of digits in the routing code is minimum and is independent of the impulse train required to position the selectors. A two-digit code will provide for access to 100 exchanges and a three-digit code for access to 1000 exchanges. A two-digit code for routing pulses followed by the first and second numerical selectors and final selector for four-digit exchange numbers caters for 10,000 line exchanges. (c) Busier routes can be allocated to first selector levels and the less busy routes on the second selector levels or subsequent selectors. (d) Along with the gradual growth of the system, the modifications of routing arrangement can be easily done without any alteration of the dialing code. (e) Stepping of selectors is done by mechanically generated impulse trains sent out by the translation equipment.

The trunking flexibility of a translator scheme is illustrated in Figs. 5.37(a) and 5.37(b) for *A* and *B* area, respectively.

For *A* area, as shown in Fig. (a), the heavy traffic routes are accommodated on the levels of the 1st selectors whilst the two separate groups of second selectors provide accommodation for all the remaining direct routes from the exchange, whilst for *B* area all the direct routes can be provided by the 1st selectors and by one group of second selector. The scheme provides complete flexibility in the allocation of levels. But there is some standard to be maintained for levels for provision of a degree of uniformity. One level at each exchange is to be allocated to the tandem junction group, when no direct route to that exchange is available. The switches provided for routing the traffic to the desired exchange are called code selectors and the switches that connect to the particular number in an exchange are called the numerical selectors.

Digit Storage: In director system of working, the four-digit numerical selectors are prefixed by a two-digit or three-digit routing code providing a six-digit or seven-digit number for the subscribers of the exchanges in a big city. Thus the calling sub dials first routing code consisting of two or three digits and then dials with an interdigital pause of less than 500 msec the numerical codes. So the correct routing of the call cannot be determined until the second digit in case of two digit routing code or the third digit in case of three digit routing code is received by the translating equipment, which may then have to transmit up to six consecutive impulse trains to position the various selectors. As this requires some time and cannot be done in the short interval between the receipt of the last code digit and transmission of the 1st numerical digit, it is required to store the numerical digits and retransmit them in their original order after the routing of the call has been established to the required exchange.

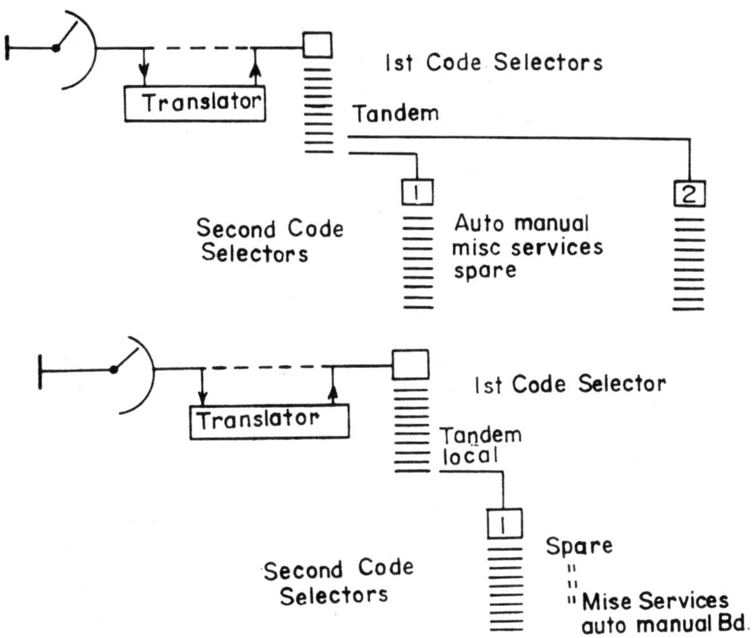

Fig. 5.37 **(a)**: Translation scheme for *A*-area
(b): Translation scheme for *B*-area

The basic trunking arrangment of a director exchange is shown in Fig. 5.38

Here, the translation equipment is divided into two parts, namely, *A*-digit selectors and directors to make it possible to design a director which would receive the routing code digits. A two-digit routing code followed by a four-digit numerical code provides access to 100 exchanges of 10,000 theoretical lines each, while a three-digit routing code followed by a four-digit numerical code, provides access to 1000 such exchanges.

Thus, in a director system with three initial digits as routing code, when a calling subscriber lifts his phone, his u/s hunts for and seizes a free first code selector, with which is associated an *A*-digit hunter. *A*-digit selectors are placed in a pool common to the whole exchange and are terminated on the banks of *A*-digit hunters. When a first code selector is seized, the *A*-digit hunter automatically searches for and selects the first free *A*-digit selector when dial tone is returned to the calling subscriber and on receipt of the first (*A*) digit, the *A*-digit selector steps to the appropriate level. At the end of the impulse train, the wiper automatically cut into the bank and search over the level to seize the first free director in that group. The *B* and *C* digit from the subscriber's dial position a two-motion selector called the '*BC* switch' in the director circuit. The destination of the call is now fully determined by the positioning of the wipers of the *BC* switch in the director circuit and the bank contacts of this selector are cross-connected as required to the banks of a sender switch, so that, at the appropriate time, the required series of impulse trains are transmitted to step the various code and tandem selectors. Then when the numical codes are received by the director and are being recoded, the director sends out the routing digits as determined by the cross-connections between the *BC* switch and the sender switch banks.

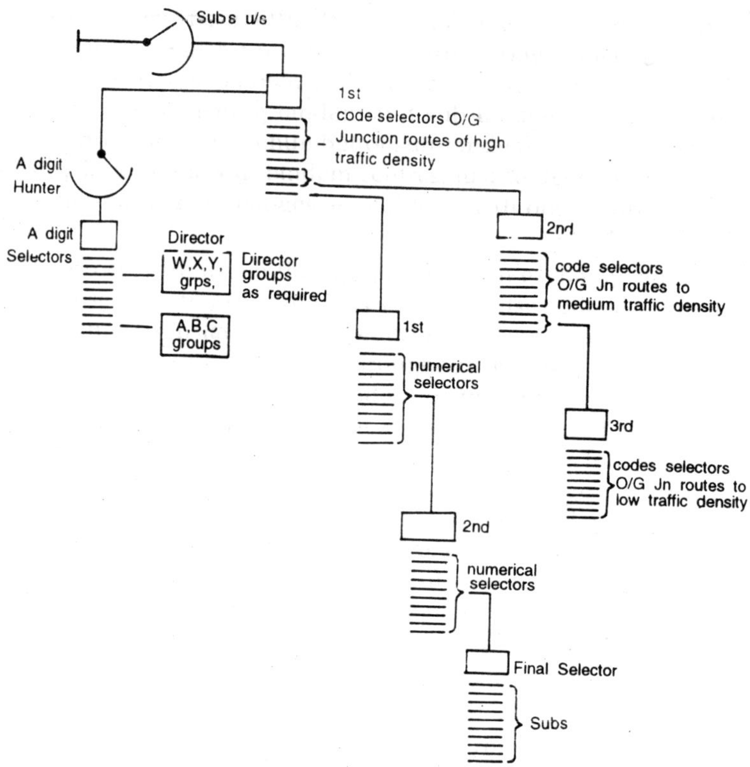

Fig. 5.38: Basic trunking arrangement of director exchange

The outgoing impulse trains are passed back through the *A*-digit selector and the *A*-digit hunter to the first code selector to the required level and subsequent impulse trains position the second code and the third code and possibly tandem selectors to switch the call through to the required exchange. After the full transmission of the routing code, the director re-transmits the numerical code to step the selectors to the desired number. If the call is meant for the local exchange, it goes straight to the first numerical selector and thence to second selector in the required group and finally to the final selector in the local exchange.

To minimise the number of switches at each code selector stage, the levels are usually allocated so that the routes of high traffic density are transmitted on the banks of first code selector, whilst the routes with less traffic are placed on the banks of second and third code selectors.

After the transmission of the numerical codes, A-digit selector and the director are released to attend to the next call, and their average holding time is about 20 sec. The calling subscribers transmission bridge is in the first code selector and metering is applied over a fourth wire to the calling sub's u/s circuit, or receipt of a line reversed condition from the final selector.

A-digit hunter is a 25-point uniselector, A-digit selector is a 100-outlet group sector, code selectors are same as 200-outlet group selectors and first and second numerical selectors are 200-outlet group selectors and final selectors are same as used in non-director exchange, except for omission of metering.

For seven-digit director system, a three-digit routing code is used. The first A digit is received by a digit selector stepping to proper level to seize the first free director to receive the second B and the third C digits and transmit a series of impulse trains to establish the call through the code selector and tandem selector stages. It is necessary that the director should store the numerical portion of the called rules number, till the routing digit are completely transmitted. Thus the director system has (a) a two-motion selector with six banks (BC switch), (b) four uniselectors for the storage of numerical portion of called subscribers number, (c) a digit distributor by means of which the incoming impulse trains are transmitted on the correct sequence to BC selector and four numerical impulse storage uniselectors, (d) a send switch controlling the number of impulses to be transmitted by the director, (e) a translation field to cross-connect the BC selector banks with send-switch banks to obtain the required translation.

Six-digit director scheme has two routing codes followed by four numerical codes, where on lifting the phone, the subscriber gets connected through his uniselector to a special type of group selector, which does not stop by the subscribers-impulses but repeat the subscribers impules to the director. The director seized by the director hunter of the first code returns the dial tone to the calling subscriber and the two code digits dialled by the subscriber are utilised to put the wipers of the BC switch in position to route the call to the required exchange after translation of the code digits. The transmission of the routing digits may be from 1 to 6 trains of impulses, as per requirement of the exchange concerned. The first routing digit received from the director steps up the wipers of the first code selector and if there are more than one translated digits, a second code selector is seized via the banks of the first code selector. The second and the subsequent code selectors are the normal group selectors.

During the time the routing digits are being sent out, the numerical digits are being received and stored and retransmitted after the transmission of the translated routing digits is completed. Thus the director equipment is required only during the setting up of a call and released after sending out the numerical digit to remain ready for the next call.

Tandem working: The tandem exchange is a switching exchange centrally located from the traffic point of view for junction working only, without having any local numbers. Instead of providing direct junctions between an exchange and each other exchange in the area, incoming and outgoing junctions are provided to a tandem exchange and the traffic is routed over these junctions to and from several exchanges in the area, others effecting a saving in cabling and switching costs, labour and space from that required for direct junctions between those exchanges. Thus tandem group of junctions are comparatively large, but carry more traffic than direct junctions.

5.9 Cross-bar Exchange System

In a cross-bar exchange, the switches directly respond to the pulses from the subscriber's dial. The selectors have no moving wipers and all connections are made by the pressure of the control springs requiring a movement of less than a millimeter. The cross-bar switch consists of a special form of multiple relay with selective operation of contact springs arranged in vertical circuit paths and horizontal circuit paths, together with a mechanism for cross-connecting any of the vertical circuits to any of the horizontal circuits by the operation of the electromagnets. The precious contact material consists of 70 per cent silica, and 30 per cent palladium, which suppresses the contact noise.

There are several types of cross-bar systems. In the B.P.O. type , there are 20 vertical circuit paths and 10 horizontal circuit paths, together with a mechanism for cross-connecting any of the 20 vertical circuits to any of the 10 horizontal circuits by the operation of electromagnets. In the Penta-conta type of cross-bar system developed in Belgium, the basic component is the multi-switch, which is used as a four-wire 52-outlet switch. The multi-switch is seldom used singly as such in the switching network. This type has been found to be more advantageous from various aspects and the cross-bar system has been in extensive use in India, surpassing all other types. The two stages of multiselector switches form the basic selector unit as shown in Fig. 5.39.

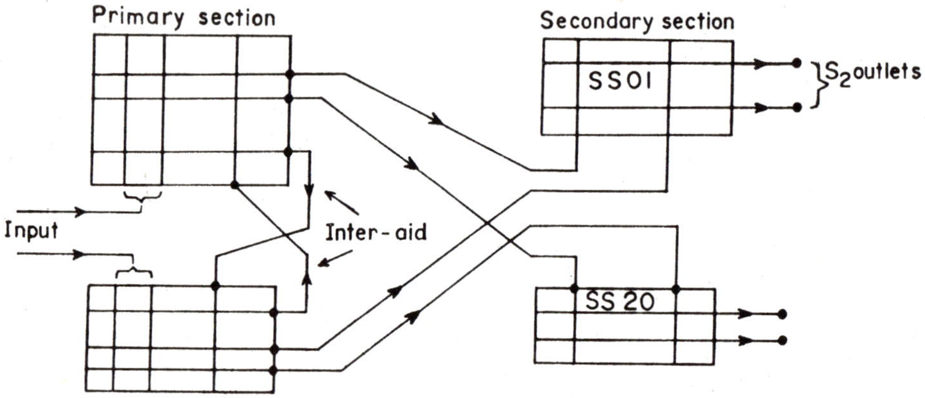

Fig. 5.39: Links between Primary and Secondary Sections of Multi-switch

This selection unit consists of primary and secondary sections, in which the vertical selectors of the primary section constitute the inlet of the stage: the horizontal multiple of the secondary section constitute the outlets. Thus 40 outlets from each of the primary section are distributed as inlets amongst the 20 secondary section at the rate of two each, called links and the remaining 12 outlets from each of the primary section are used for mutual aid, i.e. they are connected to the inlets of other primary sections. Thus a call appearing at the inlet of any primary section can reach an outlet in two ways. It can reach through (i) a direct link between the primary and secondary sections containing the outlet or (ii) another primary section which has a direct link available to the secondary section. In this case, the call proceeds through the mutual aid link between the two primary sections. Hence the two-staged network acts as an outlet switch with access to all the links between the primary and secondary sections, provided it satisfies the following tests fulfilling the conditions. Firstly, that the secondary section has at least one free outlet in the required direction, and secondly that at least one free link is available between the primary and the secondary sections. When there are a number of such secondary sections, one is chosen by a marking technique with pre-selection.

Thus, subscribers' lines via M.D.F. are connected to the terminal selector (multi-switch) frame, which is a 52 outlet switch with facility of terminating 50 directory numbers and two extra numerical numbers. At most 14 horizontal and 16 vertical frames can be provided. Subscribers originating a call via a call finder in primary section frame gets access to common control equipment, which also directs the call to the desired number in T.S. frame through a fifty selector provided in the P.S. frame. Thus, there are a maximum of 20 T.S. and 8 P.S. frames in a standard line-unit of 1000 lines. Each T.S. frame has access to each P.S. frame by a pair of links. For 8 P.S. frames, there are 16 verticals in each T.S. frame.

Line unit: A line unit consists of line and cut-off relays and terminal selector frame, call finders and 50 selectors frame, line marker frame and common 1000 relay group in marker frame. The marker and the common 1000 relays give a priority for choice of T.S. and P.S. frame and mark the required outlet, as shown in Fig. 5.40 in the block schematic of elements of line switch.

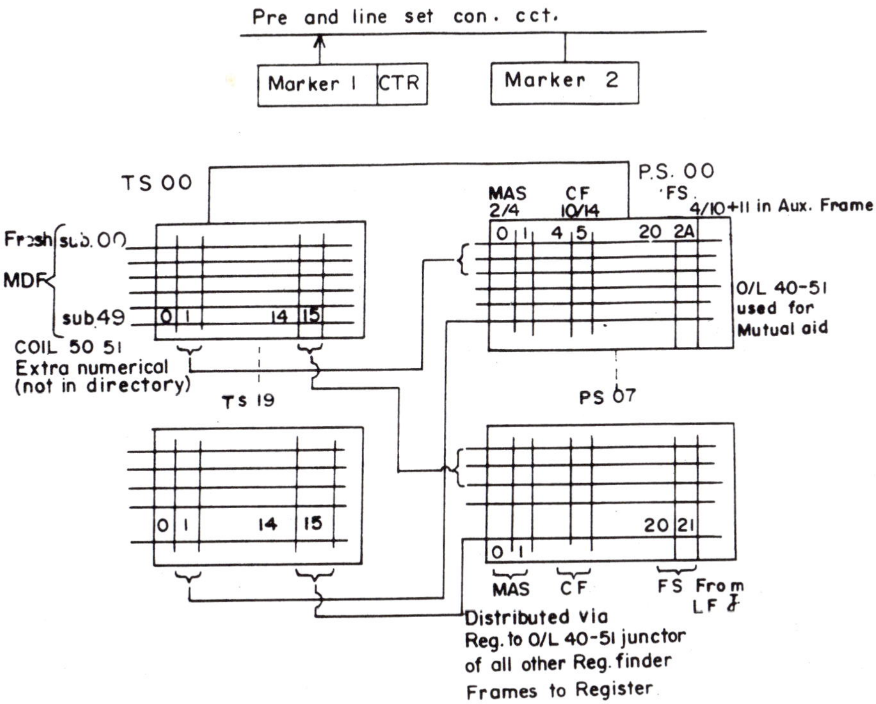

Fig. 5.40: Block Schematic of Elements of LINE UNIT

The T.S. frame handles both originating and terminating traffic and in each case the frame is seized under a line marker. When several subscribers of the same frame call simultaneously, the mutual exclusion arrangement permits one call at a time only, and if several frames call simultaneously, the marker gives priority to one only by means of distribution, leaving others to wait for their turn. If several primary sections require the same terminal multi-switch simultaneously, the marker chooses one by priority arrangement.

There is a facility for supplying busy tone to subscriber from T.S. bays and supervising false call and CSH condition of subscribers receiver, and for isolating all subscribers except a few from the exchange during emergency.

The call finders and fifty selector frames form the primary selection stage along with mutual aid selectors plus the associated fifty selectors in an auxiliary frame. When call finders are used, the primary selection gives connection between terminal frame and a free register via reg. junctor and subsequent access to group selection unit, and when fifty selectors are used, the primary section connects between the group selection unit and the terminal frame, giving access to the called line.

The function of the marker circuit is to control the operation of selection unit and the common relay unit is provided one per line unit and is equipped with one of the two line marker frames.

Group selection unit: A group selection unit represents a selection stage under which incoming traffic is distributed to various directions as shown in Fig. 5.41.

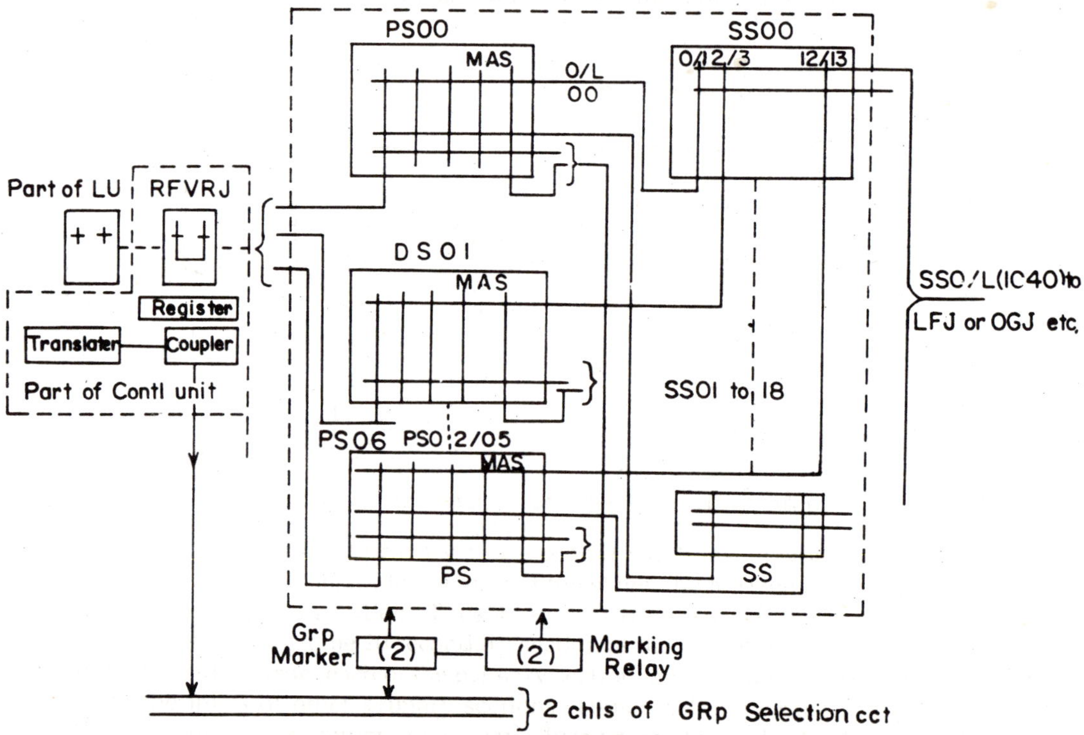

Fig. 5.41: Group Selection Unit

It handles all sorts of local, terminating and outgoing traffic and is known as originating group unit. A separate group unit known as terminating group unit is provided, where the incoming traffic from outlying exchanges is sufficiently high. During peak hours if all LFJs are busy, TGU provides for handling overflow traffic for local calls via overflow LF junctor.

A group selection unit has two multi-selector stages called primary and secondary, with a number of sections to cope with the traffic. A maximum of seven primary with 20 secondary sections forms a standard group unit. Thus there are 14 selectors in a secondary section and 40

O/L of primary section are used for direct connection to 20 secondary section and the remaining 12 O/L of PS are used for mutual aid. The primary sections of the OGU are associated with call finders of line unit via register junctor, while the primary selectors in TGU are associated with incoming junctors. For an incoming call, since the battery feed, ringing current and tone, etc. are supplied by ICJ, the outlets of TGU are connected to A.V. relays, which simply connect through the switching and speech wires to the called line unlike OGU, where outlets are connected through to the called line via LFJ, which gives the battery feed to calling and called party and ringing the called line. Each group is associated with a pair of markers which control the selection of outlets in the required direction. The marker has provision for priority distribution of PS and SS frames with the help of magnetic counter. A block schematic showing interconnections of TGU and OGU is shown in Fig. 5.42.

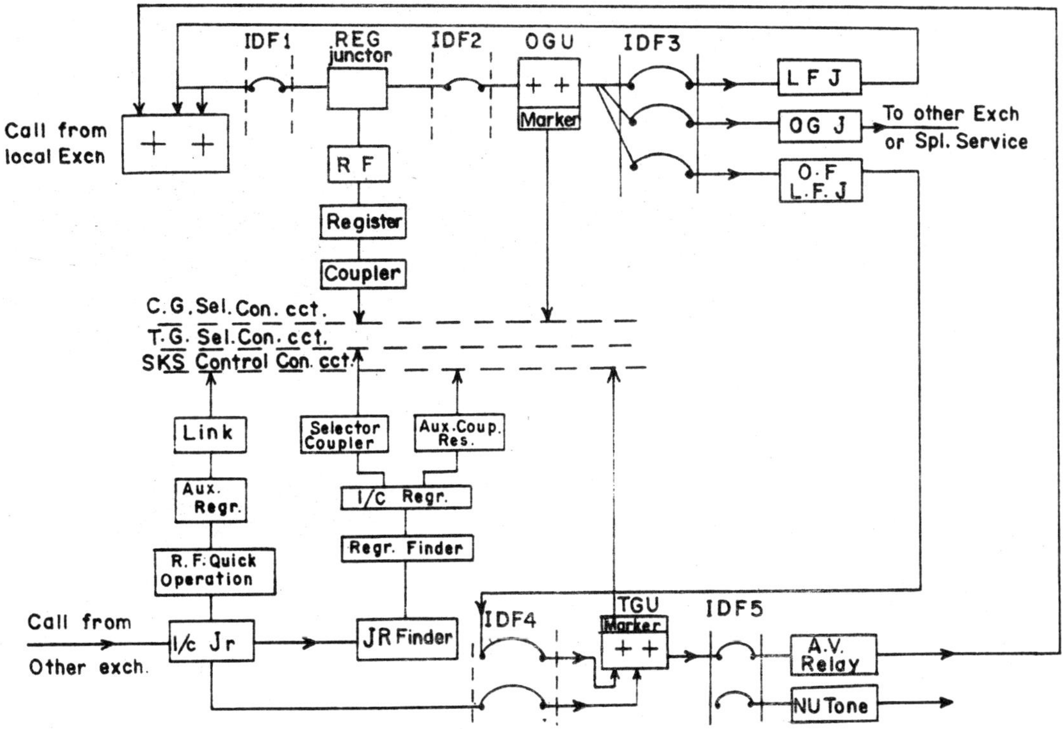

Fig. 5.42: Schematic showing Interconnections of I/L or O/L of TGU and OGU

A group unit is also connected to the register via register junctor and register finder, which provides communication between calling and called subscriber just like a line unit. The register finder frame consists of a multi-selector having 14 verticals and 56 outlets — 28 on the left and 28 on the right. The bank multiple of the RF is divided in the middle, so that 7 verticals have access to 28 register junctors on each side. The verticals on both sides are parallel in pair and each of the seven such pairs is connected to a register, providing access to 56 register junctors. On receiving information from register, the translator gives approximate direction code, which the marker receives via coupler and controls the establishment of call.

The register finder quick operation (RFQO) provides several facilities. For I/C calls from distant SXS exchanges, it searches for and keeps connected in advance a free I/C auxiliary register by means of a distributor and connects the I/C junctor to an auxiliary register, which stores maximum of first two digits received from SXS exchange during the interval when main I/C register is connected to the same I/C junctor and is ready for reception of the remaining digits.

The I/C auxiliary register can regulate a maximum of two predetermined digits and indicates to the main register the number of digits already stored and transfer the same to the main register when seized. It releases under the control of the main register.

The register is used only during establishment of a call and a group of about six registers can serve the traffic through nearly five primary sections. So the register finders (which are multi-switches) are interposed between registers and call finders which are connected to the outlets of register finders and registers are connected to verticals. Selection couplers connect the register and the information path. The senders get access through links to the information path, which is used for the exchange of information between the different common control equipment during establishment of a call such as registers or translators and the markers, O/G junctors, senders and receivers. These information paths are provided in pairs, each called a connecting circuit consisting of 20 wires divided into four groups of five wires each. The digits received from the subscribers' dial in decimal is stored in a suitable form, so that each digit is translated into a 2 out of 5 codes method. The progress of a call through the switching path from the calling subscriber to the called subscriber is shown below in Fig. 5.43.

In the progress of a call from a calling subscriber to the called subscriber in the same Pentaconta type exchange, the following three stages of selection are required: (a) the pre-selection stage, (b) the group selection stage, and (c) the line selection stage. For an exchange with second group unit, a second group selection may be required

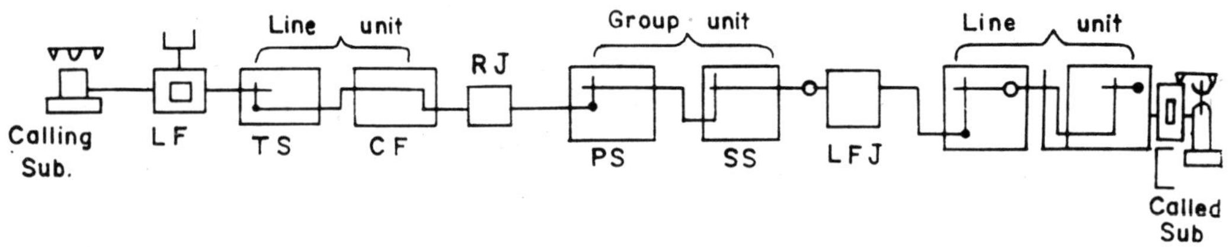

Fig. 5.43: Switching Path of a Local Call in a Cross-bar Exchange

[abbreviations: LF = Line finder, TS = Terminal selector, CF = Call finder, RJ = Register junctor, P.S. = Primary selector, SS = Secondary selector, LFJ = Local feed junctor, FS = Fifty selector].

(a) *Pre-selection*: The connection of a calling subscriber to a free register is called a pre-selection, the trunking diagram of which is given in Fig. 5.44.

A subscriber's loop is formed as soon as his receiver is lifted, causing the terminal section relay to operate and put it in the calling condition. When the digits for the number of the called subscriber are being dialled, a series of priority chains are gone through for the selection of a free register, as also the linking units and selectors, i.e. the register finder, the primary section, terminal section, call finder and terminal selector. The line marker which controls a part of this selection is also picked up on the basis of alternating priority.

(b) *Group selection stage*: The subscriber dials the digits and the pulses are received in the pulse containing chain of register. During the interdigital pause, these are transferred to the memory relays and in the same manner all the digits are stored. On receipt of two or three digits, the register engages a selection coupler, which takes test and double test of the primary section of the group unit to ensure that the primary section is not busy to set up another call. The selection coupler then connects itself to a local translator where the first few digits are transferred and analysed to form the code of the required direction.

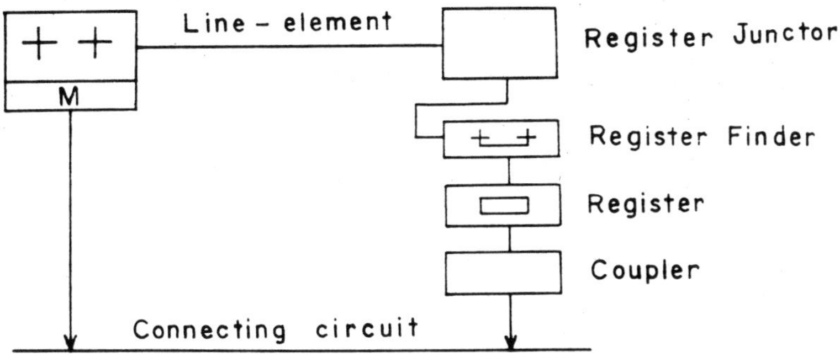

Fig. 5.44: Trunking Diagram of Pre-selection

The marker connected to the primary section connects to a group selection connecting circuit, where the selection coupler is also connected. The translator now transmits the code of the required direction in 2 out of 5 code basis to the group marker through *C* and *D* wire groups of the connecting circuit. Here also the group marker, controlling the relevant operations is picked up on the basis of alternating priority. Basically, there are two types of priority chains — fixed priority and rotating priority between the two markers.

(c) *Line selection*: The call now appears at a fifty selector of the concerned line unit through the LFJ and on receipt of all the digits of the called number, the register connects it to a selection coupler which after due tests, connect to the line marker to seize a connecting channel on *A*-wire. The last three digits of the subscriber are then transmitted in 2 out of 5 code basis to the line marker and the category of called subscriber is then transmitted to register which connects by putting earth to *B*-wire when the verticals operate and the 'la' and 'b' relays of the called subscriber also operate and when the called subscriber answers and a metering pulse is sent from the local feed junctors.

Outgoing call from one Penta conta exchange to another Penta conta exchange: An outgoing call progresses in the same manner as a local call up to the group selection stage, whence the category of the chosen outlet is sent to the register, which recognises whether the outgoing junction is connected to a penta conta or step-by-step exchange or a special service position. If it is a Penta conta exchange, the register connects itself to a multi-frequency (M.F.) sender through auxiliary folder and to an outgoing translator through a selection coupler, informing the calling subscriber's category and the first four digits. On analysis of the digits the translator orders the sender to connect itself to a link which informs the category of the calling subscriber, and the rank of the digit from which sending is to begin, through selection coupler.

Thus, the connecting circuits are released and the sender demands from the register the digits to be sent first. On receipt of the digit, the sender transmits the same through MF signal over 'a' and 'b' wires to the distant exchange, which on acknowledging the receipt demands the next digit to be sent. The operations are repeated till all the digits are transmitted; when the distant exchange intimates the called subscriber's category and the register checks the receipt, the sender is released.

Outgoing call to a step-by-step exchange: The process is similar to the above, except that a decimal sender is chosen and the stored digits are again converted into pulses and transmitted to the distant exchange.

Incoming call from one Penta conta exchange to another Penta conta exchange: The switching path of an incoming call from one Penta conta exchange to another Penta conta exchange is given in Fig. 5.45.

On completion of transmission of the called subscriber's digits from a distant calling subscriber's Penta conta exchange , these appear at the incoming junctors of the called subscriber's exchange. These junctors are connected to the inlets of the group units as well as to the outlets of the junctor finders. The incoming junctors get access to the special incoming registers through the junctor finders, which also contain multi-frequency receivers. On receipt of sufficient number of digits, the register seizes a selection coupler, which begins testing and double testing of the primary section. The incoming registers get access to the same local transmitters. The progress of the incoming call to the called subscriber is exactly the same as that described above for an outgoing call from a Penta conta exchange to another Penta conta exchange above. At the end of selection, the identity of the called subscriber is also sent back to the originating exchange initiating the call.

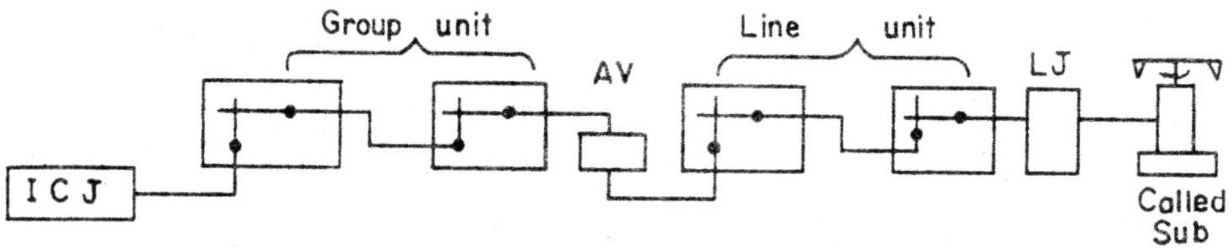

Fig. 5.45: Switching Path of an Incoming Call

[ICJ = Incoming junctor, AV = Availability relay, LJ = Line junctor].

Incoming call from a step-by-step exchange: During the time lag between the seizure of the junctor and an incoming register, an auxiliary register connects itself to the incoming junctor and stores the first two digits as received. When the main register is seized, those first two digits are transferred from the auxiliary register to the main register. The call then proceeds further as described above, but if the auxiliary register is not seized before the first pulse arrives, then the busy tone is fed to the calling subscriber from the incoming junctor itself by a special arrangement.

5.10 Private Automatic Branch Exchanges (PABXs)

Like the private manual branch exchanges (PMBXs) used in manual and auto exchange areas for distribution of incoming calls and gaining access to the public exchange, a private automatic branch exchange (PABX) is also installed in automatic areas for communication between various departments with access to public exchange. Such a PABX may be either attended type or unattended type according to the requirement of the business houses. The equipments of PABXs are of three types: (a) uniselector switching type, (b) two-motion selector type, and (c) relay switching type.

For small organisations using only 25 or 40 extensions, uniselector type of PABX is used. For big organisations where 600 to 1000 or more extensions are in use, two-motion selector type of PABXs are used. The relay type of PABXs follow the principle of Relay system of Telephony. Its advantages are that the absence of mechanical selectors minimises the amount of maintenance labour and expenditure. The system is silent in operation and may be utilised where a Strowger equipment is found unsuitable. The most common type of Relay PABX is a 50-line unit.

The diagram for a uniselector type PABX is given with its main trunking arrangement in Fig. 5.46(a).

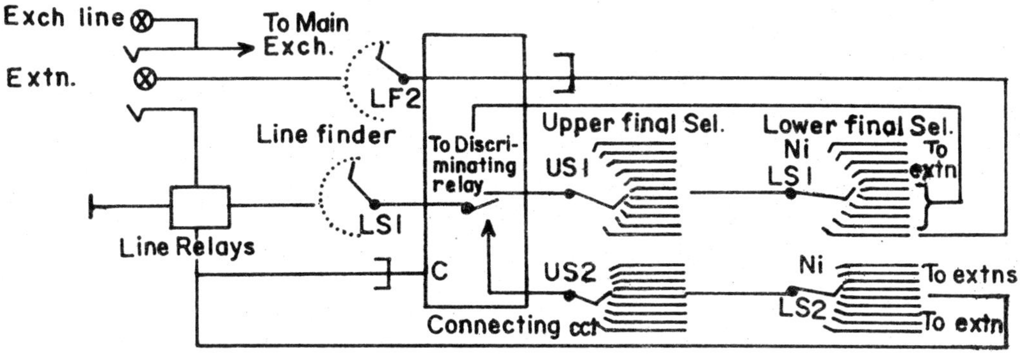

Fig. 5.46(a): Uniselector Type PABX

The unit may have a capacity up to 40 extensions and six connecting circuits and by using two such units it can work up to 70 extensions. It is designed to work on 36 V power plant. The bank of each final selector is divided into two parts which is selected by means of a wiper switching relay to give access to two different groups of lines. Call to the main exchange is obtained by dialling 0.

A two-digit scheme is adopted for extension members and the first digit is used as a discriminating signal to find out the proper final selector and the proper group of wipers on that selector and the correct half of the selector bank and the other digits are similarly used to find out the proper upper or lower selector, the first or second group of levels of the chosen selector, and first or second group of ten contacts in the selected group of levels.

The diagram of a two-motion type of PABX is given in Fig. 5.46(b).

Here, a capacity for 600 subscribers line is obtained by using a three-digit numbering scheme with a 200 point two-motion line finder, group selector and final selector. For extension to extension calls, a signal is transmitted from line circuit to start circuit and free control relay set, which preselects a line finder. From the line finder the calling line is extended to the group selector, from which the dial tone is received by the calling subscriber who dials the three digits for the called extension, i.e. the first digit is received by the group selector and the second and third digits are received by the F.S. The exchange line can be obtained either from the manual board or via level 9 of the group selector.

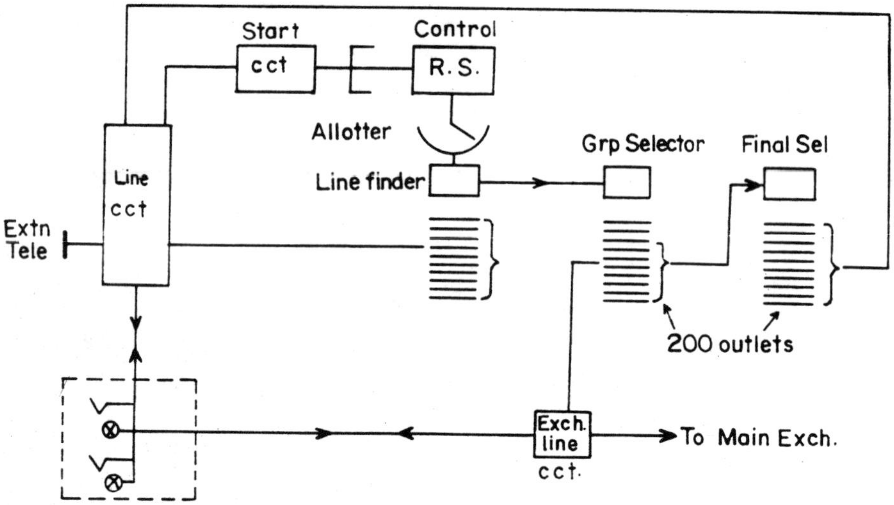

Fig. 5.46(b): 2-motion Type of PABX

5.11 Gradual Development of Trunk Exchanges and Dialling Circuits

In Section 3.3, we have described the principles of T-32 type trunk exchange, which were in exclusive use in India about four decades ago, when most of the local exchanges were of manual type. But with the gradual introduction of local auto exchanges, an improved type of trunk exchange, called T-43 trunk exchange has been gradually substituted since 1943, replacing the T-32 type trunk exchanges in most of the big cities.

In T-43 trunk exchanges, the trunk positions are divided into two main groups, viz. outgoing groups for dealing with calls from local subscribers, and incoming and transit groups for dealing with calls to local subscribers or to other trunks for transit. But in a T-32 trunk exchange where the number of positions does not generally exceed two, it is not possible to divide the positions into two groups as in T-43 type of exchange.

In T-43 type of exchange, for the incoming and transit position group, the incoming side of any trunk or junction circuit appears in the answering field of any one of these positions with calling lamps only, and this group of junctions is separate from those in the outgoing positions, except for some small group of junctions common to both type of positions. The O/G trunk multiple field is provided on a four panel basis. In outgoing positions, the trunk line multiple do not appear and the trunk circuit appears in the answering field of a position with busy lamp indication. The outgoing junction multiple is also provided on a four-panel basis.

The overall loss of a trunk circuit is kept at 6 db with transit calls built-up on a tail-eating basis in intermediate exchanges. The following standards are required for two-wire switching to maintain the overall transmission loss:

(1) Line up of channel trunk circuits to 0 db from Hybrid line to Hybrid line.

(2) Extending only the Hybrid line to exchanges and terminating the Hybrid net with a compromise network.

(3) Providing two types of line jacks for each trunk circuit, one for use with outgoing calls from outgoing positions, and another for use in incoming and transit positions, either for an incoming or an outgoing call.

(4) Controlling the overall loss of the trunk circuit by (a) switching in at the exchange originating the call, a 3 db pad in the trunk line circuit on calls emanating from outgoing positions, or (b) switching in a 3 db pad in the junction line circuit at the incoming terminal exchange, when the terminal calls are completed via incoming and transit positions.

In incoming and transit positions, the trunk line jack in the multiple field and in the answering field for any one trunk circuit are electrically same, but they differ from the line jack for the same circuit appearing in outgoing positions. At the exchange, originating a call from an outgoing position, the *K*-earth from the line jack on these positions operates a pad switching relay *P* in the trunk line circuit and introduces a 3 db pad in the trunk line circuit. On calls on the same trunk lines established from line jacks in the incoming and transit positions, the *K*-earth does not operate the pad-switching relay and hence the 3 db pad remains cut out of circuit. A diagram for trunk cord T-43 sleeve circuit is given in Fig. 5.47.

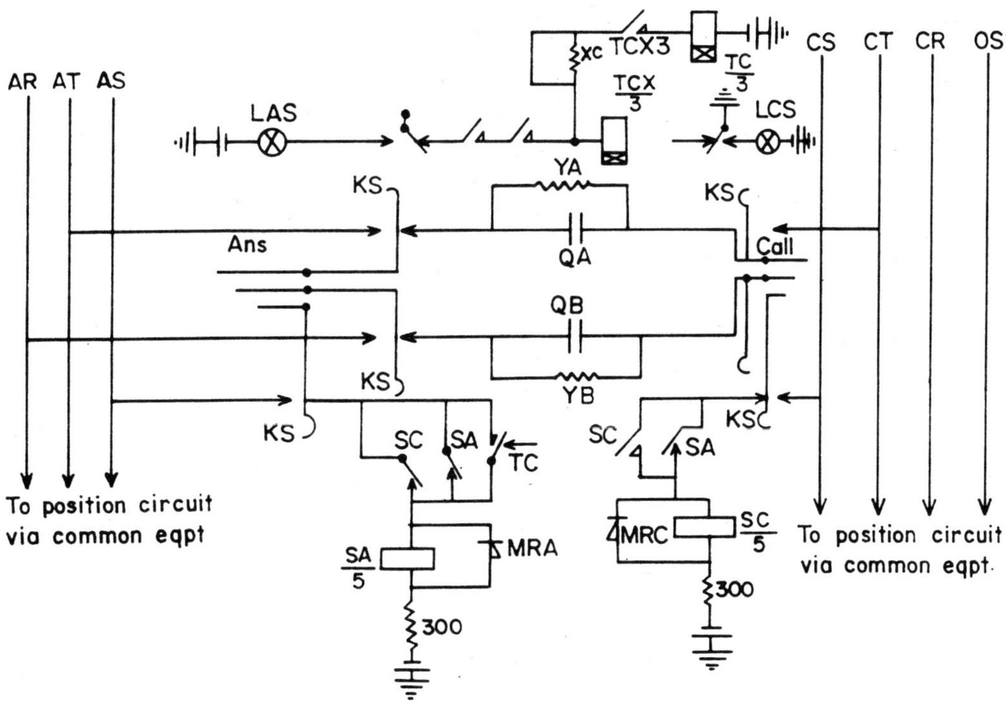

Fig. 5.47: Trunk Cord Sleeve Circuit in T-43 Exchange

(1) An incoming call is answered by the operator with the answering cord and the calling cord is used to complete the call — terminal or transit. (2) A supervisory signal presented to the sleeve of the calling cord is transmitted to the sleeve of the answering cord as an interrupted pulse signal, i.e. the backward supervisory signal is a pulse glow of the calling supervi-

sory lamp for about 200 m sec at an interval of 2 secs. (3) A supervisory signal presented to the sleeve of the answering cord is transmitted to the sleeve of the calling cord as such, i.e. forward supervisory signal is a continuous glow of the answering supervisory lamp. (4) If a clearing supervisory signal is received on the sleeve of the answering cord while the backward clearing signals are in progress, further transmission of the backward clear signals is stopped and the forward clear is transmitted through the cord circuit, i.e. clear forward signal has overriding priority over the clear backward signal. (5) When the operator at any exchange, other than terminal exchange originating the call operates, the SPEAK key of the cord circuit, transmission of supervisory signals via that exchange cord circuit is stopped.

In 1953, the T-53 type of trunk exchange was introduced with more facilities as it was operated only with keys. The trunking diagram of a T-53 type of trunk exchange is given in Fig. 5.48.

Its main advantage is the lesser wear and tear of keys than the cords used in T-43 type. In case of no answer for more than 30 secs, from any of the T-53 operators on any of the calls, a visual alarm is provided to alert the monitors to find out the operator at fault and do the needful. There being no multiple field on T-53 board, busy indication on trunk is absent and therefore to save operators from dialling unnecessarily on busy trunks, a duplicate trunk group busy lamp panel is fitted within view of all T-53 operators to indicate a lamp lighted as soon as all the trunks to a particular station are engaged. Thus the connecting link on a T-53 board consists only of: (a) Key Answer Trunks; (b) Key Call Trunks; (c) Key Call Subs; and (d) Speak Key. With this arrangement evidently two trunks cannot be called on the same link. So to obtain a trunk on a key call, subscriber is to initiate connection on a transit call by T-53 operator.

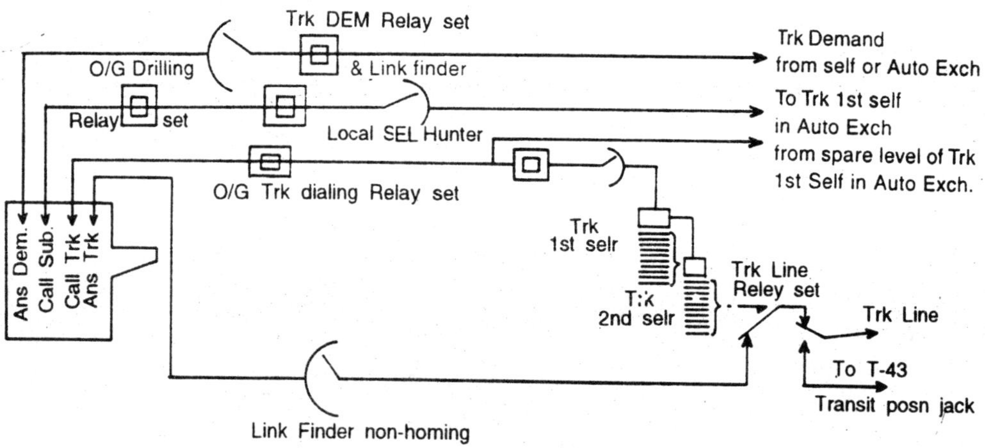

Fig. 5.48: Trunking Diagram of a T-53 Type Trunk Exchange

This is achieved by looping back a spare level (level 8) on local Auto Trunk First Selector rack to trunk selector for T-53 rack through trunk finders, etc. and thus a trunk is obtained by dialling an extra digit (8) before the trunk code on key call sub side. This facilitates connection of transit calls under initiative of T-53 operator, who can call the two distant stations, one on 'Key call trunk' side and the other on 'Key call sub side.' Thus extension of spare level from local auto trunk First selector to T-53 selector can enable anybody to obtain any station connected to T-53 exchange equipment.

In our country, the efficient technical experts at the Telecommunication Research Centre at New Delhi have, from time to time, devised different types of relay plates with facilities provided for single link operator's dialling (SLOD) and multi-link operator's dialling (MLOD), in which case the trunk circuits are terminated on trunk first selectors in auto exchanges. These trunk first selectors are equipped with one make unit of NPA contact assembly (or vertical marking bank and wipers in pre-2000 type switches) to operate and hold for levels 1 to 8 of the selectors. One extra wire is required for extending the earth from this make unit to the relay set in the trunk exchange. The development of these dialling circuits led to the introduction of subscribers' trunk dialling scheme, which was introduced in 1960 between Kanpur and Lucknow and thereafter introduced in almost all the big cities in India.

In the S.T.D. system, out-of-band signalling is used in long distance carrier channels, whereby a subscriber can himself dial from his telephone any other telephone at a distant auto telephone exchange, prefixing the trunk code number of that station, followed by the called subscriber's number of that auto exchange. For this mainly three relay sets are required: (a) the O/G relay set, (b) the terminal relay set, and (c) the I/C relay set. These relay sets work in association with one another along with a time and zone metering two-motion selector associated with O/G relay set. The facilities provided by these relay sets used for S.T.D. over carrier channels equipped with out-of-band signalling are given below.

(a) The outgoing relay set on being seized, transmits the dialled impulses to the line as well as to the time and zone metering selector, to repeat the metering pulses to the caller's meter by positive battery metering for recording the appropriate fee on the meter of the calling subscriber.

(b) The terminating relay set accepts loop/ disconnects impulses and converts the same as leg impulses to the carrier signalling equipment and on called party answering at the distant station, provides for reversal of battery and earth to the O/G relay set for metering facility.

(c) The incoming relay set provides a forward holding earth to the private wire and on a called party answering, transmit a signal on the 'send leg' at the incoming end to start metering.

(d) The time and zone metering two-motion selector associated with O/G relay set, is seized on caller dialling 0 and steps up in unison with second and third digits determining the destination and on called party answering, transmits earth pulses to the O/G relay set to be converted to positive battery pulses to operate the calling subscriber's meter. An initial pulse is given followed by a periodical pulse, according to the time pulse received from the bank of selector. This time pulse depends on the exchange to which the call is made.

The above mentioned relay sets used for long distance trunk dialling lead to the development of trunk auto exchange based on cross bar working on junction scheme.

5.12 Trunk Automatic Exchange (TAX)

The main difficulty in long distance auto trunk calls is to provide for alternate routes when congestion is found in any intermediate link, if the calls are routed through multi link. The next difficulty is the automatic charging of the calls. In some cases analysis has to be done up to five digits for determining the routing codes, for which special equipment is required for charging rates. The third difficulty is to compensate for the 2W/4W hybrid loss encountered during the switching. In addition, other facilities are required to be provided that are not used in case of a direct system. It is therefore found that instead of a direct system, an indirect system employing a common control equipment is most suitable for trunk auto exchanges. As such a common control equipment is used only during the setting up of a call, and is released after the switching operation to set up the next call, thus providing the facility, independent of the digits dialled, for re-routing a call if the direct route is busy.

Trunking scheme: The trunking scheme for trunk auto exchange is based on cross-bar working junction scheme. Here all the junctions and channels from local exchanges and distance exchanges, respectively as also from other trunk auto exchanges are terminated as incoming junctions as shown in Fig. 5.49 for the 1600 line junction of a Trunk Auto Exchange.

The incoming junctions are designed in such a way that they fit in with the line signalling of the distant end and each type of junction can have several junction classes used for separating services within one group. The control and group units are basically the same as in a local cross-bar exchange, except that five-wire switching is employed in addition to other features such as the use of through looking circuits, origin marking signal, charge determinators, pad switching, etc. It can be divided into blocks of primary units, secondary units and control unit or register block.

Primary Block: One primary block consists of a maximum of four group units, the outputs of which are grouped so as to form 3 'splits'. Split *P* is multipled over group units 1 and 2, and split *P'* of the same capacity is multipled over group units 3 and 4. Split *S* is multipled over four group units of the block. Thus a split is looked upon as a group of inlets having a common multiple on the outlet side.

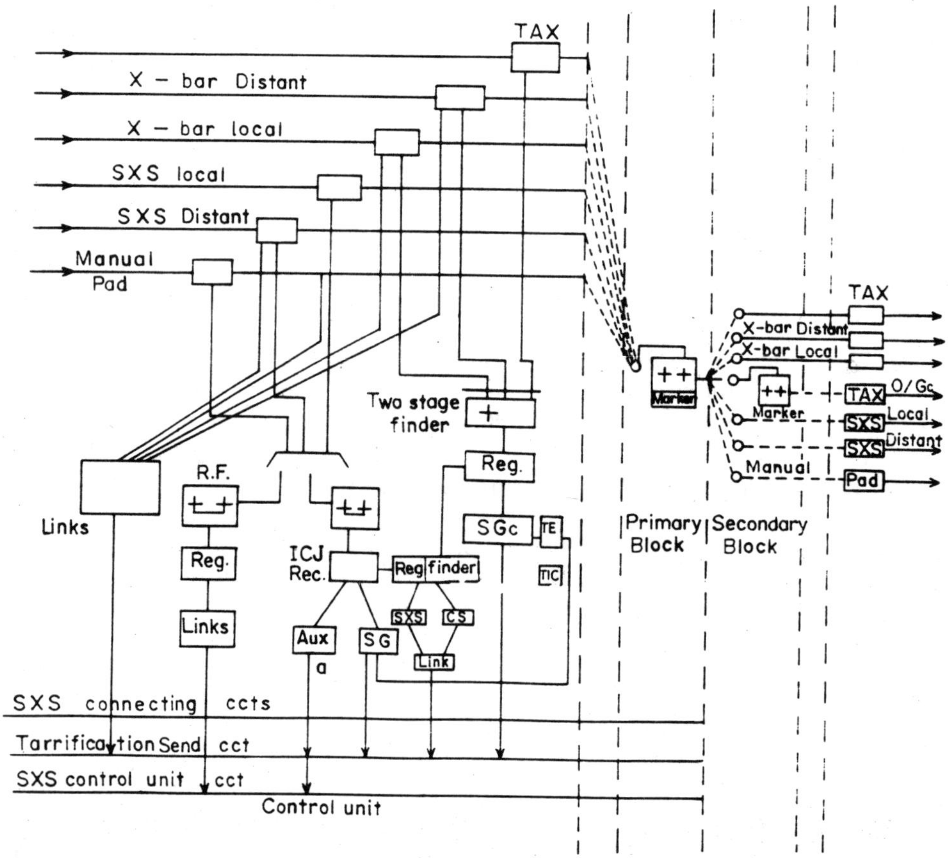

Fig. 5.49: Working Junction Diagram of Trunk Auto Exchange (1600 lines)

Secondary Block: Each secondary block comprises a maximum of two group units. One secondary block corresponds to one primary block with its outlets connected to the same *p* and *p'* splits and in some cases, a separate overflow secondary block is provided with outlets serving small groups of outgoing trunks for particular directions.

Register Block: It is a group of registers serving a certain number of incoming junctors via a double stage connecting scheme composed of junction finders and register junctors. The incoming junctors served by one register group must be connected to only one primary block, i.e. they can have access to a maximum of two '*p*' splits. But a translator has to deal with only four '*p*' splits as two translators can serve a maximum of two register groups.

Through Looking Circuit: It is used as a link between the translator circuit on one hand and the outgoing junction circuit on the other, so that the translator can detect the availability of free outgoing junctors in a certain direction. It also blocks a route for normal traffic. Thus an outgoing direction may be spread out over all group units, i.e. primary and secondary blocks. An 'all trunk busy relay' is provided per split and per direction to find out the availability of junction. This relay remains operated as long as a free junction is available in a particular direction. All the junctors of a particular direction and particular split are multipled as shown in Fig. 5.50.

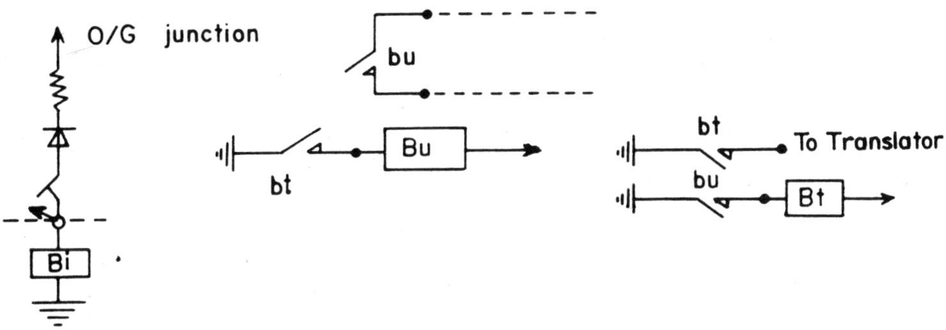

Fig. 5.50: Multipling of Junctor of a Particular Direction and Split

Here, relay *Bi* remains operated as long as one free junction in this direction is available and the information is passed to the translator. When all the junctions are found busy, the relay *Bi* is inoperative and relay *Bt* operates and gives an indication to the translator that no free outlet is available. Up to 6000 lines exchange of incoming and outgoing lines, the through looking circuit is provided in a ratio of one relay per direction per split with relay only, and for 6000 to 16000 lines, the same is provided with electronic through looking circuits.

Origin Marking Signal: It is provided to categorise the incoming junctors in separate groups requiring different facilities. The origin marking signals are DC signals and MF signals forwarded from incoming junctors to registers for digit insertion, metering and junction indication, etc. A provision is made for 24 origin markings for incoming junctions from *SXS* exchanges and 16 from cross-bar exchanges.

Selection: The following information is sent to the register finder as soon as the I/C is seized by an I/C junctor: (i) identity of primary split to which the junctor is connected; (ii) class of junction; (iii) type of line signalling, i.e. 2W or 4W line. The junctors are arranged in conformity with the signalling arrangement of the distant exchange and after the receipt of first three or four digits including digit 0, the register seizes a coupler, making a double test on the primary selector to which the I/C junctor is connected. If the double test is successful, the coupler seizes a translator and the register sends the first three or four digits to translators, which analyses and gets the information concerning allocation of at least one free outlet in one group unit in the wanted direction with the help of through looking circuit.

In case of a single selection, the translator sends the direction code to first group selection marker. In case of two-stage selection, it sends the identity to all secondary blocks to the first group selection marker to select one secondary block with easy access, i.e. without mutual aid. After completion of the first selection, the translator is seized to send the direction code to the second group selection marker. If it is found that the direct route is busy, then the through looking circuit tests the alternative routes at most three and if all routes are found busy, then busy tone is sent from I/C junctor or originating trunk auto exchange.

Signalling: For signalling with SAX (local and distant) cross-bar exchange (local, distant and TAX) and manual trunk exchange, the register-signalling for interwork with TAX is indicated below:

(a) For SAX local: DC loop signalling.

(b) For SAX distant: 'Tone-on' continuous working over outband of carrier channel.

(c) For manual exchange including manual trunk exchange: DC loop signalling.

(d) For cross-bar exchange: MF 2 out of 5 compelled signal type.

To set up a call on auto trunk, a calling subscriber dials 0 followed by the station code and then the directory number of that called subscriber. At the originating exchange, the register connects itself to the TAX and transmits all the digits including 0 stored already there, when the common control equipment including the local originating register releases. The I/C trunk register after the receipt of the first three or four digits transmits them to the translator, which determines the route to be followed and the rate at which the calling subscriber is to be charged if the call is proved successful. But if these digits are not found sufficient for complete determination of route or the rate to be charged, it asks for more digits to be sent. With the help of through looking circuits, the translator understands the existence of at least one free junction, through which it passes the information to the marker to make a free outlet and then the translator sends the information to the distant exchange or second TAX by senders, when the common control equipment except the register releases. In case the call is passed via second TAX, the digit 0, code digit required are sent to the second TAX register, which switches to the local exchange and common control equipment at the second TAX releases with its register. Then the numerical digits are sent in MF in case of cross-bar or in pulses in case of Strowger exchange by the sender of originating trunk auto exchange. After the call is put through, the register at the originating trunk auto exchange is released and thereby the call is metered at the calling subscriber's meter.

5.13 Air Conditioning of Telephone Exchange and Telecommunication Equipment Rooms

Just as human beings can work with ease and comfort for longer periods of time, when the climate is very congenial and the temperature is moderate, so also all the electrical, magnetic and electronic equipments used for telecommunication systems can work with the highest capability and efficiency with the least wear and tear, when the temperature and humidity of the rooms containing these equipments are kept within the permissible limit. Such artificial condition of the air of the room increases the working efficiency and enhances the longevity of these equipments. In the process of air-conditioning the rooms of a telecommunication building, various types of refrigerants are used. The selection of a particular refrigerant out of Arcton,

Frigen, Genetron, Isotron, Freon 12 (CCl_2 F_2) dichloro difluoromethane and Freon-22 ($CHCl$ F_2) monochlo-difluoromethane, etc. is to be made after due consideration of various factors, such as boiling point, pressure for condensation, density, non-inflammability, non-toxicity, odourlessness and stability, etc. of the refrigerant as the size of the pipes, orifices, valves, etc. depend upon the characteristics of the refrigerant used.

It has been observed that for air-conditioning the telecommunication buildings, the most suitable refrigerant is either Freon 12 or Freon 22. But the vapour-density of Freon 22 is 40 per cent higher than that of Freon 12 at 105^0F. So the sectional areas of pipes, various passages, orifices, valves, and working pressures at various points for the two refrigerants would be different for the desired temperature. So the replacement of Freon 12 by Freon 22 (or its equivalent) or vice versa in a particular plant must not be done without reference to the designer to avoid troubles.

A refrigeration cycle consists of the following components: (i) compressor, (ii) condenser, (iii) evaporator, (iv) expansion valve, (v) refrigerant piping, (vi) accessories, (vii) air handling unit, (viii) safety controls, and (ix) water circulation system, as shown in Fig. 5.51 for an air-conditioning plant with a direct atmospheric type of cooling tower.

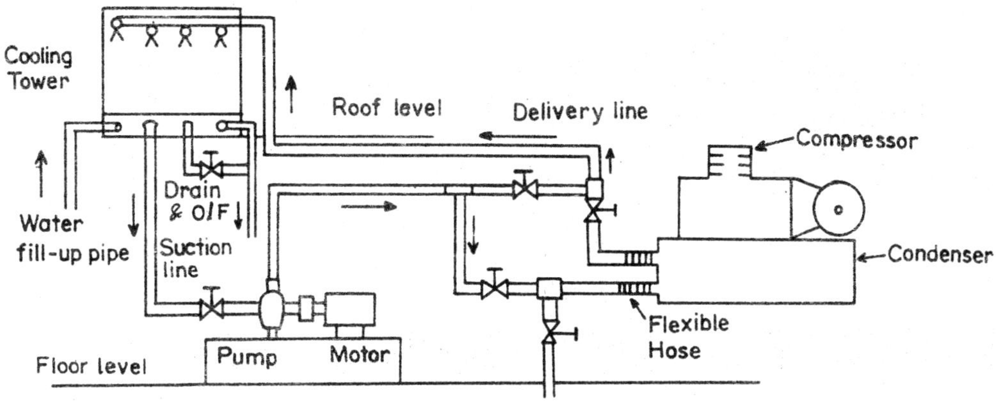

Fig. 5.51: Block Diagram of an Air-conditioning Plant with a Direct Atmospheric Type of Cooling Tower Showing Refrigeration Cycles.

The functions of each component is described below.

(i) *Compressor*: Its selection is dependent upon the capacity of the pump used. First of all the amount of refrigerant by weight to be evaporated in the cooling coil per unit of time is determined. This gives the amount of heat in BTU/hr and then the capacity of the compressor is selected to circulate this amount of refrigerant through the system. Also the temperature and the humidity conditions are required to be considered in deciding the capacity of the condenser. The system may have either forced lubrication where an oil pump is embodied in the compressor, or a splash lubrication system where the oil pump is replaced by a set of scoops fixed to the lower half bush-bearings at the big end bearings. The cool suction gas on its way to the upper side of the cylinder passes through manifolds cast into the wall of the body of the compressor. It brings about cooling of the walls heated by the compression of refrigerant and friction of the pistons.

(ii) *Condenser*: It converts the hot refrigerant gas delivered by the compressor into liquid at room temperature. The condensers may be of (a) shell and tube type, (b) air cooled type, and

(c) evapora tive type. Thus for 1 ton of refrigeration and 8 ^0F rise of water temperature from 90 ^{0}F to 98 ^{0}F, the quantity of required refrigerant =

$$\frac{\text{heat rejected / hr / ton} \times \text{no. of tons}}{\text{temp. rise of water} \times \text{sp. heat of water}}$$

$$= \frac{240 \times 1}{(98-90) \times 1} \; 30 \text{ lbs/min.}$$

(iii) *Evaporator*: Here the liquid refrigerant evaporates on release of pressure and the heat liberated is taken out from the air to be conditioned. The cooling arrangement consists of a coil of copper tubes fitted with aluminium or copper fins. The air from the outlet of the coils is cool enough to absorb the sensible heat in the room.

(iv) *Expansion Valve*: Generally, a thermostatic expansion valve with external equaliser is used in aircondition installations, where it regulates the rate of flow of liquid refrigerant to the evaporator in proportion to the rate of evaporation of the liquid refrigerant in the evaporator. The diagrams of shell and tube type condenser (a), evaporator (b), thermostatic expansion valve (c), and a solenoid valve (d) are shown in Fig. 5.52.

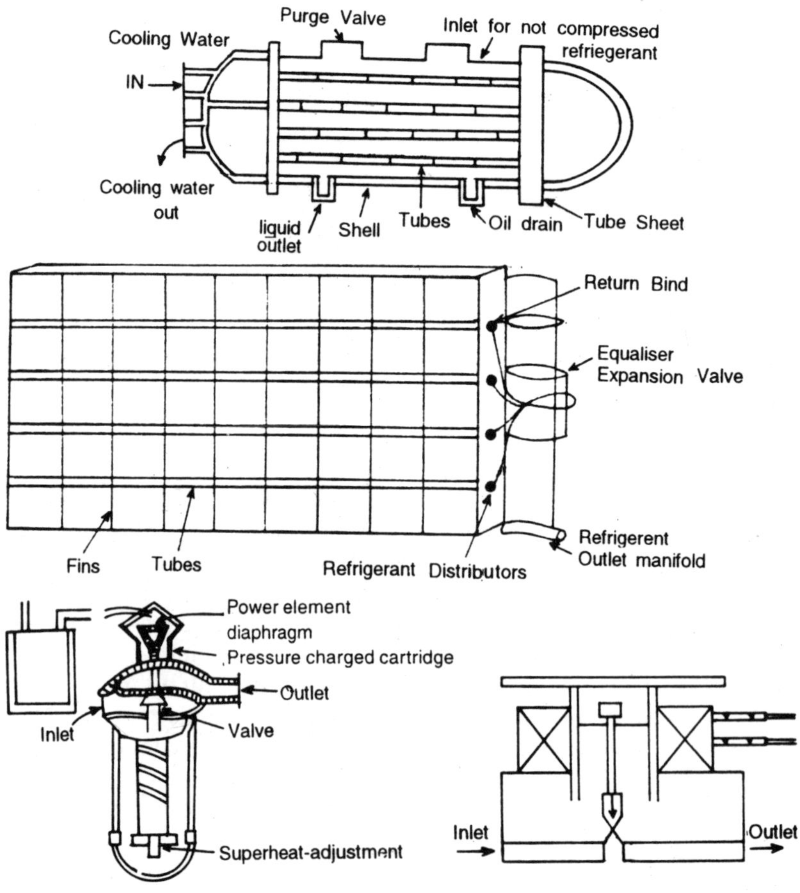

Fig. 5.52 (a) Shell and Tube Type Condenser. **(b)** Evaporator.
(c) Thermostatic Expansion Valve. **(d)** Direct Acting Solenoid Valve.

(v) *Refrigerant Piping*: For Freon 12 and 22 type air-conditioned units, the refrigerant is conveyed from unit-to-unit by means of pipes or tubings, whose materials must not corrode or react with the refrigerant and must remain leak proof. The sizes of the pipes should be selected to correspond with the pressure drop and velocity change to avoid friction loss and noise. The pipes should be supported on metal hangers and brackets to minimise vibrations and stress on joints.

The line from the condenser receiver to the expansion valve is called the liquid line. The suction line from the cooling tower to the pump should be designed for a pressure drop of 1 to 2 per sq. inch and the velocity should be normally between 1000 to 2000 feet/min, and the minimum at light loads should be 800 ft./min for horizontal runs and 1500 ft/min for vertical runs. The pipe between the compressor and the condenser is called the discharge line or hot gas line and the velocity in the discharge line varies from 1000 to 5000 ft/min with a pressure drop of 2 to 4 per sq. inch. In case of more than one compressor, the oil drain from one to the other compressor is prevented by equalising oil levels by running pipes below oil level. Similarly, for gas equalisers in crank cases, pipes are run above the oil level.

(vi) *Accessories*: (a) the solenoid valve is usually installed in the liquid line ahead of the expansion valve for automatically shutting off the flow of the liquid refrigerant to expansion valve and cooling coil; (b) the moisture should not exceed 10 parts per million by weight in the refrigerant under charge, for which hydrator may be used initially to absorb moisture; (c) the pressure gauges at different temperatures are installed for measuring the suction pressure, discharge pressure and oil pressure; (d) a liquid line filter is installed to protect the system from undesirable impurities, dust and foreign particles. The cooling coil is designed to work between 30 ^{0}F and 35 ^{0}F for a telecommunication building containing equipment.

(vii) *Air handling Unit*: It consists of a mixing chamber for return air and fresh air. The other accessories are air filters, air cooling cum dehumidifying coils, a bypass damper in case a close control of humidity is desired, a spray type humidifier in case dry climates are encountered, a set of electrical or warm water heaters for monsoon heating and an air-blower with a drive as shown in Fig. 5.53.

It may be of built-in masonry type or of pre-assembled sheet metal, in which case its capacity is limited to 20 or 25 tons only. For plants larger than 25 tons, a masonry shell type air handling unit is required.

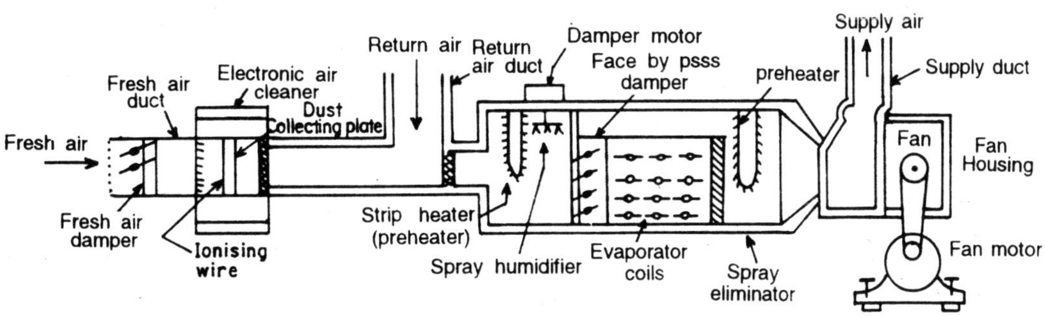

Fig. 5.53: Air Handling Unit

(viii) *Controls*: Safety controls are used to safeguard the plant. High pressure control is connected to the discharge side of the compressor to see if an electrical switch opens and the compressor automatically cuts out when the discharge pressure exceeds a pre-determined value of 176 to 180 lb/sq inch for F-22 type, i.e. in both cases about 25 lb/sq inch more than the normal operating pressure. The low pressure control is connected to the suction side of the compressor, to cut off the compressor when the suction pressure exceeds a pre-determined value of 15 lb/sq inch below the operating suction pressure. The oil failure control operates when the difference is more than 10 lb. The safety relief valve is mounted at the top of the horizontal shell and the tube condenser, so as to blow off the hot gas when a very large discharge pressure builds up and the high pressure control fails to function. After the blow off, the pressure comes down to the setting of the relief valve and the spring-loaded device automatically closes again, bringing about relief of the high pressure and minimising the loss of refrigerant. The standard settings of the relief valves for F-12 and F-22 types are 235 lb/sq inch and 300 lb/sq inch, respectively. Similarly, the temperature is controlled by the thermostats and the humidity is controlled by humidistat on rise and drop of temperature and humidity.

(ix) *Water circulation*: When the water is not recirculated in the case of 100 gm per hour/ton of refrigeration, the water regulating valve is provided at the inlet of the condenser which operates to allow the condenser water to run off to the drain. But when the water is recirculated, the water regulating valve is not used and the water from the outlet of the condenser is again it is provided at the inlet of the condenser which operates to allow the condenser water to run off to the drain. But when the water is recirculated, the water regulating valve is not used and the water from the outlet of the condenser is again pumped into the cooling tower to bring down the temperature to the limit, within which pumped into the cooling tower to bring down the temperature to the limit, within whichit can be recirculated for the next cycle of operation. In a cooling tower the water circulation is made at the rate of 200 g per hr/ton of refrigeration, by which the water is cooled by nearly 8 to 10 degrees. An atmospheric type of cooling tower is of 10 to 15 ft high with an area depending on refrigeration capacity. Galvanised iron louvres sloping inwards are provided on the vertical faces of the tower to prevent undue loss of water and allow air to pass through the cooling tower uninterrupted and get cooled to the required temperature.

A strainer is provided in the sump of the cooling tower at the suction line, whereby the cooled water is drawn up by a centrifugal pump set for circulation through the condenser as depicted earlier in Fig. 5.51.

6

Open Wire Carrier Systems, Underground Cable Carrier and Coaxial Cable Systems

6.1 Open Wire Carrier Systems

In our country, one channel (1+1), three channel (1+3), eight channel (1+8) and twelve channel (1+12) open wire carrier systems are worked on physical trunk lines duly transposed with repeaters at suitable distances along the route, to provide one physical outlet plus the number of channels in the system. But on underground cables including the coaxial cables, only 12 channel carrier system is worked with suitable repeaters at short distances.

The open wire carrier systems, specially the eight channel and twelve channel carrier systems, require trunk lines constructed under C-J or C-8 transposition scheme for efficient working, though, of course, ordinary single channel carrier or three channel carrier system can work on non C-8 trunk lines of high insulation, if they are properly transposed and are free from inductive effects from other trunk lines or power lines in the vicinity, causing cross-talk interference and distortion on the transmitted signals.

(1+1) Carrier System: The principle of a single channel O.W. carrier system is described diagrammetically in Fig. 6.1.

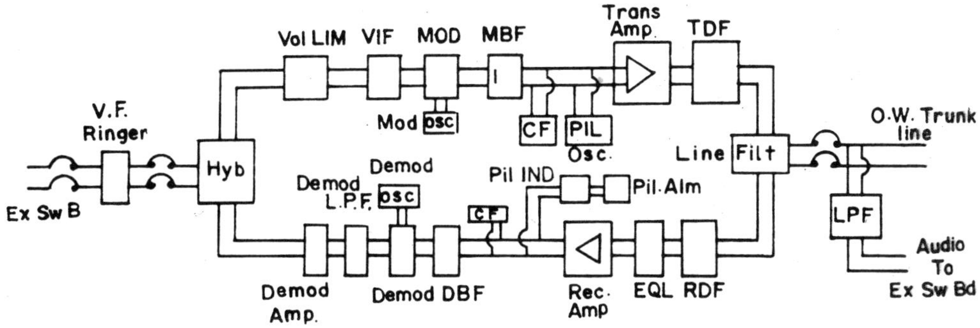

Fig. 6.1: (1+1) O.W. Carrier Telephone System

Here, the channel and the audio line are transmitted into the trunk switch board.

The Transmit Path: The outgoing speech from the exchange passes via the 2 wire - 4 wire Hybrid unit through the voice input filter, volume limiter to the modulator. The voice input filter cuts off the unwanted frequencies beyond the voice band and the volume limiter reduces the exceedingly high volume of speech levels preventing overloading in the transmit amplifier causing distortion and interference in the speech. In the modulator panel is fed a frequency from modulator oscillator and the speech frequencies are superimposed on this modulation frequency for amplitude modulation producing both upper and lower sidebands. In case of terminal *A*, the modulator band filter transmits the upper sideband and suppresses the lower sideband, while in the case of terminal *B* the reverse is the case. An amplification of about 49 ± 1 db is given by both transmit and receive amplifiers and a pilot oscillator supplying a pilot frequency fitting with the sidband transmitted in each direction is fed into the input of the transient amplifier, the gain of which is so adjusted by means of pads and potentiometer that at its output, the transmitted level of the sideband is +19 ±1 db and that of the pilot frequency is ± 0.5 db. From here the transmitted sideband goes to the open wire line at + 18 ± 1 db level at H.P. section of the line filter out point via the transmit directional filter. A compensating filter is connected across the MBF output for correcting the impedance condition for both ends of frequency range.

Receive Path: The sideband transmitted by the terminal *A* reaches the distant terminal *B* at the other end of the physical trunk line via a H.P. section of the line filter to the receive directional filter and then to the equaliser consisting of a fixed and a variable section to equalise the line losses for the high and low frequencies in the sideband, as the higher frequencies of the band undergo higher loss. Then the received sideband enter the receive amplifier, where necessary amplification is given to bring the sideband at the receive amplifier output to a level of +6 db ± 0.5 db. Also at the output of the receive amplifier, the pilot frequency is filtered out through a filter to the pilot indicator, where the filtered pilot frequency is rectified, smoothed and amplified and then put on DC meter at a level of ± 0.5 db on the mid-zero mark of the indicator scale to specify the relative condition of the physical line loss. From the output of the receive amplifier, the received sideband goes to the demodulator via demod band filter and is demodulated by a demod frequency exactly equal to the mod frequency giving out the original audio band, which on passing through the demod low pass filter enters into the demod amplifier, where it gets necessary V.F. amplification to give a level of -6 db at the Hybrid point.

From here, the audio band goes to the local switch board via the ringer panel. This has got suitable relays in the transmit and receive side operation for ringing current conversions as described in V.F. signalling under Sec. 3.1 B(6). When the trunk board operator requires the distant end operator, he sends a 17 ~ ringing current, which is converted into 500/17 ~ in the V.F. ringer panel and goes through the carrier channel to the other end, where it is again converted to 17 ~ by rectification and operation of relay to glow the switchboard lamp to draw the distant end operator to respond.

Whenever the pilot level goes beyond ± 1 db, on the pilot indicator, an alarm both visual and audible goes on to attract the attention of the technical staff on duty to do the needful adjustment of the pilot level indicator, by adjusting the gain of the receive amplifier and thereafter take necessary action by contacting the next repeater section for the verification of the physical line, etc. and do the needful.

For narrow band transmission, the speech band width is from 300 c/s to 2600 c/s and for broad band transmission, the speech band width is from 300 c/s to 3400 c/s.

The modulation frequency in single channel narrow band transmission is taken as 6000 c/s, so that the U.S.B. is 6300 c/s to 8600 c/s and the L.S.B. ranges from 3400 c/s to 5700 c/s. As one sideband only is transmited, the other sideband along with the carrier is suppressed in each direction.

(1+3) Carrier System: The principle of three-channel carrier systems working on open wire trunk lines is shown in the schematic diagram in Fig. 6.2.

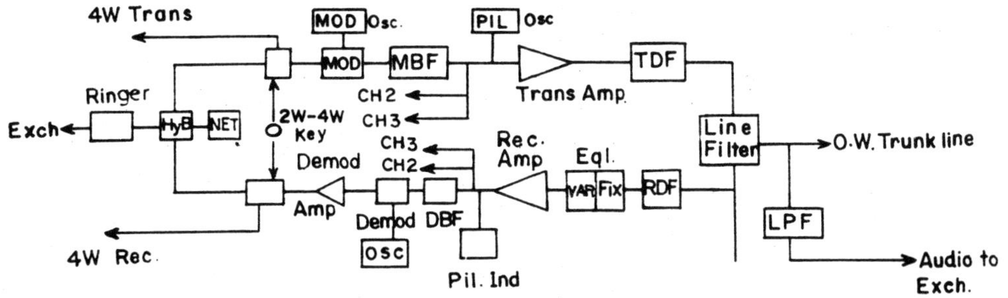

Fig. 6.2: (1+3) O.W. Carrier Telephone System

The carrier frequencies and the sidebands in *A-B* and *B-A* directions are shown below. Here only one sideband is transmitted in each direction and the other S.B. along with the carrier frequency is fully suppressed in each channel.

Type	Direction	Ch. No.	Carrier (kc/s)	Sideband	Range (kc/s)
SIS	*A-B*	1	12.9	13.2	15.5
		2	9.4	9.7	12.0
		3	6.3	6.6	8.9
	Pilot		12.8		
	B-A	1	24.4	21.8	24.1
		2	20.7	18.1	20.4
		3	28.4	25.8	28.1
	Pilot		28.5		
SIT	*A-B*	1	14.3	14.6	16.9
		2	10.9	11.2	13.5
		3	7.7	8.0	10.3
	Pilot		14.2		
	B-A	1	23.7	24.0	26.3
		2	19.8	20.1	22.4
		3	27.7	28.0	30.3
	Pilot		23.6		

The function of each panel is identical for each channel, with the 1+1 system already described and the only difference is in the carrier frequency and the sideband range. The outputs of the three MBFs are joined in parallel across a compensating filter to correct for the impedance conditions at both ends of the frequency range before they enter into the transmit amplifier, the output of which passes to the line filter via a transmit directional filter.

The outgoing signals of the three-channel system are separated from the incoming signals by a receive directional filter joined in parallel with the line side of the transmit directional filter. In case of terminal *A* the TDF is a LPF while the RDF is a HPF, while in the case of terminal *B*, the reverse is the case.

Only in the case of synchronisation of the modulated oscillation frequency of channels of terminal *A* with the demod oscillation frequency of terminal *B* and vice versa, both the side-bands are sent in *A-B* direction on channel 2 and in *B-A* direction in channel 1 and 3 with the following procedure.

To synchronise channel 1, the MBF of channel 1 is patched out at terminal *B* and 800 c/s tone at 0 db level is sent at Hybrid line point of channel 1. At terminal *A*, the DBF is patched out of channel 1 and at demod amp outright hand points of channel 1, the beats of the tone are listened while the air condenser of demod oscillation channel 1 is adjusted if necessary, by altering the tapping of variable condenser till the beats heard are less than one per second. The air condenser is then locked up and the DBF of channel 1 normalled. Then demod amp out-right hand side of channel 1 is patched to the mod in right hand side of channel 1 to give a loop back and the testing staff at terminal *B* are informed accordingly. The terminal *B* staff then normal the patching of MBF in channel 1 and listen for beats at demod amp out point of channel 1, by adjusting the air-condenser settings in the demod oscilation of channel 1 till the beats are less than one per second. The air condenser is locked and all the patchings at both ends are normalled and the speech and rings are verified between both ends.

A similar procedure is followed in case of channel 3. But in the case of channel 2, the MBF is patched out at terminal A and DBF is patched out at terminal *B* and the remaining procedure is similar.

To check the high frequency line-up, 800 c/s tone at 0 db level is sent in turn at the hybrid line point of each channel and measured at the first repeater station receive amplifier out in the corresponding direction. If these levels are within +18 ± 1 db, the loss due to the first section of the physical line between the terminal and the first repeater is taken as within limit. But if it is not within +18 ± 1 db, then sections 0.4, 0.4 and 0.2 db/kc in variable equaliser are put by trial and error method to bring it within this limit. This H.F. line up test is finished from one repeater to the next repeater till we reach the distant terminal and adjustments for the variable equaliser sections are made at each repeater. At the other end terminal Rec Amp Out point, the adjustment of Rec Amp potentiometer and input pads are made to adjust the receive level at +6 ± 0.5 db. Then the same line-up is taken for the reverse direction and adjustments are made as in the previous direction.

After finishing the H.F. line up test, a V.F. Line-up test is taken by sending the 800 c/s tone at the hybrid line point of one terminal at 0 db level and measuring the same at the distant terminal hybrid line point after adjusting by altering the input variable potentiometer stepping of demod amplifier to bring the level at – 5.5 ± 0.5 db. This test is repeated for all channels and the system is put for service after verifying the speech and ring from the local exchanges at both ends.

(1+8) Carrier System: In the eight channel O.W. carrier system, a single stage of group modulation is employed only in the *A-B* direction after channel modulation by the particular carrier frequencies, while the signals are transmitted in the *B-A* direction after channel modulation by those particular frequencies. Here also only one sideband is transmitted and the other side-

band along with the carrier frequency is suppressed fully before transmission. The frequency allocation for each channel in both directions and the transmitted sidebands and signal frequencies for each channel in each direction are shown below. Here broad channels with bandwidth 300 Hz to 3400 Hz are used.

Direction	Ch.	Ch carrier kHz	Grp carrier kHz	Transmitted S Bank, kHz L.S.B.	Trans. Sig. freq. kHz
A-B	1	36	176	136.6-139.7	135.7
	2	42	"	130.6-133.7	129.7
	3	48	"	124.6-127.7	123.7
	4	54	"	118.6-121.7	117.7
	5	60	"	112.6-115.7	111.7
	6	66	"	106.6-109.7	105.7
	7	72	"	100.6-103.7	99.7
	8	78	"	94.6- 97.7	93.7
B-A	1	36		36.3-39.4	40.3
	2	42		42.3-45.4	46.3
	3	48		48.3-51.4	52.3
	4	54		54.3-57.4	58.3
	5	60		60.3-63.4	64.3
	6	66		66.3-69.4	70.3
	7	72		72.3-75.4	76.3
	8	78		78.3-81.4	82.3

A block schematic for (1+8) O.W. carrier system working with a three-channel carrier system on the same physical line is shown below in Fig. 6.3.

Here, signalling is provided with in-built out of band arrangement separately for each channel, so that it can be used for C.B. signalling or for transmission of dialled impulses. A frequency of 4.3 kHz is used for signalling in all channels and is injected into the signal injection circuit by the corresponding signal modulation panel at the output of the voice input filter to be modulated along with the speech signal in the channel modulator. As the signal modulation

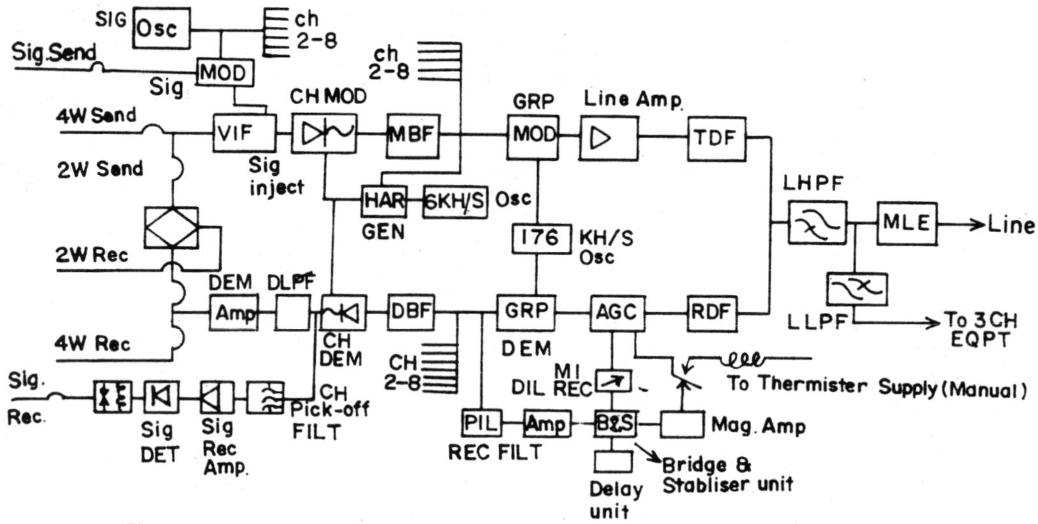

Fig. 6.3: (1+8) Channel Carrier System Working with (1+3) Channel System on the Same O.W. line

panel works with 'tone-on' in the space condition, the tone is cut off when earth is extended for signalling. It can be seen that corresponding to the idle or active condition of the signal send leg, a 'tone' or 'no tone' condition is maintained on signal send path. In the receive path at the other end, the demodulated signal frequency is picked-off by the signal pick-off filter, amplified, detected and the detected signal frequency operates a telegraph relay, which sends earth signal via receive leg to the switch board to glow the call lamp.

In the channel transmit path in the case of terminal *A*, the output of the channel modulator band filters joined in parallel in the range of 36-84 kHz are group modulated by a group carrier frequency of 176 kHz, and the L.S.B. of the group modulated output in the range of 92-140 kHz passes via the group modulation filter to the line amplifier which gives a gain of 45 db. The output of the line amplifier is adjusted to +17 db per channel, which is then fed into the O.W. line via transmit directional filter, line H.P. filter and the miscellaneous line equipment (MLE) panel containing the building out network, the loading unit and the miscellaneous network unit.

In the case of terminal *B*, the channel modulated signals in the range of 36-84 kHz enter the line amplifier direct and are transmitted to the line via TDF, LHPF and MLE panel.

In the channel receive path, the incoming carrier sidebands pass through the MLE panel, line high pass filter and enter through RDF to the AGC panel. Group modulation is done only at terminal *B*. So the output of AGC panel is in the range of 36-84 kHz in case of terminal *A* and in the range of 92-140 kHz in the case of terminal *B*. This also comes in the range of 36-84 kHz after the group demodulation is performed and L.S.B. separated by DBF, the output of which are separately channel demodulated by the demodulator of the respective channel by the corresponding channel demod frequencies and the audio output is separated by demod LPF and amplified by the demod amplifier where its level is suitably adjusted and passed to the switch board through 2W-4W terminating set.

The pilot frequency in *A-B* direction is 92 kHz and in *B-A* direction it is 84 kHz. An auto gain control arrangement keeps the output of the group demod amplifier in the *A-B* direction and the AGC amplifier output in the *B-A* direction more or less constant over a substantial margin of ± 5 db variation of incoming levels, beyond which the pilot bell rings to draw the attention of the technical staff on duty for taking necessary action. After the pilot frequency is picked off from the output of group demod or AGC panel by pilot receive pick off filter and fed to the magnetic amplifier via Bridge and Stabilizer Unit which rectifies the pilot signal to a DC voltage and applies the same to the thermistor heater, an increase in the pilot level decreases the thermistor bead resistance as a result of the increase in the thermistor heater current. This increases the negative feedback and thereby reduces the gain of the amplifier and vice versa, when the pilot level decreases.

A delay unit introduces delay in the operation of AGC to avoid hunting in the AGC action. The gain of the AGC amplifier is variable from 15 to 35 db with a normal value at 25 db over a frequency range from 35 to 150 kHz.

In the miscellaneous line equipment (MLE) panel adjustment is made by strapping B.O. network sections to match the capacitance C in pF of the lead-in cable from the O.W. line terminal pole to the MLE panel so that B.O.N. sections 0.05, 0.1, 0.2, 0.4 when added to the fraction $C/3000$ makes a total of 1.0, 0.8, or 0.6 depending on the loading unit used.

H.F. line-up is taken like three-channel system and the levels measured on loss measurement should be adjusted as follows:

At AGC out/Grp demod Amp out	-8 ± 1 db
At demod amp out	-2 ± 1 db.
On V.F. line-up at Hybrid line	-5.5 ± 1 db.

(1+12) Channel O.W. Carrier System: A block schematic of (1+12) channel O.W. carrier system is shown in Fig. 6.4.

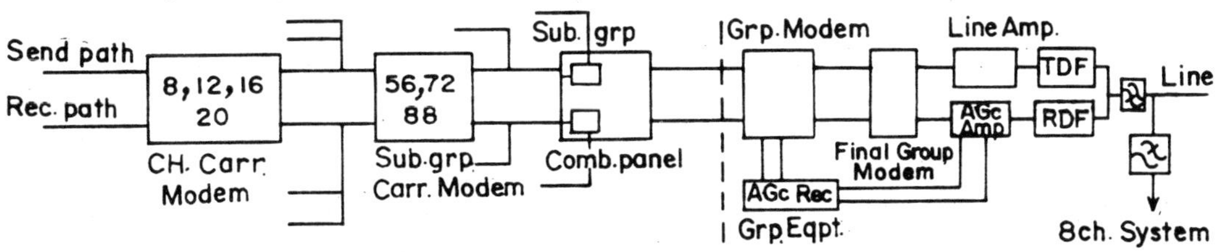

Fig. 6.4: Block Schematic of (1+12) Carrier System on O.W. line

Transmit path: The 12 channel carrier system is worked on O.W. trunk lines transposed on C-8 or C-J scheme. The system works on U.C.P. (Uniform Common Practice), wherein a sub-group is formed with every four channels with the carrier frequencies of 8, 12, 16 and 20 kHz. The lower side band of 4-20 kHz band width after channel modulation is sent for subgroup modulation with subgroup carrier frequencies 56, 72 and 88 kHz respectively. The upper sideband of 60-108 kHz band width after subgroup modulation is sent for group modulation. In the first stage of group modulation with carrier frequency 340 kHz, the U.S.B. (400-448 kHz) is sent for second stage of group modulation with frequency so chosen that the desired sideband after modulation is of such bandwidth that the (1+12) channel system can work on the same trunk line along with (1+3) or (1+8) channel carrier system without any interference. If no other system works on the same C-8 trunk line along with (1+12) channel carrier system, the first group modulator output of U.S.B. (400-448 kHz) is fed to the final group modulator and is modulated with 308 kHz and the L.S.B. (92-140 kHz) is fed to the line amplifier which is a 4 watt broad band (36-160 kHz) amplifier, where it is amplified and sent through the transmit directional filter to the line filter at a level of +18 ± 1 db to the trunk line through the matching transformer for lead-in cable.

Two pilot frequencies, flat and slope, are sent in each direction to enter at the line amplifier input point and at the output of the line amplifier the level is adjusted at ± 0.5 db in each direction for controlling the gain of the AGC amplifier. These are 92 and 144 kHz in the *A-B* direction and 40 and 80 kHz in *B-A* direction.

Receive path: In the receive path the signals enter through line filter, the receive directional filter, fixed equaliser, variable equaliser to the AGC panel amplifier, at the output of which the two pilot frequency signals, flat and slope, are separately picked off through pilot pick off filter and amplified and rectified to be shown in the indicator.

The controlled received signals at the AGC out are successively demodulated twice via group DBF to get the original subgroup 60-108 kHz which on subgroup demodulation produces 4-20 kHz band for four channels, which are again separated by channel band pass filters and separately demodulated with channel demod frequencies of 8, 12, 16 and 20 kHz to get the original V.F. band of signals sent from the far end.

The original group 60-108 kHz is made into two groups, one 36-84 kHz for one direction and the other 92-140 kHz for the other direction. As the original group 60-108 kHz is only partially superimposed on the groups sent in both directions, the exact replica of the sent signals are not obtainable in single stage modulation, for which double modulation process is taken. In one direction, the group carrier frequencies 340 and 308 kHz modulate 60-108 kHz band to give 92-140 kHz band and in the other direction 340 and 484 kHz group carrier frequencies modulate the 60-108 kHz band to give 36-84 kHz band.

The two pilot frequencies, flat and slope, are sent from each terminal station and help the functioning of AGC by decreasing the amplification when the pilot level increases and increasing the same when the line pilot decreases with the help of a marginal relay and keeps the line level more or less constant.

Signalling: An inbuilt out of band signalling process is used employing signal carrier frequency which lies 450 H.P.S. higher than the upper limit of the speech frequency spectrum of the respective channel. The modulated carrier is introduced at the subgroup stage of the channel modem and permits trunk dialling and C.B. signalling to be employed. The procedure is as follows: Incoming signalling pulses from the exchange operate a static relay circuit, which applies pulses of the desired carrier signalling frequencies to the signalling circuit, which pass through the subgroup and group modulation to the line and are abstracted at the other end from the subgroup demod with the aid of a skew hybrid, the output of which is detected for operating the telegraph relay for sending earth via receive leg to the switchboard to glow the call lamp.

As an alternative, 500/20 H/S V.F. ringing procedure can also be used for two-wire terminal side.

For each bayside, an AC power unit of 230 V provides HT supply and a 6.3 V AC for LT supply. All the required carrier frequencies for different stages are derived from a 4 kHz master oscillator through frequency multiplication process with the help of harmonic generator and required filters. The signalling carrier frequencies are supplied from the signalling and generator bayside, which supplies signalling facilities to 120 subgroups with the help of requisite number of signalling equipment baysides.

6.2 Underground Symmetrical Cable Carrier System

As the railways are now electrically conducted, a heavy inductive effect is imposed on the open wire trunk lines running in parallel with the electric traction line along the side of the railway route. As such, symmetrical cable carrier systems have been installed on underground symmetrical cable laid along the railway route to get rid of such electric inductions. For this, sheathed cable or safe screened cable, or steel tape armoured cables are used. Also transformers are used to decrease and limit the increase of voltage up to 300 V even during the worst condition. The earth-returned circuits are altogether avoided. For this the CCIF has given a direction to use a 12 channel basic group of 60-108 kHz bandwidth and superimpose the same on different group carrier frequencies to get a 60 channel super group.

Frequency modulation plan: Broad band speech in the range (0.3 - 3.4) kHz is modulated with channel carrier frequencies 8, 12, 16 and 20 kHz for the four channels and the L.S.B. (4-20 kHz) is modulated by sub-group carrier frequencies 88, 72 and 56 kHz to get a basic group 60-105 kHz for 12 channels. Five such groups are separately modulated by group carrier frequencies 420, 468, 516, 564 and 612 kHz and the lower sideband 312-552 kHz is modulated by a super-group carrier frequency 564 kHz to get a L.S.B. (12-252 kHz) of 60 channels to form a super group.

This type of frequency translation is a combination of old equipment practice 9O (O.E.P.) and uniform common practice (U.C.P.) which is called the CP-4 Plan. But in CP-5 plan, there are three channel frequencies 12, 16, 20 kHz and four sub-group carrier frequencies 120, 108, 96, 84 kHz. But in both the plans, the basic group of 12 channels is of the same (60-108) kHz bandwidth and the five group and super group modulation frequencies are the same in both the cases as also the signals of the sixty channels of the supergroup output are sent in both cases within a band width of 12-252 kHz.

The frequency modulation plans of CP-4 and CP-5 are shown in Figs. 6.5(a) and (b), respectively.

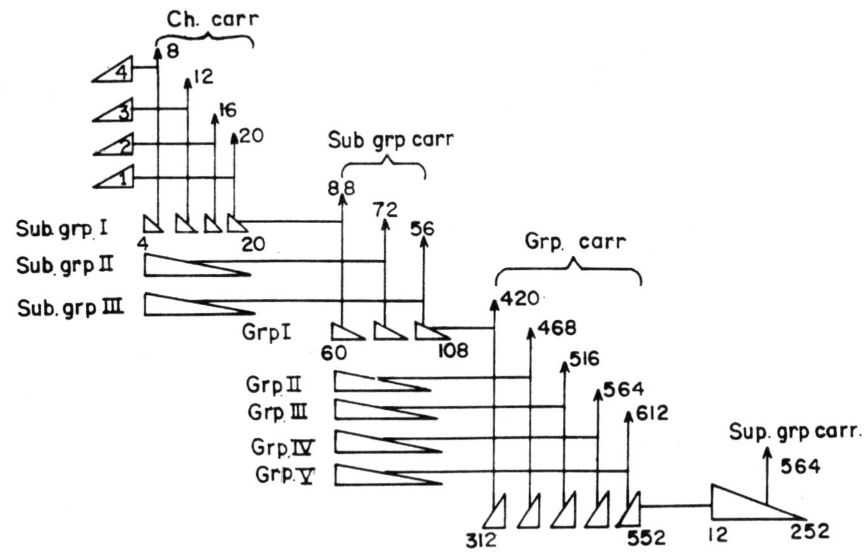

Fig. 6.5(a): Frequency Modulation Plan CP-4

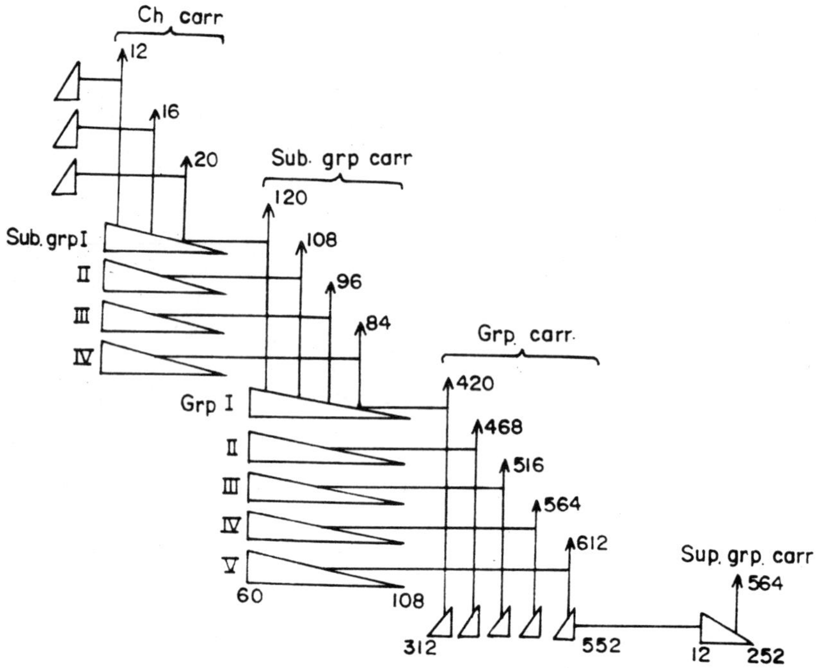

Fig. 6.5(b): Frequency Modulation Plan CP-5

A block schematic for channel and subgroup modulation and demodulation is shown in Fig. 6.6.

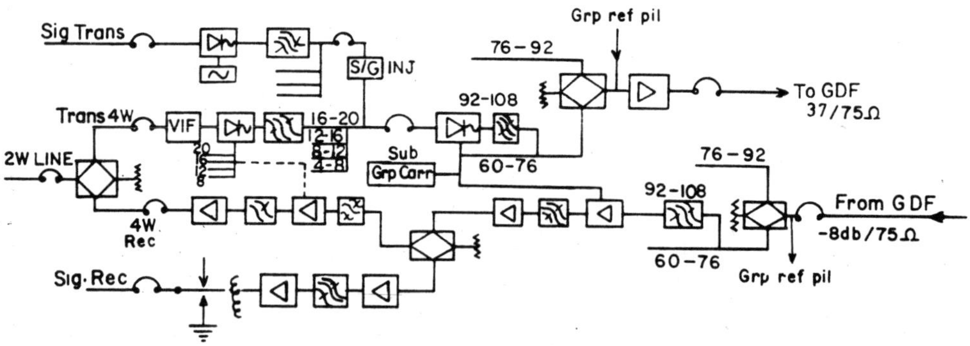

Fig. 6.6: Channel and Sub-group Modulation and Demodulation

In the transmit path, the broad band, (0.3 - 3.4) kHz speech is modulated by channel carrier frequencies 20, 16, 12 and 8, and the LSB outputs are separately modulated by sub-group carrier frequencies 88, 72 and 56 kHz and the U.S.B.s are contained through a Hybrid transformer to obtain a basic group 60-108 kHz of 12 channels, which is amplified and sent to GDF at a level of 37 db at an impedance termination of 75 ohm. In-built out of band signalling using

3.825 kHz frequency for signal modulator is separately injected through signal injecting circuit at the output of the MBF in the channel panel. The signal channel can also transmit dialling pulses.

A group reference pilot frequency of 60 kHz is fed in the transmit path after the subgroup combining hybrid transformer by a separate hybrid transformer.

In the receive path, a basic group 60-108 kHz of 12 channels from GDF at -8 db at 75 ohm termination impedance is separated into different sub-groups by a hybrid transformer. The sub-groups thus separated are separately demodulated by the respective sub-group carrier frequencies and after necessary amplification are fed to the skew hybrid, where the receive signal frequencies are separately picked off, rectified and by the operation of the telegraph relay earth is sent to the exchange switch board to glow the call lamp. The speech band separated by the skew hybrid is demodulated by the respective channel demodulator carrier frequency and the LSB is amplified by demod amplifier and sent to the exchange via hybrid panel.

The group reference pilot frequency is picked off by a hybrid transformer before the sub-groups are separated by a second hybrid transformer.

The block schematic for group modulation are demodulation is shown in Fig. 6.7.

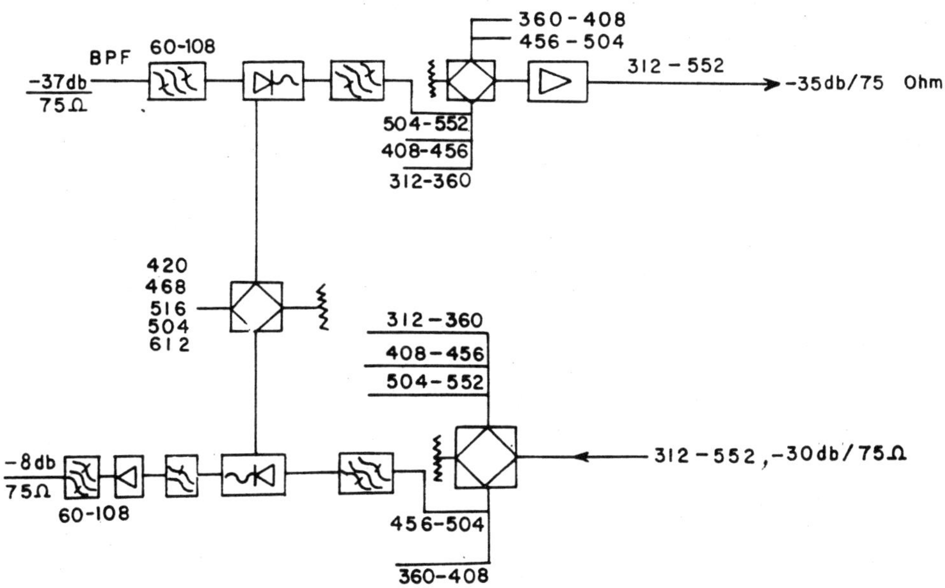

Fig. 6.7: Group Modulation and Demodulation

In the group modulation and demodulation equipment, the five basic groups 60-108 kHz are translated to a basic super-group in the range of 312-552 kHz in the transmit path and in the receive path the basic super-group in the range of 312-552 kHz is translated back to five basic groups 60-108 kHz separately.

Thus, in the case of, say the fifth group, where it is modulated with a group carrier frequency of 512 kHz and the LSB is taken out through a band pass filter. Similarly the group carrier frequencies of 564, 516, 468 and 420 kHz are used to translate groups 4, 3, 2 and 1, respectively to produce L.S.Bs 456-504, 408-456, 360-408 and 312-366 kHz. These sidebands are combined by a hybrid transformer by feeding in parallel the output of band pass filters of groups 1, 3 and 5 to one arm of the combining hybrid and the outputs of BPFs of groups 2 and 4 are fed in parallel to the other arm of that hybrid. The combined output in the range of 312-552 kHz is the basic super group, which is amplified to give a level of -35 db at 75 ohm impedance for further transmission.

In the receive path, the super-group signals 312-552 kHz at a level of -30 db at 75 ohm impedance enter the hybrid transformer, whence the different groups are separated by the corresponding group separating filters. The fifth group 504-552 kHz is demodulated by a group carrier frequency of 612 kHz and the L.S.B. 60-108 kHz is filtered out and amplified and sent to the channel equipment at a level of -8 db at 75 ohm impedance. Similarly the other groups are demodulated with group carrier frequencies 564, 516, 468 and 420 kHz respectively and the LSBs are picked up by filters and in each case the basic group 60-108 kHz of 12 channels are separately amplified and sent to the channel equipment at -8 db at 75 ohm impedance.

The sub-group modulation and demodulation process is shown below in block schematic of Fig. 6.8.

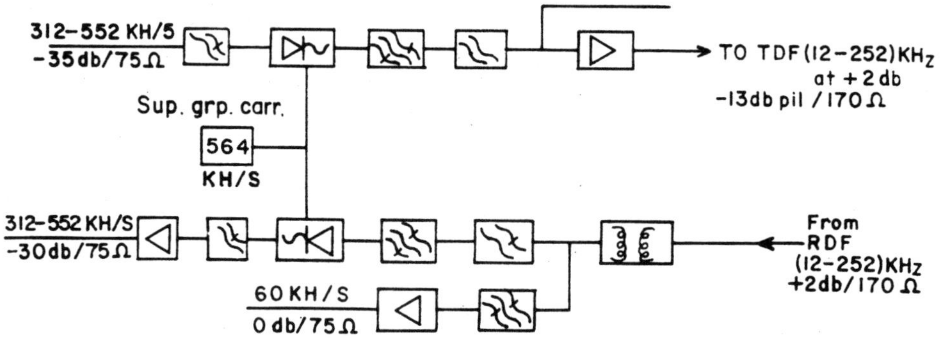

Fig. 6.8: Super Group Modulation and Demodulation

The super group modulation equipment converts the basic super group of 60 channels in the range of 312-552 kHz to the line frequency range of 12-252 kHz in the transmit path and amplifies it to a level of +2 db/170 ohm impedance for transmission into the cable. In the receive path it accepts the line frequencies in the range of 12-252 kHz from the terminal line amplifier and demodulates it to the basic super group frequencies.

In the send path, the basic super group of sixty channel 312-552 kHz is modulated with a super group carrier frequency of 564 kHz and the L.S.B. is filtered through a LPF and the output is amplified in the transmit amplifier, at the input of which a pilot frequency of 60 kHz is injected for regulating and synchronising. The pilot is 15 db lower than the sideband frequencies at the line point, where its level is -13 db/170 ohm at TDF.

In the receive path, the line frequencies at 12-252 kHz range are amplified by the receive amplifier in the 'terminal regulated repeater' and taken to the super group demodulating equipment. The synchronising pilot of 60 kHz is extracted by means of a crystal filter at the input of the super group demodulating equipment and is separately amplified by an amplifier to give an output level of 0 db/750 ohm impedance. The receive line signals are then separately demodulated with a super group carrier frequency of 564 kHz and the LSB 312-552 kHz is amplified for 22 db to a level of -30 db/75 ohm, which is then sent to the group demod equipment through RDF.

Terminal and intermediate repeaters are used to amplify the received level to a fixed value. The intermediate repeaters are installed at an interval of about 12 miles. There is always one attended intermediate repeater at a suitable position on the route and the rest are unattended. At every terminal repeater and at every third repeater an AGC amplifier with regulator is used and the rest of the repeaters are unregulated. In each direction of any repeater, the amplifier is joined with a pilot indicator, whose level in case of unregulated is beyond ± 4 db from the centre 0 scale gives an indication at the terminal supervisory panel for necessary action by the staff on duty. The block schemtics for unregulated and regulated repeaters are shown in Figs. 6.9(a) and (b), respectively.

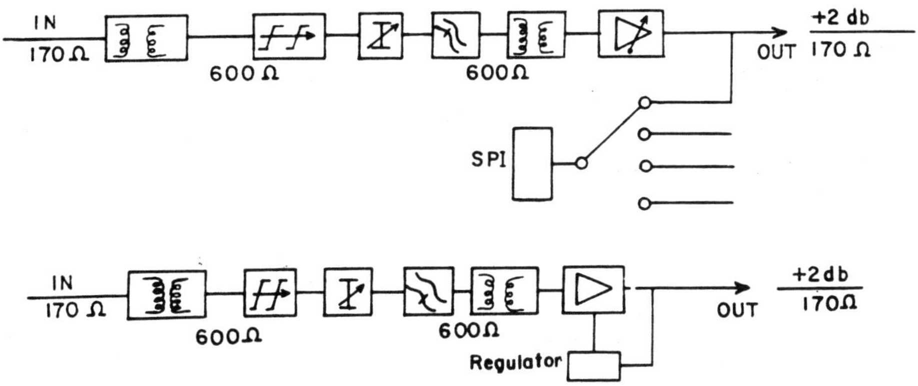

Fig. 6.9: (a) Unregulated repeater,
(b) Regulated repeater

The power supply arrangement in underground symmetrical cable system is shown in Fig. 6.10 below.

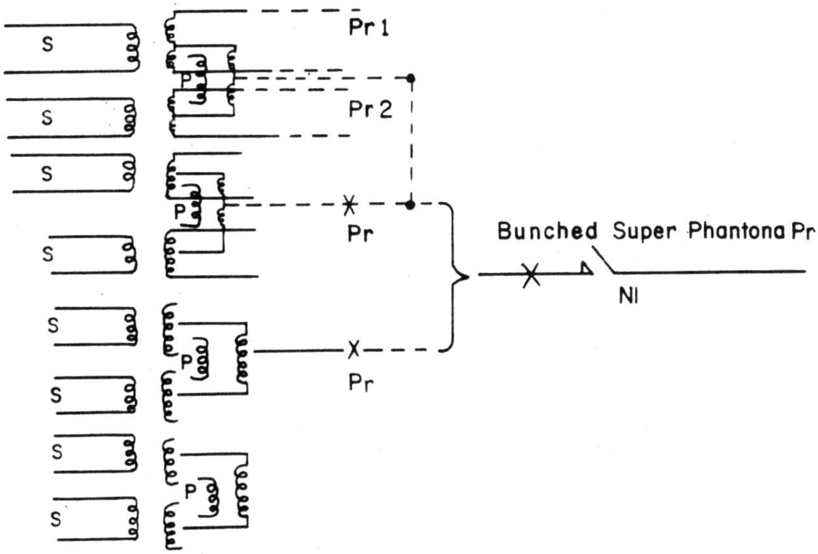

Fig. 6.10: Power Supply in U/G symmetrical cable.

The signals are transmitted in the side circuits simultaneously with the supply of power in the bunch or super phantom bunched circuits. The side transformer circuits are designated with *S*, where signal sidebands are transmitted and the phantom circuit transformers are denoted with *P* where power is fed.

6.3 Coaxial Cable Carrier System

This system can work either on CP-4 or CP-5 frequency translation plan to get the basic group of 12 channels from 60 to 108 kHz by filtering the sidebands through LP filter. Below is shown in Fig. 6.11 the frequency translation plan of 960 channel coaxial system with CP-5 Plan.

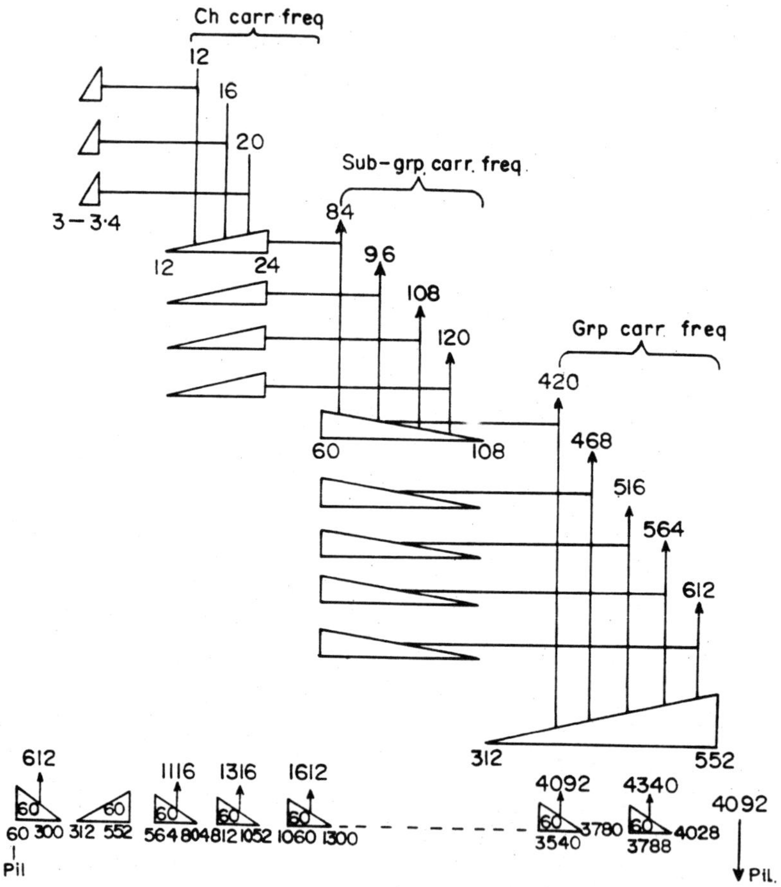

Fig. 6.11: Frequency Modulation Plan for 960 Channel Coaxial System

The five basic groups of 12 channels 60-108 kHz are each separately group-modulated by carrier frequencies 420, 468, 516, 564 and 612 kHz to get a super group of 60 channels from 312 to 552 kHz.

For 960 channel system, 16 such supergroups are taken and leaving the second, the remaining 15 supergroups are modulated further with carrier frequencies in kHz as follows:

Ist	super group	Carr freq. 612 kHz	
2nd	"	Nil	"
3nd	"	1116	"
4th	"	1364	"
15th	"	4092	"
16th	"	4340	"

Two pilot frequencies 60 kHz and 4092 kHz are sent to supervise the line level and to control the regulation of amplification of line amplifiers as also to equalise the line loss for different frequencies by automatic hunting procedure at the repeater station. With the 60 kHz pilot frequency, the value of the Mop-up equaliser is altered to correct the slope of the transmitted band of frequencies. Apart from these two pilot frequencies (60), 308, 556, 808, 1056, 1304, 1552, 1800, 2048, 2296, 2544, 2792, 3040, 3288, 3576, 3784, (4092) kHz are used as representative test frequencies for each super group. The channel and sidegroup modulation are done in two ways:

In CP-4 plan, speech frequencies 0.3 to 3.4 kHz is modulated with four channel frequencies 8, 12, 16 and 20 kHz and the LSB 4-20 kHz taken separately to sub-group-modulated by three subgroup carrier frequencies 56, 72 and 88 kHzs and by a HPF, the USB of 12 channels from 60-108 kHz is taken out as the basic group. In a CP-5 plan, speech band of 0.3 to 3.4 kHz is modulated separately by three channels of carrier frequencies 12, 16 and 20 kHz to form a sub-group which is separately modulated by four sub-group carrier frequencies 84, 96, 108 and 120 kHz to give a basic group of 12 channels from 60-108 kHz bandwidth. The block schematic for channel and sub-group modulation and demodulation is shown below for CP-4 in Fig. 6.12(a).

In CP-4 plan, the speech level in the transmit path at a level -4 db/600 ohm at 4 wire access bay enters the limiter and after channel modulation and sub-group modulation goes out at GDF at -37 db/75 ohm at sub-group out point.

In the receive path, the basic group of 120 channels from 60-108 kHz enters the sub-group demod at a level of -8 db/75 ohm and after sub-group and channel demodulation goes to 4 wire rec. point at a level of -2 db/600 Ω termination to switch board. Similarly, in CP-5 plan, the 4 wire trans level is -4 db/600 ohm at 4 wire access bay and -36.5 db/150 ohm at group trans amplifier out, whereas in the receive path the level at super-group demod is -30.5/150 Ω and after sub-group and channel demodulation, the receive level is -2 db/600 ohm at 2 wire line at hybrid out point to the switch board. A pilot frequency is injected at trans amplifier input and extracted at group pilot extradition hybrid in the receive path. A signal frequency of 3.825 kHz in each channel is modulated separately with channel carrier frequency and transmitted with the speech band and in the receive path the same is extracted by a signal pick-up filter and converted from AC to DC in the signal detector and sent to signal receive instrument as shown in block schematic Fig. 6.12(b).

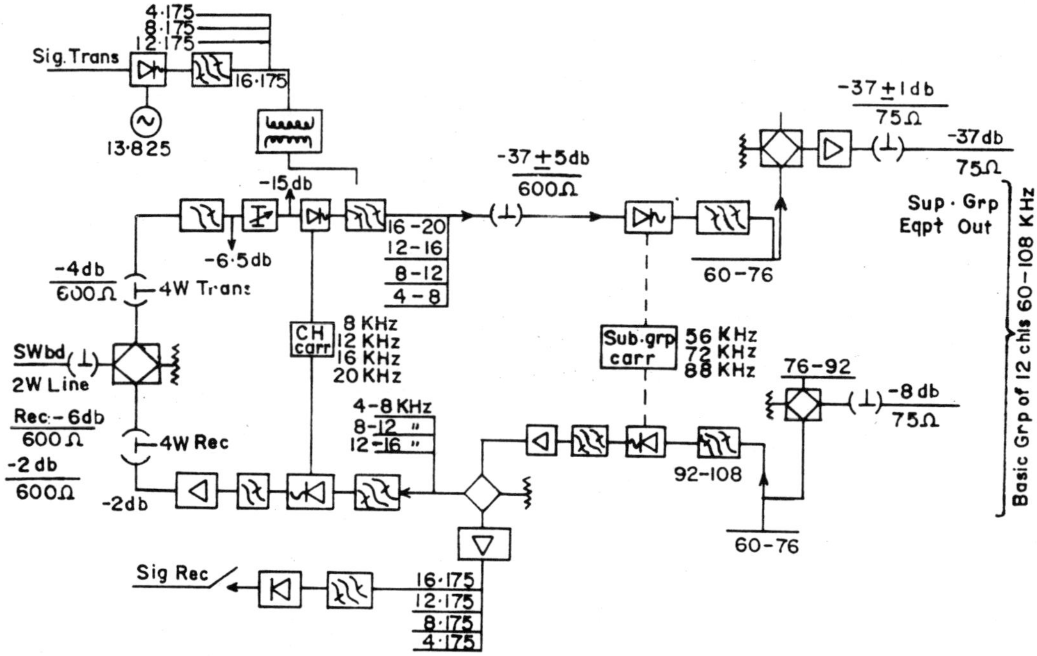

Fig. 6.12(a): CP-4 Plan for Carrier Channel and Sub-group Modulation and Demodulation

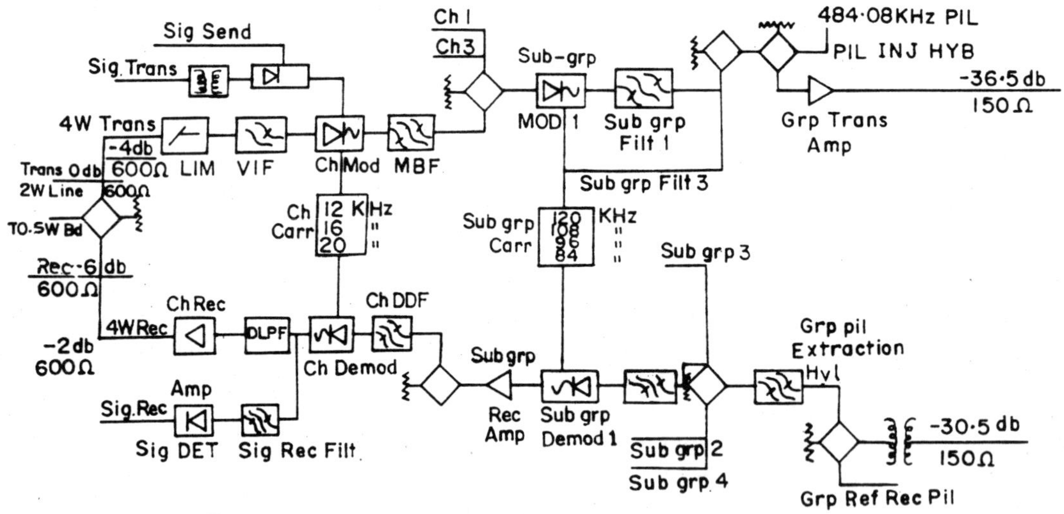

Fig. 6.12(b): CP-5 Plan for Channel and Sub-group Modulation and Demodulation

The group modulation and demodulation is shown in the block schematic given below in Fig. 6.13.

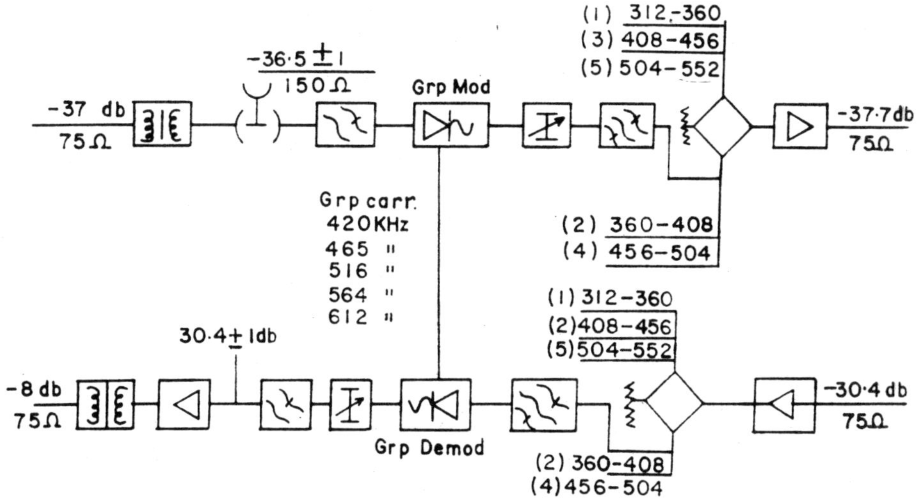

Fig. 6.13: Block Schematic of Group Modulation and Demodulation

In the transmit path, the basic group of 12 channels of 60-108 kHz before entering the group modulator from the GDF at a level of -37.5 db/75 ohm terminating impedance is converted to -36.5 db at 150 ohm impedance as the input impedance of group modulator is 150 ohm and its output impedance is 75 ohm. Thus each group is separately modulated with group carrier frequencies of 420, 468, 516, 564 and 612 kHz and the LSB is separated by a LP filter and all LSB's are combined by a combining hybrid by entering odd groups through one side and even groups through the other to get a basic supergroup of 60 channels of bandwidth from 312 to 552 kHz. This is amplified to give a level of $\frac{-37.7 \pm 1 db}{75\ ohm}$ at SGDF.

In the receive path, the basic super group at $\frac{-30.4 \pm 1}{75\ \Omega}$ db level is amplified by the receive amplifier to a level of -10.4 db/75 ohm and is again separated to groups by a group separating hybrid. Then each group is passed through a group DBF after which they enter a group demodulator separately for each group. Here they are separately group demodulated by group demod carrier frequencies 420, 468, 516, 564 and 612 kHz. Here the group demod input impedance is 75 ohm and the output impedance is 150 ohm. Thus demodulated group of 12 channels with sideband 60-108 kHz is separated by a L.P.F. and then amplified and the termination impedance is again changed by a transformer to get a level of -8 db at 750 ohm termination at group distribution frame (GDF).

Super group Modulation and Demodulation: The block schematic for sub-group modulation and demodulation is shown in Fig. 6.14.

Here, the inter-channel cross-talk from the side channels is eliminated by using decoupling hybrid coils. By this procedure, the super groups are united together before transmission and the united super groups are separated after reception. As they are united in two stages, so also are they separated in two stages. The first basic super group band of 312-552 kHz is modulated and demodulated with a super group carrier of 612 kHz. The second basic super group band is sent without any modulation. From the third to the 16th super group, the bands are respectively modulated and demodulated by super group frequencies 1116, 1364, 1612 ... ... 3844, 4092 and 4320 kHz. These super group frequencies from the third to the 16th super group are odd multiples of 124 kHz master oscillator output. The modulator and demodulator carrier frequencies are pushed through a hybrid coil as shown, so that sufficient decoupling is provided between the modulating and demodulating circuits. Thus they are united and separated in three stages through hybrids as shown.

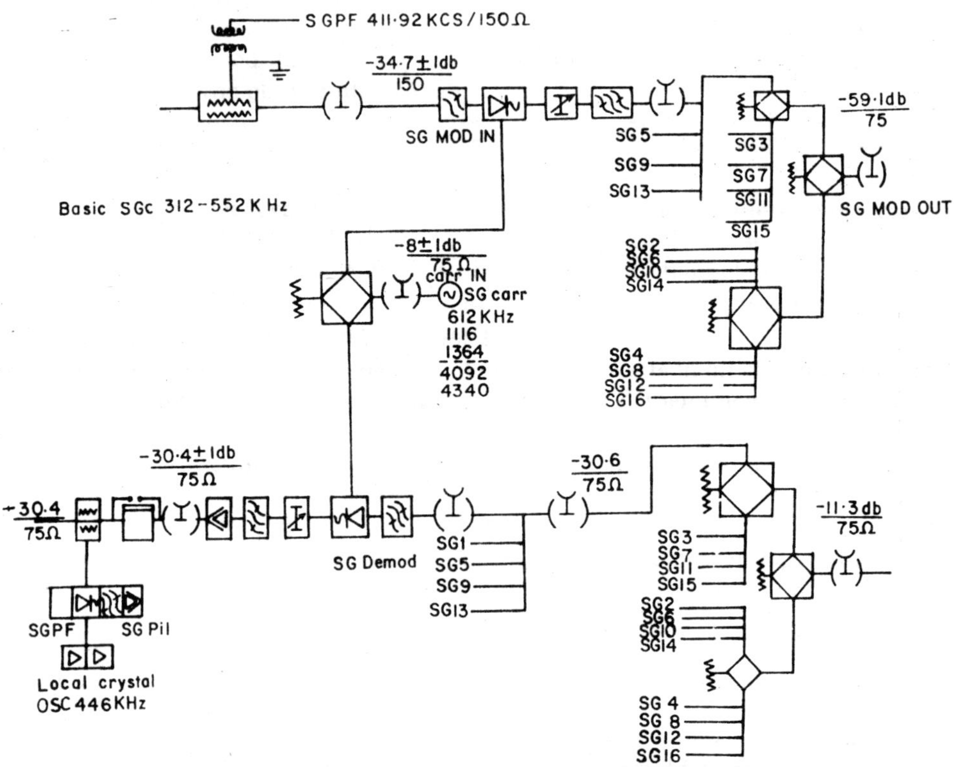

Fig. 6.14: Block Schematic of Super group Mod and Demod

In the send path, the signal level of the frequency band at the output of the modulator is 60-4028 kHz at -59.1 db at 75 ohm impedance termination and in the receive path the signal level of the frequency band at the input of super group modulator is -11.3 db at 75 ohm terminating impedance for the incoming band 60-4028 kHz. These are also demodulated after being separated by three hybrids. The demodulated outputs are separated by passing them through low pass filters and the basic super group bands 312-552 kHz at 75 ohm termination is fed to the G.D.F.

7

AC Telegraphy with Teleprinter and Telex Service

We have described in detail the various types of direct current telegraph circuits in Chapter III. Now we shall describe the alternating current telegraphy with working principle of teleprinters and telex system.

7.1 Historical Development of AC Telegraphy

Various efforts were made for the transmission of AC signals from as early as 1924, by transforming the DC signals into AC in the transmit station in the form of voice frequency tones to the distant station in the carrier channels. At the distant station, these tones were rectified and reconverted into DC signals to operate the receive relay to reproduce the transmitted signals into their true forms. After various experiments specially by B.P.O. and the Siemens and Halske Company of Germany from 1924 to 1931, the CCITT fixed the range of voice frequency for multi-channel AC telegraph system from 300 to 2600 Hz in 1931, since when Great Britain, Germany and U.S.A. almost simultaneously produced the multi-channel voice frequency telegraph system in 1936. Each design was more or less within the limit prescribed by CCITT and consisted of a voice frequency oscillator, a key device by which the DC pulses could be converted into AC tones at the transmit station by start-stop principle as per telegraph codes, which were modulated with a modulatory frequency and transmitted to the distant station by a carrier speech channel along with the tones combined through send filters of other VFT channels after proper modulation and amplification.

At the distant receiving station, these tones are separated by corresponding receive band filters, detected and rectified to DC pulses by rectifiers to operate the receive relay for reproducing the send signals. The diagram for the circuit used is shown in Fig. 7.1.

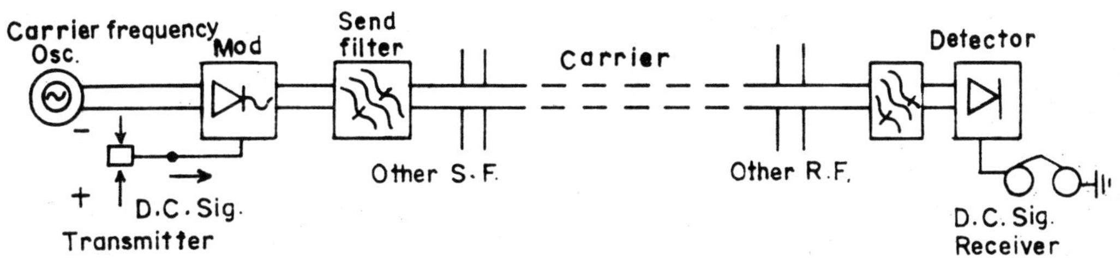

Fig. 7.1: A Channel of Multi-channel VFT System

CCITT recommended that the series of frequencies formed by an even multiple of 60 Hz should be used with the lowest at 420 Hz. Thus working on four-wire basis, each VFT channel frequency with a channel bandwidth of 120 Hz for each VFT channel was employed for an amplitude modulated VFT system. Thus in a carrier channel of narrow bandwidth from 300 to 2600 Hz, as many as 18 VFT channels could be accommodated in AM VFT system. The channel bandwidths are as given below: 420-540, 540-660, 660-780, 780-900, 900-1020, 1020-1140, 1140-1260, 1260-1380, 1380-1500, 1500-1620, 1620-1740, 1740-1860, 1860-1980, 1980-2100, 2100-2220, 2220-2340, 2340-2460, 2460-2580 Hz.

Standardisation of frequency and bandwidth of each VFT channel as mentioned above led to the limitation of speed of telegraph signals transmitted by a teleprinter in VFT channel. For this CCITT standardised the working of teleprinters in VFT channels to a speed of 50 bands equivalent to 66 words per minute in 1929.

7.2 Power Levels

The selection of power level for each VFT channel has to be considered very carefully, as on one hand the power level is desired to be as high as possible, on the other hand, the power level should not be high enough to disturb the adjacent channels in the form of cross-talk interference, nor should it produce overloading effect on line amplifiers and distort the wave form of the signals due to production of harmonics.

Now if the maximum instantaneous power W is in watts in all channels for a system of N channels, working on a line of characteristic impedance Z, then the maximum voltage E across the impedance is given by $W = \frac{E^2}{Z}$ or $E = \sqrt{WZ}$ If e = the maximum permissible voltage per channel, then $e = \frac{E}{N} = \frac{\sqrt{WZ}}{N}$ and the maximum permissible power w per channel is given by

$$w = \frac{e^2}{Z} = \left(\frac{E}{N}\right)^2 \times \frac{1}{Z} = \frac{WZ}{N^2 Z} = \frac{W}{N^2} \text{ watts.}$$

Now CCITT has provided 5 mw as the maximum power for a VFT system. Hence for 18 channel VFT system, the maximum power per channel should be $\frac{5 \times 10^{-3}}{18^2}$ watts or 15 mW, so that it can safely work with negligible interference with other side channels.

7.3 AM VFT System

The send filters are provided at the output of the modulator of each channel: (i) to restrict the sidebands of the modulated carrier to the portion of the frequency spectrum for that channel, (ii) to prevent transmission of unwanted frequencies, and (iii) to offer a high impedance to line at the operating frequencies of other channels.

Similarly, the receive filters are connected to the receive lines via 1:1 transformers, separating the line from the filters whose functions are: (a) to accept with minimum attenuation the transmitted band of frequencies, and (b) to offer high impedance to the frequencies beyond the pass bands to minimise cross-fire between channels.

Compensating filters parallel to the output points of send filters and input points of receive filters are provided to equalise the frequency-attenuation of end channels. A block schematic diagram of 18-channel AM VFT system is shown below in Fig. 7.2.

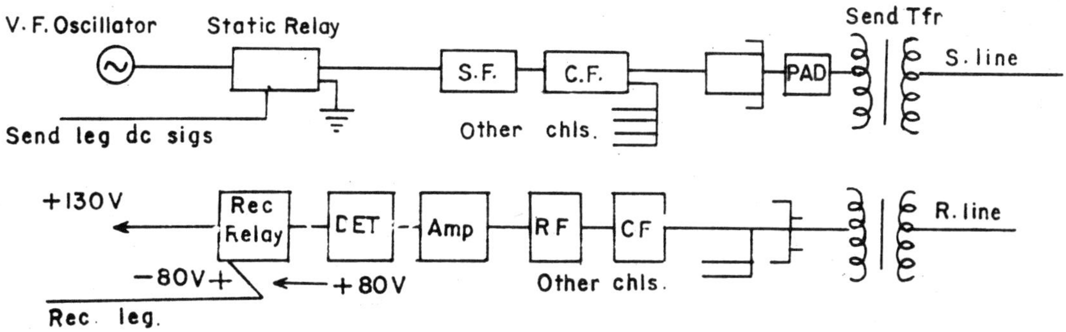

Fig. 7.2: Block Schematic of 18 channel AM VFT System

The static relay modulator consists of a copper oxide or selenium rectifier circuit as shown in Fig. 7.3 below.

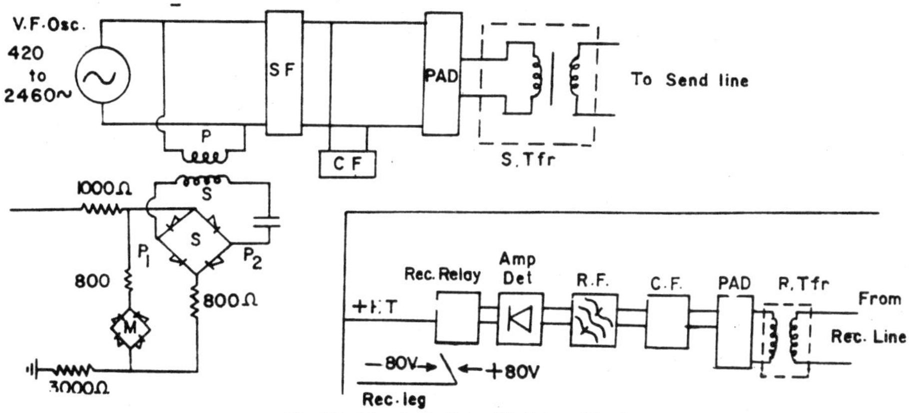

Fig. 7.3: The Static Relay Modulator Circuit

When telegraph signals are sent in the form of -80V for spacing signal and +80V for marking signal from the telegraph transmitter, then Voice Frequency Oscillator output is short-circuited for spacing signals and remains open for marking signals with the result that the telegraph DC signals are converted into AC tones for marking signal and no tone for spacing signal. Rectifier M works during the marking signal when rectifier S remains inactive and during the spacing signals, the reverse action takes place. The rectifier S is joined as a bridge circuit on the secondary coil of the transformer at points P_1P_2 and a condenser is put in the middle, so that no damaging DC can pass to the secondary coil and damage the send frequencies by creating distortion. As the rectifiers are mutually identical, the marking and spacing currents are also equal to each other. So at the output of the static relay, only marking and spacing carrier currents are present, which are brought to the same level by send filter and compensating filter by means of pad and then sent to the line via send transformer.

In the receive path at the distant receive station, the various bands of different channels are separated by the help of separate receive filters and amplified to the desired level. Then with the help of separate detectors, they are converted to DC signals from AC tone and received by the receive relay for recording the message transmitted from the sending end. In this way the messages are exchanged from one end to the other.

As the tone on-off system as described above works on the principle of single-current system, it has got its disadvantages also. The auto-gain-control (AGC) system on the receive amplifier detector controls the receive signal level between the limits ± 5 db to ± 10 db and the time constant for the auto-gain control is only 5 sec. For this the sensitivity of the detector remains constant during the tone off condition. So if the received tone is faded or increased all of a sudden, the same cannot be quickly corrected due to the long time-constant of AGC. For this during such flicks, the received signals cannot be detected correctly and therefore mutilation occurs during such periods with single tone system. To get rid of such troubles, the two-tone system of working is followed in long-distance system as also in wireless, radio and microwave links. In this method, two adjacent VFT channels are required, of which in one channel marking tone and in the other spacing tone is sent, so that the receive relay works on the double current on the marking and spacing channel as shown in Fig. 7.4.

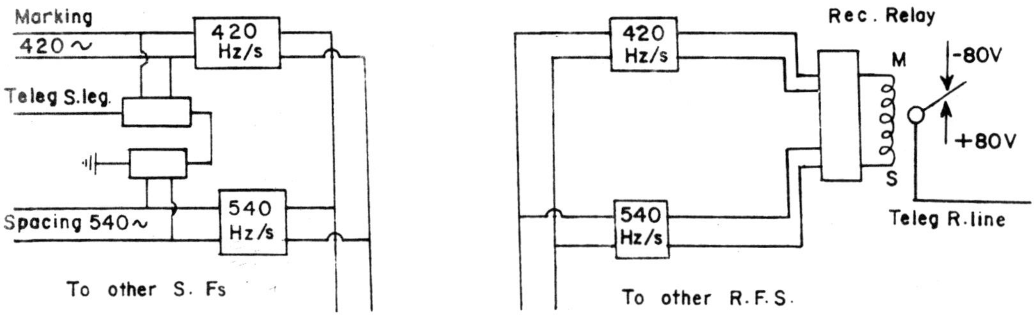

Fig. 7.4: Two-tone VFT Channel Circuit

As one tone or the other remains always present in this process, so one AGC can function successfully well at all times. So a smaller time constant is used in this process for AGC, so that the intensity of prolonged flicks period can be decreased to a minimum. This is why this process is used in radio links.

7.4 FM VFT System

Up to 1956, only AM VFT was recommended by CCITT. In 1956, the CCITT recommended for FM VFT system working with a maximum of 24 channels on double tone principle with the following instructions: (a) the space signal should be of negative current and mark signal should be of positive current, and (b) the frequency for send line for space signal should be of higher order than that sent for mark signal. Based on the above recommendations, a deduction of send power level is made for FM VFT system.

Let there be C carrier frequencies with an amplitude of V each. Then the maximum average amplitude = CV. If this signal is to be sent free from inter-channel distortion due to inter-channel modulation, then the transmitter should have a maximum power handling capacity $P \propto$

$(CV)^2$. Hence the power available for each channel = $\frac{P}{C^2}$. If we assume that the noise present in the common path is uniformly distributed in frequency, then the noise in each telegraph channel is proportional to its bandwidth. Hence the noise in each telegraph channel before demodulation is n, so that n is given by the equation

$$n = \frac{\text{total noise power in common path} \times \text{Band width of telegraph channel}}{\text{Total band widths}}$$

Before demodulation, the signal-to-noise ratio for a channel would be proportional to $\frac{P}{C^2} \times \frac{1}{n}$. From this, it is observed that the signal- to-noise ratio for each channel decreases if the number of channels is increased. The block schematic of a 24 channel FM VFT system is given in Fig. 7.5.

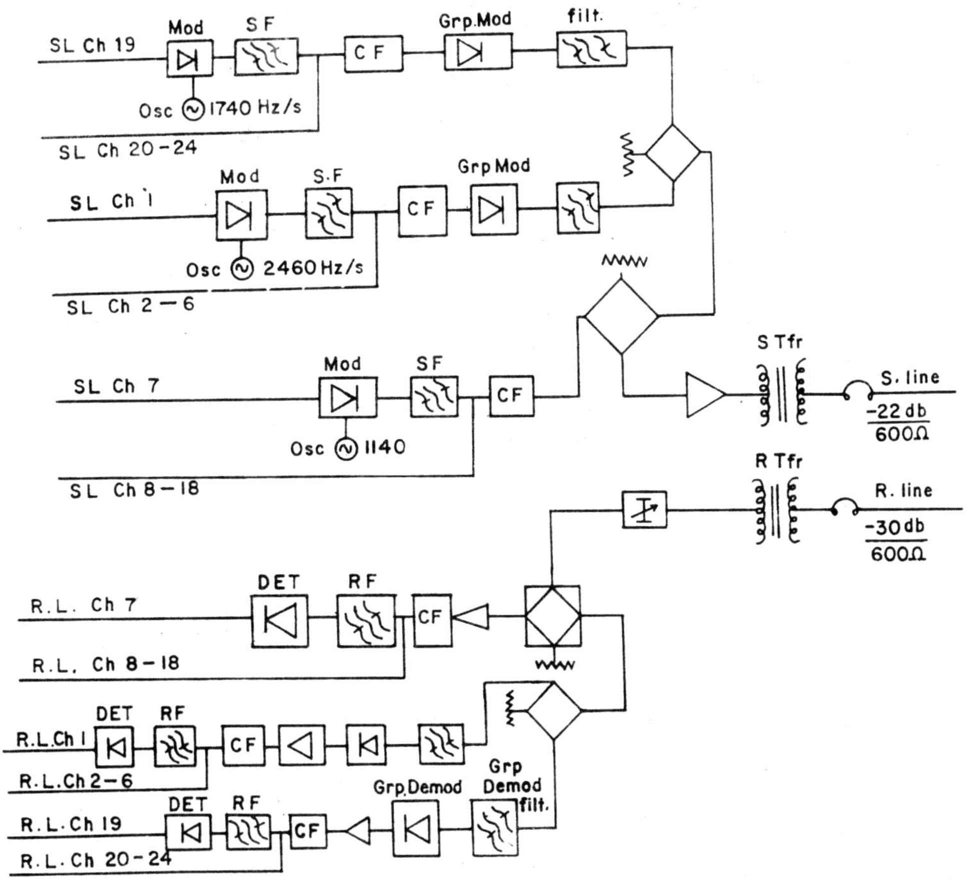

Fig. 7.5: Block Schematic of 24 channel FM VFT System

The frequency allocation of 24 channel FM VFT system is given below:

Ch No.	MidCh Freq. Hz	On Ch. panel Sp. -80V	Hz Mk +80V	Grp. Carr freq.		On line Sp.	Mark
1	2460	2430	2490			450	390
2	2340	2310	2370			570	510
3	2220			2880		690	630
4	2100				810	750	
5	1980				930	870	
6	1860	1830	1890		1050	990	
7	1140	1170		1110		1170	1110
8	1260	1290		1230		1290	1230
9	1380	1410		1350		1410	1350
10	1500	1530		1470	Nil	1530	1470
11	1620	1650		1590		1650	1590
12	1740	1770		1710		1770	1710
13	1860	1890		1830		1890	1830
14	1980	2010		1950		2010	1950
15	2100	2130		2070		2130	2070
16	2220	2250		2190		2250	2190
17	2340	2370		2310		2370	2310
18	2460	2490		2430		2490	2430
19	1740	1710		1770		2610	2550
20	1620	1590		1650		2730	2670
21	1500	1470		1530	4320	2850	2790
22	1380	1350		1410		2970	2910
23	1260	1230		1290		3090	3030
24	1140	1110		1170		3210	3150

The group modulated frequencies 2880 Hz and 4320 Hz are derived by frequency division process from the crystal oscillator 8640 Hz. Channels 7 to 18 form the basic group with a single stage basic modulation and go straight on the line without any group modulation. But channels 1 to 6 and 19 to 24 are formed by the two-stage modulation. Channels 1 to 6 get the first stage basic modulation corresponding to channel 18 to 13 and then are modulated by a group modulating frequency of 2880 Hz. Similarly channels 19 to 24 get first the basic modulation with frequencies corresponding to channel 12 to 7 and then get a group modulation with a group modulating frequency 4320 Hz. So on the line the space frequency is always 30 Hz above and the mark frequency is always 30 Hz below the mid-channel frequency, as per recommendation of the CCITT and in the receive path -80V is given for space signal and +80V is given for mark signal. When there is a disconnection of line, a space signal is given on the line in the receive path.

7.5 Simultaneous Speech and Duplex Telegraph Circuits

On a C-8 trunk line, or on a carrier channel simultaneous speech + Duplex telegraph, one or more circuits can be worked whenever required, specially during the emergency periods. It is observed that about 95 per cent of our speech bandwidth remains within 300 to 2600 Hz and for broad band the same is 300 to 3400 Hz. Though the fine quality of speech is lost to some extent, but for commercial purposes the pitch, intelligibility, and exactness of human voice remain more or less the same, so that we can clearly understand and recognise that a particular person is speaking. So if by a telegraph filter, a bandwidth from 1540 Hz to 2030 Hz is separated for working a single telegraph channel, from a narrow band channel or speech plus three or four FM duplex telegraph circuits from a broad band channel, then there is no appreciable difference in the quality of commercial speech and the working of the telegraph circuits.

Speech + AM Duplex Telegraph Circuit: A block schematic for speech + single channel AM duplex telegraph circuit working on a trunk line or carrier channel is shown in Fig. 7.6.

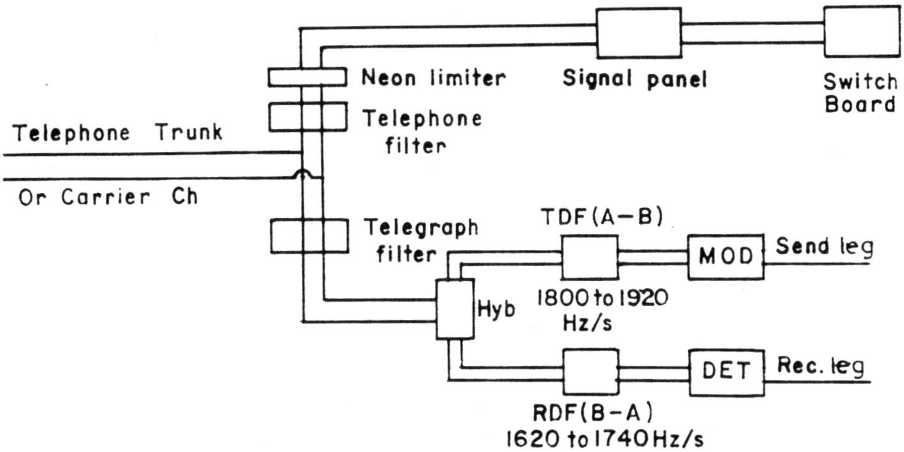

Fig. 7.6: Block Schematic for Speech and Single Channel AM Duplex Circuit

Like AM duplex telegraph circuits on narrow band channel, DC telegraph signals are converted into AC signals by static modulation in the transmit path, while in the receive path, the AC signals are reconverted into DC signals to operate the receive relay.

In the *A-B* direction, the telegraph carrier oscillation frequency is 1860 Hz while in the *B-A* direction, the same is 1680 Hz. Higher sideband is sent in *A-B* direction, while lower sideband is sent in *B-A* direction. So in *A-B* direction 1860 ± 60, i.e. 1800-1920 Hz band is sent at *A* station and in the *B-A* direction 1680 ± 60, i.e. 1620-1740 Hz band is sent at *B* station. Thus at *A* station, the received band is 1620-1740 Hz and at *B* station the received band is 1800-1920 Hz.

For open wire line, a ringing current of 17 Hz is used but for a carrier channel a ringing tone of 500/20 Hz must be used.

Speech and FM Duplex Telegraph Circuit: In a broad band channel, *S* + 3 *dx* or 4 *dx* FM VFT system can work like ordinary VFT systems.

S + 3 Duplex FM VFT System: The schematic diagram for a speech + 3 channel duplex FM VFT system is shown in Fig. 7.7.

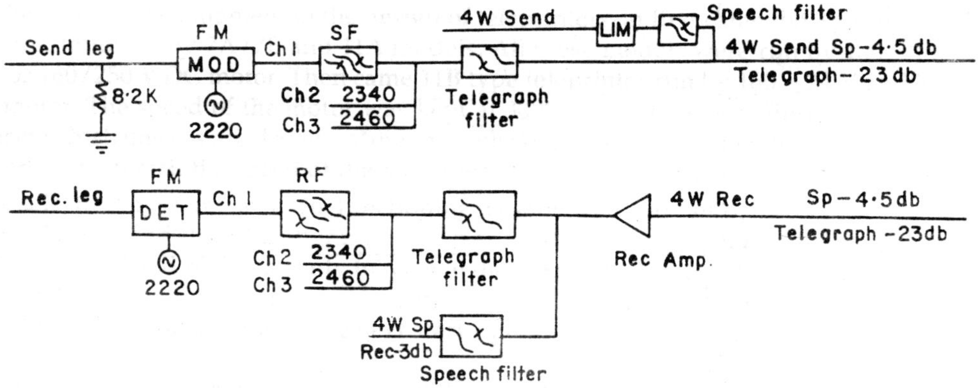

Fig. 7.7: (Speech + 3 Duplex) FM System

The frequency distribution is shown below:

Channel No.	Mid ch. freq. in Hz	On the line Sp (-60V)	Mk(+60V)
1	2220	2250	2190
2	2340	2370	2310
3	2460	2490	2430

In the transmit path, the space signal (-60V) is superimposed on osc. modulator, and the changed osc. mod. frequency is f +30 Hz and at the time of mark signal (+60V), superimposed on osc. modulator, the changed osc. mod. frequency is f -30 Hz. Here the bandwidth of FM channel is only 60 Hz. These are passed through the send filters and combined with the outputs of other channels at the input of telegraph filters. At the output of telegraph filters, these telegraph tones at a level of -23 db are combined with the speech level of -4.5 db at 4W send path.

In the receive path, the speech channel frequency band is amplified along with the telegraph channel frequency bands and both speech and telegraph bands are separated by a speech filter and telegraph filter, respectively. The speech channel is sent to exchange at a level of -3 db at 4W receive point and the telegraph channels at the telegraph filter output at a level of -16 db enter the respective telegraph receive filters, at the output of which they are separately demodulated and reconverted from AC to DC in separate FM detectors to operate the receive relays in the respective receiver circuits.

Here frequencies below 2.19 kHz work for speech and frequencies from 2340 to 3180 Hz band is used for telegraph circuit, as shown in the frequency distribution plan given below:

Channel No.	Mid ch. freq. in Hz	On the line	
		Sp (-60V)	Mk(+60V)
1	2400	2460	2340
2	2640	2700	2580
3	2880	2940	2820
4	3120	3180	3060

In the transmit path, when space signal (–60V) is applied to telegraph signals, the mod. osc. frequency is changed to $(f + 60)$ Hz and when mark signal (+60V) is applied to telegraph signals, the same is changed to $(f - 60)$ Hz where f is the mid-channel frequency in Hz so that the bandwidth of the send filters of (Sp + 4Dx) FM telegraph channels is 120 Hz. These are combined together at the input of the telegraph filter and sent to the line along with the speech channel combined at the output of the speech filter. In the receive path the telegraph channels are separated from the speech channel at the output of the receive amplifier and are separately treated as in the case of S + 3 duplex system.

7.6 Teleprinters — the Start-stop Telegraph Instruments

In Morse telegraph codes, we have the code signals made out by various combinations of dots and dashes for letters and digits. But if the line impedance is not properly balanced for resistance and capacitance variations due to some changes in the line characteristic, then the length of dot and dash signals also vary producing some combination, indicating a signal other than that sent and thereby appear to be a different letter or digit to the telegraphist at the receiving end due to this wrong representation or misunderstanding. If, however, the line could have been instantaneously balanced, then such misunderstanding due to distortion of signals could not have happened.

But as the line is subject to disturbances at any time, it is not possible to balance the line for resistance and capacitance unbalance instantaneously. Many telegraph engineers developed various methods of balancing the line such as the Baudo system, Wheatstone automatic system and Marre system, etc. But the effect of distortion could not be significantly diminished due to the non-synchronous action of send and receive instruments, due to the frequent unbalancing of line impedance. So to decrease the intensity of such conditions, a lot of research work was carried out in which the receive instrument was stopped after the reception of every letter or digit and the same was started again before sending a letter or digit. For this a start signal was sent from the sending instrument before sending a signal code and a stop signal was sent after the code signal. Thus a code signal was made of five units of marking and spacing currents preceded by a start signal of spacing current of one unit and followed by a stop signal of marking current of 1.5 unit.

Thus developed the start-stop system of telegraph working, called the teleprinter, in which both the send and receive instruments are stopped at a fixed point of the same phase. While the receive instrument is stopped, the send instrument goes on continuously sending the stop signal (marking current) till the next code signal preceded by a start signal sent followed by a stop signal. Thus the chance of distortion due to lack of synchronisation between send and receive instruments is considerably diminished and the chance of mutilation due to unbalance of less impedance is considerably decreased.

In the gradual development in the design of teleprinters, at first came the 3A/3R type with DC motor, followed by 7A/7B and 11A models. All these models were run by 100/110 V DC motor or 160/250 V DC motor. Then came 11B type teleprinter run by 160 ± 24 V and 220 ± 30 V DC motor. The speed of the motor was kept at 3000 revolutions per minute. Both the send and receive instruments are driven by the same electric motor, controlled by a centrifugal type of governor, by which the speed of the motor was kept within ± 5 per cent.

The selecting mechanism starts functioning as soon as a start signal is received in the receive instrument and stops functioning as soon as the five unit code signals followed by a stop signal is received and remains ready to receive the next signal. Thus whatever letter or digit is formed by the selecting mechanism by the five unit code signals is printed by the typewriter of the receive instrument as soon as the next start signal is sent. In this way the start-stop system works.

According to the CCITT recommendations, each letter or digit code signal is sent in 150 m sec of which 20 m sec is for the start signal, 20 msec for each unit of 5 unit code taking 100 m sec and 30 m sec for the stop signal. Also the difference of speed of send and receive side motors should remain within 0.75 per cent limit. Generally ± 80 V DC supply is given for signal currents, deriving the same power from AC power unit or secondary battery. The teleprinter motor consumes 0.4 ampere current per hour, which is supplied from the positive and negative terminals of ± 80 V battery, i.e. at 160 V.

The motor speed of 3000 revolutions per minute is kept constant, controlled by a series-wound or shunt-wound motor as shown below in Fig. 7.8(a) and (b), respectively.

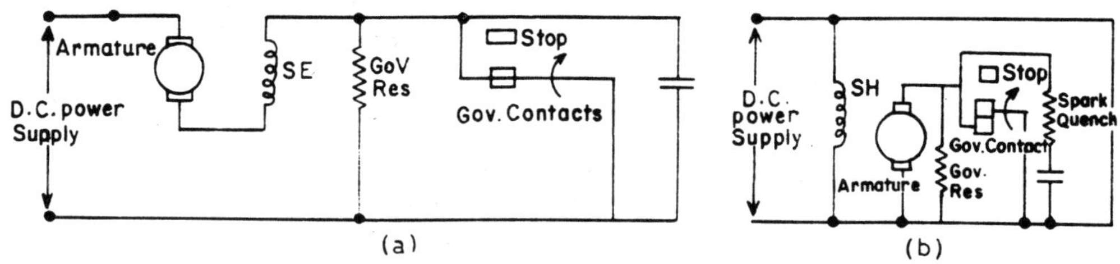

Fig. 7.8 (a): Series-wound Motor Normal Governing Circuit
(b): Shunt-wound Motor Armature Governing

In Fig. 7.8(a), in the idle condition, the governor resistance is short-circuited by the governor contacts in a series-wound motor armature and field coil circuit. When the motor runs at a high speed, it generates a centrifugal force which works on the contact arm and tries to separate it. When the strength of this centrifugal force becomes greater than the tension of the spring, then the contact is separated and the governor resistance is joined in series with the motor circuit. Then current decreases and consequently the speed also decreases until the strength of the centrifugal force falls below the spring tension, when the governor resistance is again short-circuited and the current begins to increase and repeat the previous cycle.

Similarly, in the shunt-wound motor system, the governor controls the speed by separating the closed contacts as soon as the speed exceeds the standard controlled speed limit and thereby brings the armature in series with the governor resistance, when the armature current is decreased and the speed decreases until the governor contacts are again closed, short-circuiting the governor resistance. Then the speed again increases and repeats the previous cycle. A spark-quench circuit is provided to bye-pass the spark caused by the make and break of the governor contacts.

There are about a couple of thousands of parts in a teleprinter, which are mainly assembled in three units, viz. the teleprinter sub-assembly, the keyboard unit, and the printing machine.

Teleprinter Sub-assembly: The following items are associated with the T.P. sub-assembly — main base, combination head, type head, receiving cam unit, type hammer unit, control lever unit, operating electromagnet, starter control lever unit, motor governor, main shaft unit, starter, switch, ribbon bracket, cover, etc.

The Keyboard Unit: It is connected to the sub-assembly by two screens and consists of keyboard assembly, transmitter and answer back unit.

The Printing Machine. It is either of page-printing or tape printing type and the two parts are changeable as per necessity. The selecting mechanism works in synchronisation with the received code signals, so that they can compose the letters or digits in correct sequences. In the printing machine, while the T.P. motor is running, the type head spindle is rotated continuously via the gear wheel.

To get a clear idea of the functions of these parts in each unit, we shall describe briefly the principle of work done in 7A/7B models of teleprinters. As the 7A model was introduced in 1930, it does not tally with the recommendations of CCITT. Its receive mechanism was designed for receiving at 7½ units and not 7 units. For this 7A model receive cam sleeve makes one revolution in 140 m sec instead of 150 m sec as recommended by CCITT. The speeds of various shafts are shown in Fig. 7.9(a) for model 7A teleprinter.

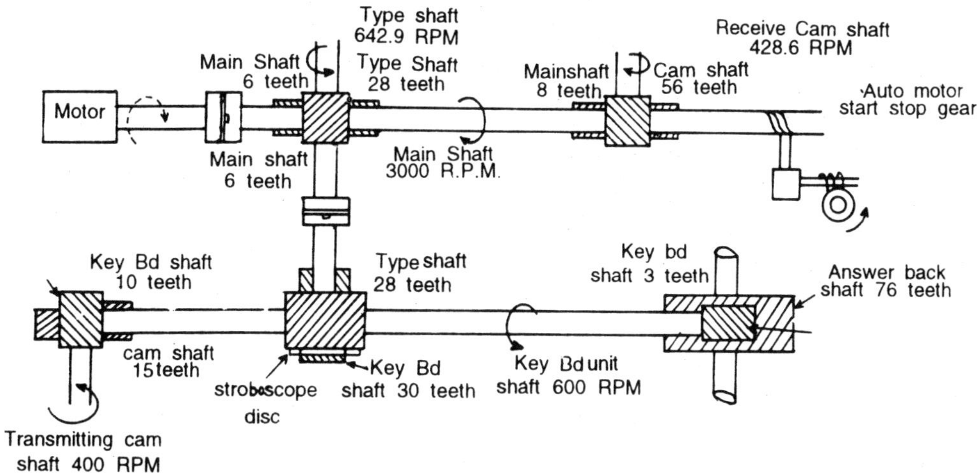

Fig. 7.9(a): Shaft and Gear Details of 7A Model Teleprinter.

In 7B model teleprinter, the signals are transmitted at 7 ½ units in 150 m sec and is made up of start element 20 m sec, five code elements 100 m sec for 20 m sec each and stop-element 30 m.sec. To conform to CCITT standards, the receiving mechanism is designed to receive either 7 unit or 7 ½ unit signals. The receiving cam shaft makes one revolution in 130 m sec before the end of 140 msec signal and in case of 7½ unit transmission 20 m sec before the end of 150 m sec signal. The speeds of the various shafts and gears of 7B model teleprinter is shown in Fig. 7.9(b).

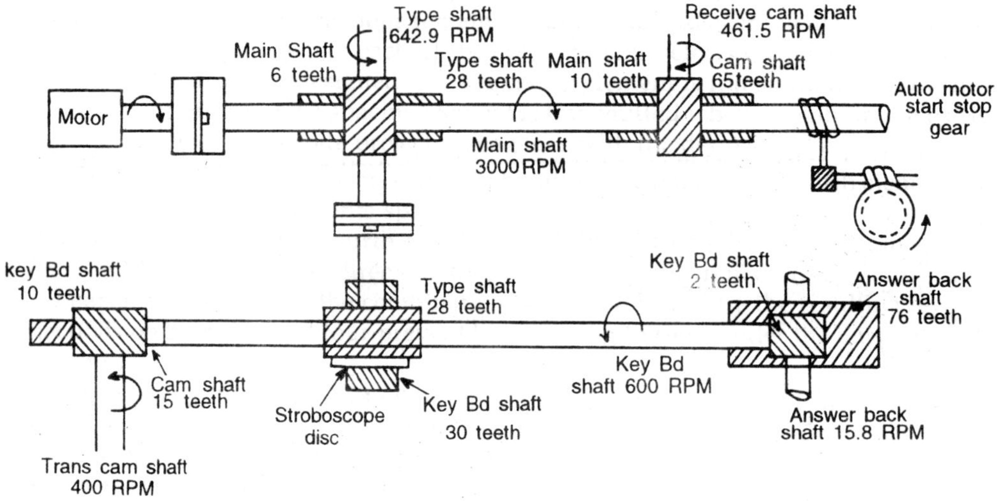

Fig. 7.9(b): Shaft and Gear Details of 7B Model Teleprinter

The maximum deviation of transmitted signals due to line variation that a receiver can somehow receive is 50 per cent of the transmitted signal element. So for a 20 m sec signal unit, the maximum displacement permissible is 10 m sec. In practical design of teleprinter, the margin is kept at 40 to 45 per cent of transmitted signal element, i.e. 8 to 9 m sec (early or late) for a 20 msec signal unit. So to ensure a high degree of margin in the receiver, the operating time of the clutch driving the receiving cam is to be kept as short and uniform as possible, to reduce the errors in the timings of the selecting instants, as the receiving cycles of each character has to be finished before the arrival of the next start signal. So it is obvious that the duration of the receiving cycle should be less than that of the transmitter, so that the receiver is in stop position when the next start signal arrives. For this, account must be taken for the operating time of the receiving clutch, speed errors in the transmitter and the receiver and the effect of signal distortion in the line.

Another factor to be considered is that some teleprinters employ transmitters having 140 m sec character duration for working with those of 150 m sec character duration in the transmitting cycle. This is overcome by designing the receiver to complete its cycles in 6 ½ unit signal element, as it permits a time displacement of 20 msec between successive start signals without endangering the stopping of the receiver by the stop signals.

The smooth working of a teleprinter circuit depends on three mechanisms — the transmission mechanism, the receiving and selecting mechanism, and the printing mechanism.

Transmission Mechanism: In the keyboard, the key bars are free to move vertically between two pairs of guide plates. A pair of roller is fitted to each key to reduce friction. The keys also pass through slots in each of four racks, which prevent lateral movement. With five unit codes, only 2^5 or 32 arrangements can be made. But as five space signals cannot form a letter, only 31 arrangements can be made, out of which 26 arrangements are for 26 English letters and the remaining five arrangements are used for letter shift, figure shift, carriage return, line feed and space signals. The teleprinter keyboard works like a typewriter. In each row there are 10 keys. When the transmitter is on the letter-shift position, the T.P. transmitter sends letters and letters are printed in the distant end receiver and when it is on the figure-shift position, it sends digits and the distant end receiver prints digits only till such time the transmitter is put back on the letter-shift position for sending letters. Thus the work goes on.

The keybar and locking bar mechanism of 7B model T.P. is shown in Fig. 7.10.

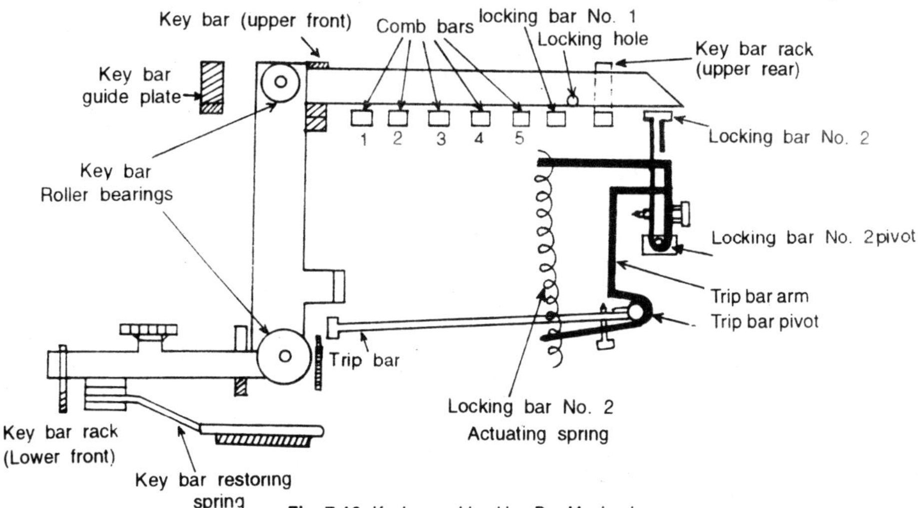

Fig. 7.10: Keybar and Locking Bar Mechanism

There are six bars supported by the racks at each end, so that they lie immediately below the upper horizontal extensions of the key bars. The first five bars are called the combination bars, each of which controls one element of every signal combination to be transmitted and the sixth bar is called the first locking bar. Each of the combination bars and the first locking bar has a spring fitted to the right, pulling the bars to the right, but the bars are normally kept held to the left by the resetting levers for transmitting cam sleeve, trip and locking bar, as shown below in Fig. 7.11.

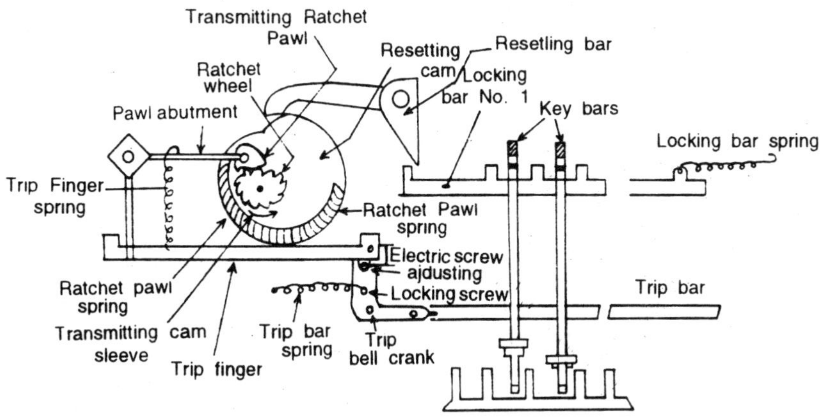

Fig. 7.11: Transmitting Cam Sleeve, Trip and Locking Bar Mechanisms

In this position the vertical extensions of the combination bars are immediately below the tails of the identical trans-unit selecting levers. There are a number of square projections identical with the spacing elements of the signal code, so positioned that with the normal position of combination bars, the projections remain just below and adjacent to the left hand side of the identical key bars. Thus while each of the combinations selects one element of the code elements, the keyboard is so designed that a marking element is transmitted with the movement of the combination bar and a spacing element is transmitted when a combination bar is held in its normal position, as its code projection engages with the depressed key bar.

For example, the letter Y has its code combination as first, third and fifth elements marking, and second and fourth elements spacing. So projections are provided on the corresponding positions of the second and fourth combination bars, so that they may remain stationary but not on combination bars one, three and five, so that they may remain free to move. Thus when *Y* key bar is pressed down, the combination bars are released by the resulting levers and the projections on bars two and four prevent bars moving to right under the tension of the spring attached to them.

The locking bar No. 1 moves to the right during the transmission of each character and one of the hooked teeth on its upper edge engages with the hole on the depressed key-bar thus locking the key in the depressed position until the character is completely transmitted, as shown in Fig. 7.10. During the period of transmission of the character and before the release of the first key bar, the operation of a second key bar is prevented by the locking bar No. 2. It is situated just behind the upper horizontal extension of the keys and is supported on a spindle moving on bearings fitted at each end and an L-shaped locking bar arm is mounted as an extension of the spindle, whilst a trip bar is fixed to the spindle of the trip bar assembly. The two arms are attached to a spiral spring holding them together. Normally, the locking bar No. 2 is held away from the chamfered end of the key bar by the thrust of the trip bar arm.

As soon as a depressed key bar moves the trip bar downwards, the trip spindle is rotated and the action of the trip bar arm and the tension spring causes locking bar No. 2 to move forward to a position above the chamfered end of the depressing key bar, but beneath the ends of the key bars not depressed. So now no other keybar can be depressed till the release of the first key bar after the completion of the transmission of the code combination. The trip bar is rigidly attached by means of four rods to a spindle resetting on bearings at each end of the keyboard assembly. The trip bar is also connected to the locking bar No. 2 by the locking bar and trip bar arm as shown in Fig. 7.10 and at its left hand end the trip bar is coupled to the trip bell-crank as shown in Fig. 7.11. So on depression of a key bar, the projection on vertical portions moves the trip bar downward causing movement of the trip bell-crank to begin transmission of the code combination.

The transmitting cam sleeve rotates at a speed of 400 RPM. So in one revolution it takes 150 m sec, i.e. in 20 m sec it rotates about 48°. The start element of space signal is sent during the interval from 12^0 to 60^0 during which period, none of the cam tracks are cut away. The five code signals are transmitted, being cut away by cam tracks No. 2 to 6 each for 48^0 arcs, successively between 60^0 and 360^0. Besides the selecting lever cam tracks, the send and receive switch levers are also operated from a cam, which lies between the rear selecting lever track and the pawls, whilst the resetting lever is controlled from a cam on the rear end of the cam sleeve, which also serves the disengagement of the trip finger from the pawl abutment connected to the trip lever by means of a spiral spring.

In modern practice, the striker-type transmitter is commonly used, which is similar to the old link-type transmitter, except that a timing cam with seven indentations cut accurately in its periphery is fixed to the cam sleeve, so that the indentations are in correct phase relationship to the operating position of each signalling cam track. The schematic of the striker type transmitter is shown in Fig. 7.12.

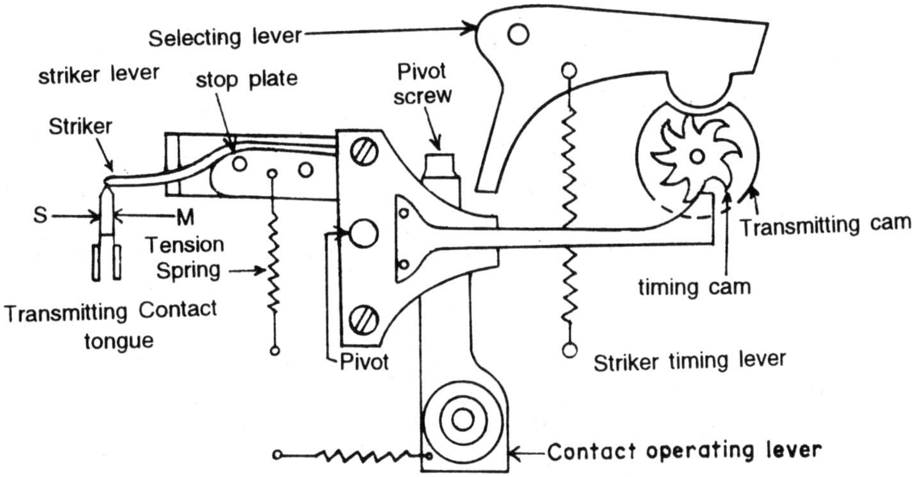

Fig. 7.12: Schematic of Striker Type Transmitter

Here, the selecting levers along with the transmitting cam sleeve and keyboard combination bars are used to determine only the sense of the signals to be transmitted, i.e. mark or space and a striker operated from the timing cam is used to effect the movement of the transmitting tongue from the mark to space and vice versa.

The contact operating lever operates in conjunction with the striker and the striker timing lever which determines the position to be occupied by the contact tongue. The nose of the striker timing lever is held in engagement with the periphery of the timing cam by the spring tension. The striker lever has a slotted guide controlling the vertical movement and the timing lever in conjunction with timing cam first raises and then lowers the striker seven times during transmission of each code combination, accurately timed to correspond to the change over periods of code elements.

After the transmission of code elements, the start-stop selecting lever rides back to its recess and the tongue moves to mark contact when stop signal element is transmitted. The pawl abutment that was released during the rotation of the cam sleeve engages the pawls after the cam has completed one revolution and forces them out of engagement with the ratchet. The cam sleeve then comes to stop and is ready for the transmission of next character and so on.

7.7 Receiving and Selecting Mechanisms

In the receiving circuit of a T.P., the signals operate a polarised receive electromagnet and the movements of its armature initiates several steps. (a) It sets the selecting and printing mechanism in motion, (b) it selects the character required to be printed, (c) it brings the mechanism to a standstill position at the finish of a signal, and (d) it controls the automatic start-stop switch of the motor. The electromagnet consists of two U-shaped permanent magnets mounted side-by-side with like poles adjacent and the laminated pole pieces embracing the coils. The armature is pivoted at the centre and is enclosed by the coils. The electromagnet link is fitted by means of a spring clip to the end of the armature extension, which extends beyond the stop-plate as shown in Fig. 7.13 below.

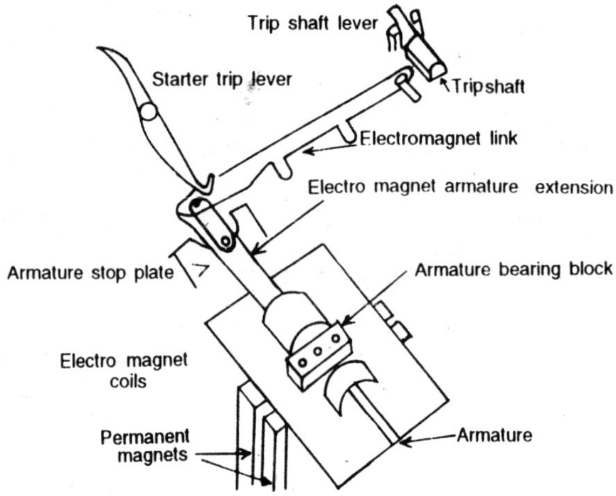

Fig. 7.13: Receiving Electromagnet Link

The link attaches the armature to the trip shaft and also engages with the starter trip lever. A bias adjusting spring is used to give a spacing bias when electromagnet is operated from single current signal but for a neutral bias under double current signalling conditions, no bias adjusting spring is fitted. As the electromagnet armature moves in response to the incoming signals, the trip shaft is either on the mark or space position depending on whether the received signal is marking or spacing. The trip shaft performs two functions: (i) it controls the start-stop of the receive cam sleeve by make or break contacts of the pawl abutment, and (ii) it controls the deflections of the finger-setting blade corresponding to the incoming signals and transforms the signal codes into an identical five comb-setting fingers. Also the receive cam sleeve is driven by means of a clutch consisting of a ratchet and pawl similar to those employed for transit cam sleeve. On receipt of the first spacing impulse, i.e. start signal, the electromagnet armature moves causing the trip shaft to turn in such a direction that the link attached to the pawl abutment moves it from the path of the ratchet pawls. The spring keeps the head of one of the pawls into engagement with the rotating ratchet and the cam sleeve begins rotation. The cam sleeve makes a complete revolution for the reception of a signal combination. A retention lever is provided to hold the receive cam sleeve in a standstill position, when the clutch is out of engagement. The receive cam tracks are shown below in Fig. 7.14.

The receiving cam sleeve is a hollow steel cylinder with five cams cut in its surface in the form of groves as shown in Fig. 7.14 above, with the cam track shapes at their respective phase positions controlling the sequence of movement of levers actuated via cam rollers. These rollers ride in the grooves, controlling the different operations which are required for the selection and printing of the particular character. The traversing cam is coupled o the traversing link, which is given a reciprocating movement by the shape of the operating cam track.

As soon as the receive cam sleeve rotates, the traversing link is operated causing the finger-setting pin to traverse the upper ends of the five comb-setting fingers and from its normal

t

position opposite the third finger, it moves first to the position opposite the first finger and after passing each of the five fingers in turn, it finally goes to the position opposite to the third finger and rests. By this time the cam sleeve completes one rotation and comes to stop.

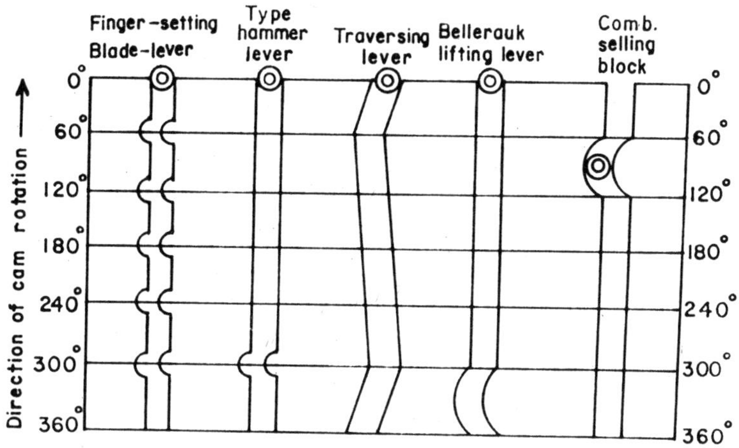

Fig. 7.14: Receiving Cam Tracks

Printing Mechanism: The receive combs are used to select the bell-cranks. These bell cranks make the type head to be positioned, so that a character corresponding to the incoming signal combination is properly printed. The indentations cut into the periphery of each receive comb are so arranged that for each one of the 32 combinations of positions to which the five combs are moved by the comb-setting fingers, certain indentations on the five combs are brought into alignment and a slot is given beneath each pair or adjacent bell-cranks, depending upon the setting of the combination of the character.

One bell crank of each pair corresponds to 'letters' and the other to 'figures' for the same signal code. The sixth combination, also called the shift comb determines which bell-crank of a pair shall drop into the prepared channel. This is similar to other receiving combs, except that it is not provided with an extension and the indentations being equally spaced, the distance apart, or pitch is equal to twice the basic pitch of the receive combs. Therefore for each combination received, only one bell-crank is operated, the position of the shift comb determining whether the bell-crank will be brought into action in the letters or figures group.

The bell-cranks are mounted on the periphery of the receive combs and are carried on an annular ring. The shift comb and the lever of combination head unit, the exploded and assembled views of the type head clutch are shown respectively, in Figures 7.15(a), (b) and (c).

The clutch spring and the scissors spring are connected together at their looped ends and are inserted into the clutch drum in such a way that the free end of the clutch spring is pointing opposite to the direction of rotation of the type-head. Pin No. 1 on the stop plate engages with eye No. 1 formed by the clutch and scissors spring and Pin No. 2 on the latch cam engages with eye No. 2 of the scissors spring. Pin No. 3 and 4 on the stop plate engages in bushes on the typehead.

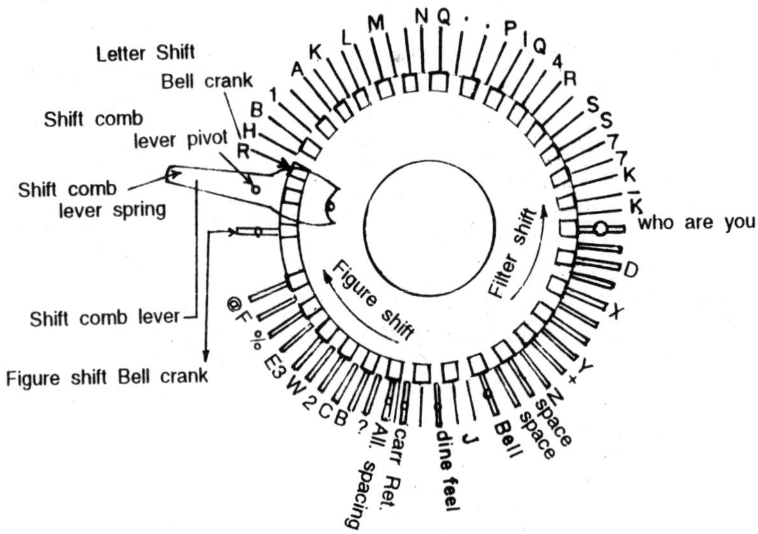

Fig. 7.15(a): Shift Comb and Lever of Combination Head Unit

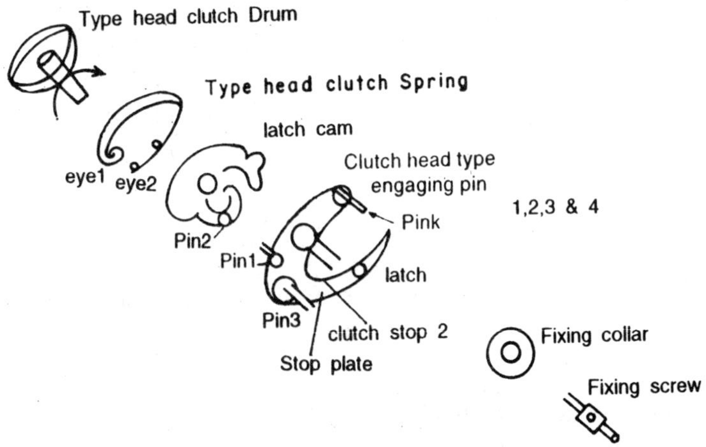

Fig. 7.15(b): Exploded View of Type Head Clutch

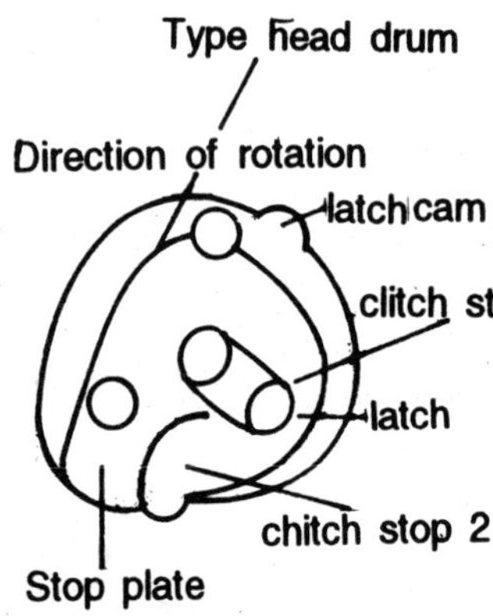

Fig. 7.15(c): Typehead Clutch (assembled)

When a selected bell-crank moves into the slot prepared for it, its extremity obstructs the path of rotating clutch Stop No. 1 and stops its movement and thus the speed of the stop plate and the typehead comes to rest. After a small movement of the stop plate and typehead, the clutch Stop No. 2 abuts against the back edge of clutch No. 1 and the type head is brought to rest. Simultaneously, the clutch latch moves behind the latch cam and locks the clutch on the desired bell-crank and remains ready for printing when the type hammer is operated during the next revolution of the cam sleeve. The type hammer is operated from the receive cam sleeve and is made to strike one of the types according to the position in which the typehead has been brought to rest. The type moves forward carrying one edge of the type retaining plate with it. When the type hammer is withdrawn after the strike, the retaining plate restores the type to its normal position under the influence of its springs. Thus printing goes on during the succeeding revolution of the cam sleeve.

An ink ribbon is fed forward each time a character is printed and the feed direction is reversed when the ribbon reaches its end. The carriage operating mechanism controls the movement of paper carriage and consists of a letter-feed mechanism, carriage return mechanism and, line feed mechanism. The carriage is mechanically returned by the carriage return mechanism, after a carriage return signal is given, or manually returned by using a carriage return key fitted on the left. By an automatic start-stop switch the T.P. motor is automatically started at the time of reception of signal at the beginning of transmission and stopped at the end of transmission.

An answer-back unit enables the operator to understand that the T.P. machine at the distant end of the circuit is ready for reception. It is brought into operation from the transmitting end by sending 'who are you?' signal, which is secondary to letter *D*. The 'who are you?' signal is sent when the *D* key is depressed after a "figure shift" has been sent. The schematic of letterfeed, carriage return and line feed mechanism and answer back mechanism are shown respectively, in Figures 7.16 (a), (b), (c) and (d).

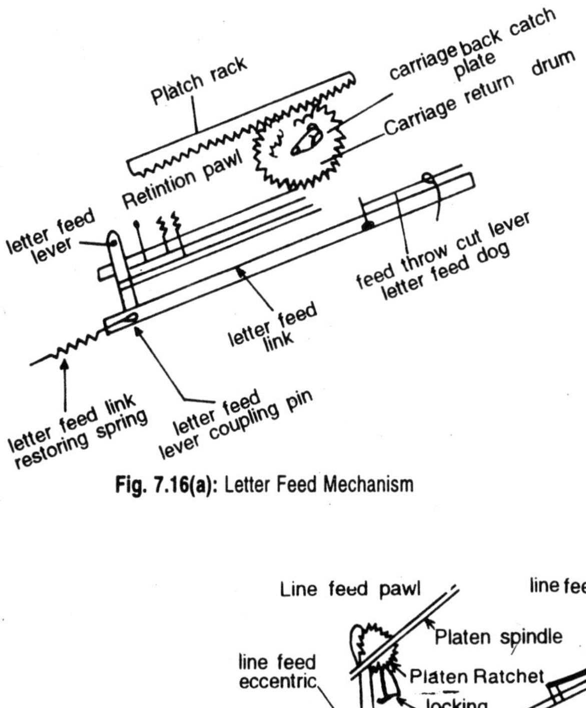

Fig. 7.16(a): Letter Feed Mechanism

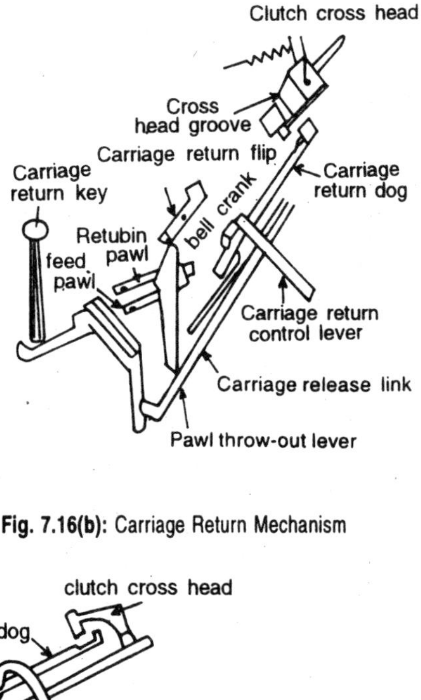

Fig. 7.16(b): Carriage Return Mechanism

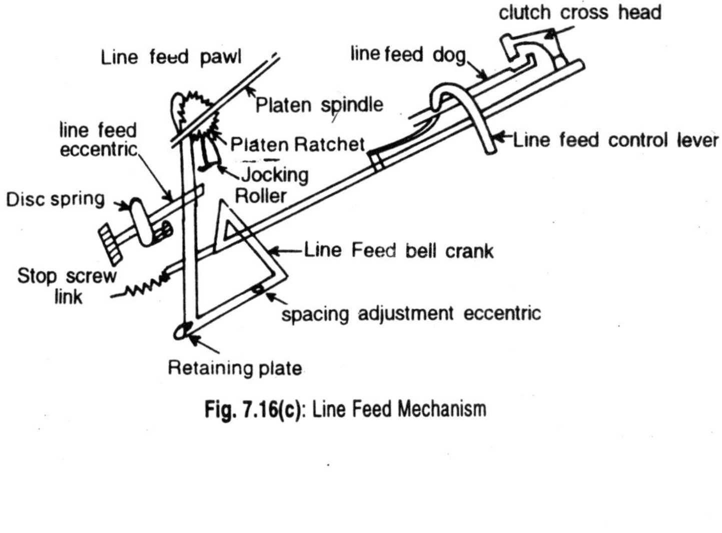

Fig. 7.16(c): Line Feed Mechanism

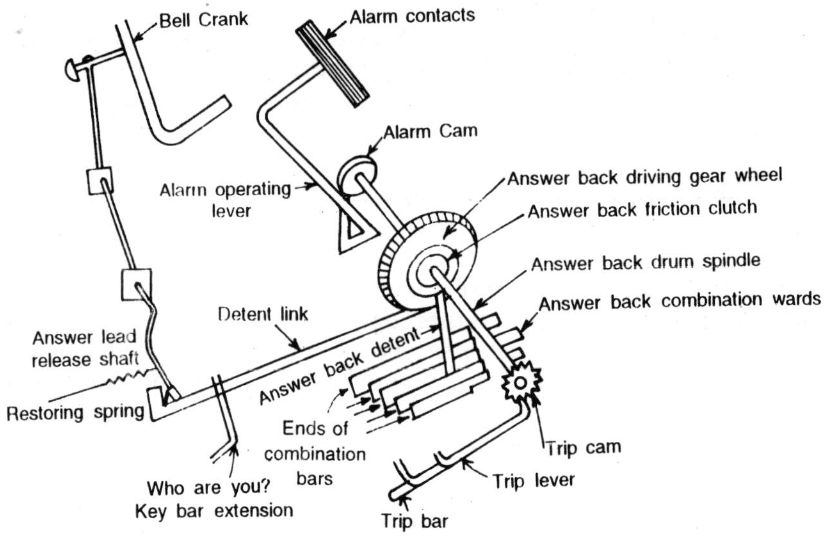

Fig. 7.16(d): Answerback Mechanism

Here, the answer back drum spindle is mounted at right angles to the keyboard combination bars and its forward ends carry the word barrel containing the answer-backwards, each corresponding to one character determined by the projections on its outer edge, engaging with the ends of the combination bars. A projection prevents the corresponding combination bar from moving, thus transmitting a space signal and where the combination bar is allowed to move due to the absence of a projection, a mark signal is transmitted.

We have described in detail the working mechanism of 7B type teleprinters. The other models, namely Model No. 11, etc. work with slight modifications of this mechanism.

7.8 Automatic Switching of Telegraphic Circuits

With the increase of demand in the use of T.P. and the consequent increase of T.P. traffic, a method of automatic switching of teleprinter circuits was applied under the name of teleprinter automatic switching system (T.A.S.S.) in the continental countries. In designing such an exchange for automatic switching, the trunking and grading of the selectors are done on the basis of telegraph traffic, just as the same is done in an automatic telephone exchange, on the basis of telephone traffic. Since the group selectors are very expensive, separate selectors are not provided for each station line and instead, the switching centre end of a station line is terminated on a station line circuit (S.L.C.) comprising of a uni-selector and small relay group. The S.L.C. connects the station line to a group selector for an outgoing call. It also connects the station line to the final selector for incoming calls. The routing of call on the same switching centre for a called number, say 498, of T.P. set is shown in Fig. 7.17(a).

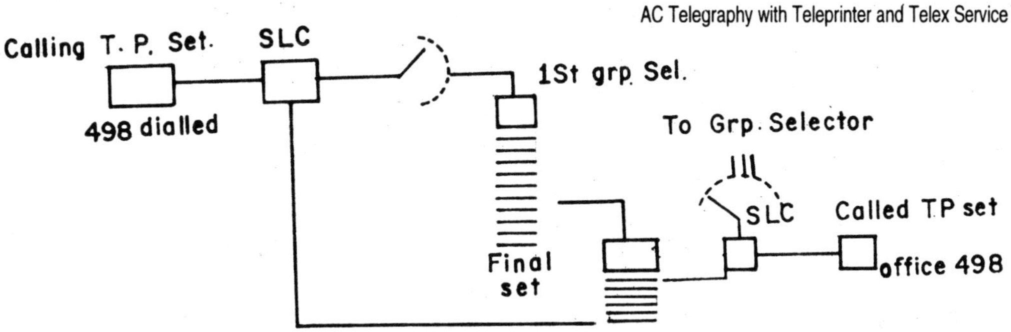

Fig. 7.17(a): Auto-switching of a T.P. Set Within Same Switching Centre

The routing of a call between offices within two different switching centres is shown below in Fig. 7.17(b) for two switching centres A and B.

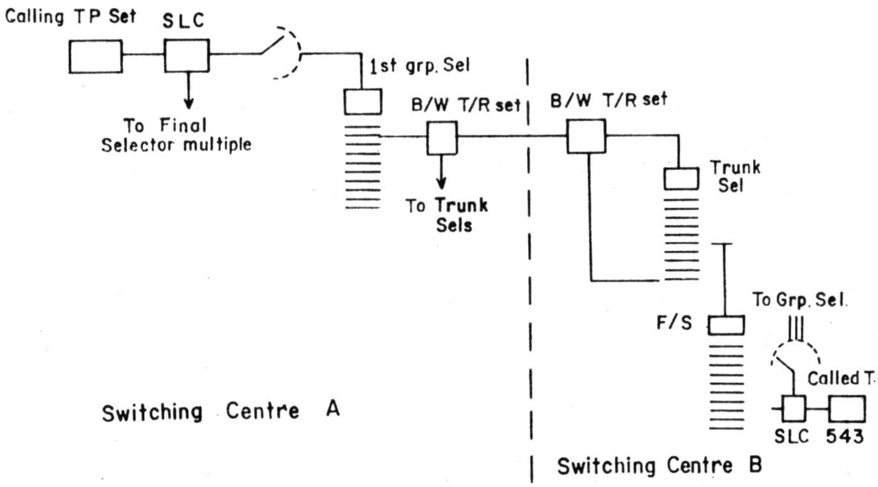

Fig. 7.17(b): Automatic Switching of a T.P. Set on a Distant Switching Centre

Suppose a T.P. set number in centre *A* wants to dial a number, say 534, of a T.P. set in centre *B*. Now the trunk route from *A* to *B* centre may be designated by an additional digit, say 7. Then it is necessary to dial 7534 to get the desired TP set. The connection to trunk circuits is made by means of a bothway trunk relay set at the switching centre A and at the switching centre *B*, the trunk circuit is connected via a second bothway trunk relay set to an incoming trunk selector. Such a system for automatic switching of telegraph traffic was later on converted to telex exchange for telex service.

Telex Service: A subscriber's switched telegraph service enabling the users to communicate directly with one another by means of a start-stop teleprinter machine is called telex service. This telex service was introduced in India during the late sixties when most of the big cities were having automatic telephone exchanges connected with each other with subscribers' trunk dialling facilities through coaxial or microwave channels. The work of installing telex exchanges began in India from the year 1963 and within a couple of years it was installed in cities like Bombay, Delhi, Madras, Kanpur, Calcutta, Ahmedabad, Nagpur, Secundrabad, Coimbatore and other places. At present, there are more than 200 telex exchanges in India working with lines from 20 to 1000, with a total number of telex subscribers of more than 50,000.

To originate the call for telex transmission over a telephone network, the teleprinter signals must be within the frequency range required for the telephone system or for long-distance channel connecting different auto-exchanges on S.T.D. system of working. Single tone V.F. currents are utilised for transmission purposes, a tone being transmitted for spacing element and no tone for marking elements and idle periods. At the receiving end, the V.F. currents are rectified to operate the electromagnets of the T.P. and a local record is kept at the send station via leak circuit.

A separate telex directory giving details of procedures for obtaining any connection is given to telex subscribers. The telex code numbers of all the telex subscribers served by the telex service in India are given in this telex service directory. Generally, the exchanges are divided into zones or areas, e.g. Bombay, Calcutta, Delhi, Madras, etc. respectively with codes (011), (021), (031), (041), etc. Every area or zone is again sub-divided into districts for which some additional code numbers are added to the zone or area codes. The telex service works for every subscriber with a telephone fitted with a dial and a teleprinter unit with page printing arrangement and certain other auxiliary equipments. A voice frequency converter provides a single tone V.F. current to be modulated by the teleprinter transmitting contacts and to convert the receive V.F. signals into double current impulses suitable for operating the T.P. electromagnet.

To operate a call, the subscriber dials the number of the wanted sub, who on answering, is requested to switch on to T.P. and both the calling and the called sub change over the switch to the T.P. position and replace the telephones. This extends the line to the T.P. circuit at each end and the calling party depresses the figure key for 'who are you?' signal of the keyboard, thus actuating the answer-back unit of the distant T.P. and gets a verification that the correct sub has been connected. The T.P. messages are then transmitted in each direction alternately. The connection of 1500 Hz V.F. converter are made under a steel table, on which is placed the telex service T.P. and telephone with other accessories.

In closed circuit working, a current of 40 mA is required. The telegraph batteries are of +60 V and -60 V, so that a loop resistance of 3000 ohm is found for normal work to satisfy this (i.e. $120/40 \times 10^3 = 3000\ \Omega$). Here, of course, the line coil of polarised relay is taken as 70 Ω and that of the dial unit as 120 Ω (i.e., two 240 Ω coil in parallel).

In the send path, the polarised relay in the sender's T.P. sends the make/break oscillations in the form of double current pulses or oscillations via relay contact. The output from the oscillator circuit is connected to the Tongue (T) and Space (S) contact of the teleprinter as shown in Fig. 7.17(a) below, in the oscillatory circuit of 1500 Hz of V.F. telegraph converter.

In the receive path, a rectifier valve responds to the received signals followed by two amplifier valves in push-pull arrangement, and the T.P. electromagnet is operated from the output of this circuit as shown in Fig. 7.17(b) in the push-pull amplifier circuit. The rectifier circuit combines the function of a rectifier, a voltage limiter, and an echo-suppressor.

At the beginning of a spacing signal, the rectifier anode current rises in the primary winding of the transformer including emf in each secondary winding, in the direction of the full arrow as shown, which in effect increases the negative grid potential of valve amplifier 1 and reduces that of valve amplifier 2. As a result, the magnetic flux in I-C coil of the T.P. electromagnet is reduced, while that in *O*-C coil is increased. The armature was neutrally adjusted and it remained in marking position. As an effect of the spacing signal, the armature moves to the spacing position due to the increase of current in the spacing (O-C) coil and decrease of current in the (*I*-C) coil.

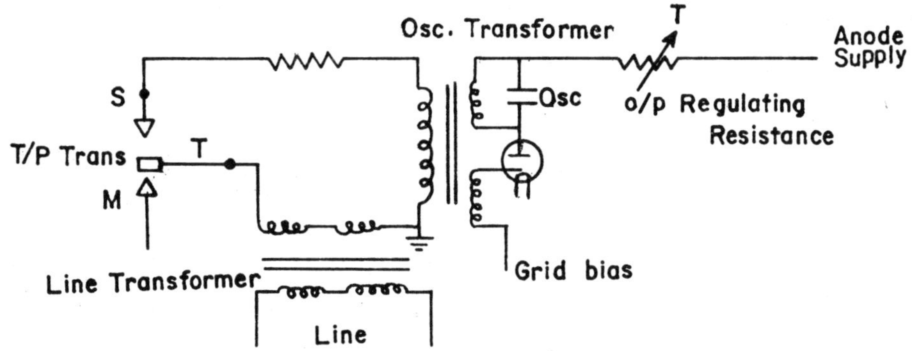

Fig. 7.17(c): Osc. Circuit of 1500 Hz of V.F. Telegraph Converter

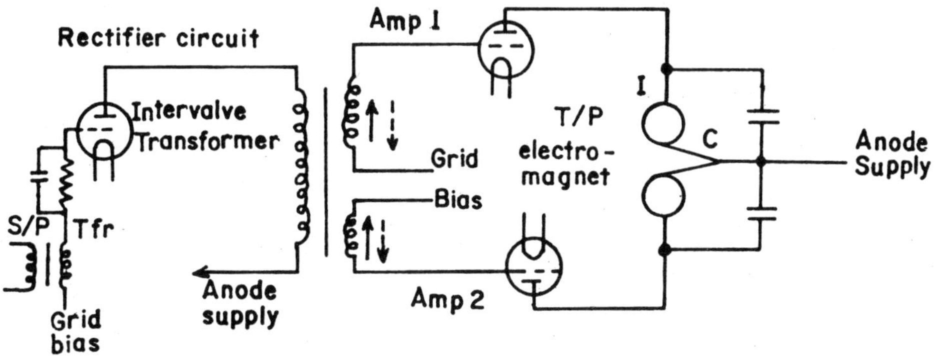

Fig. 7.17(d): Push-pull Amplifier Circuit of 1500 Hz

On cessation of spacing signal, the rectifier anode current falls and an emf is induced in the direction of dotted arrows in the secondary winding. Now the grid of the amplifier 1 is less negative and that of amplifier 2 is made more negative, for which anode current in *I-C* coil increases and in *O-C* coil decreases, for which the armature moves to the spacing stop.

The working principle of a long distance extension of telex circuit is shown below in Fig. 7.18 for working via V.F.T. system and telex exchange.

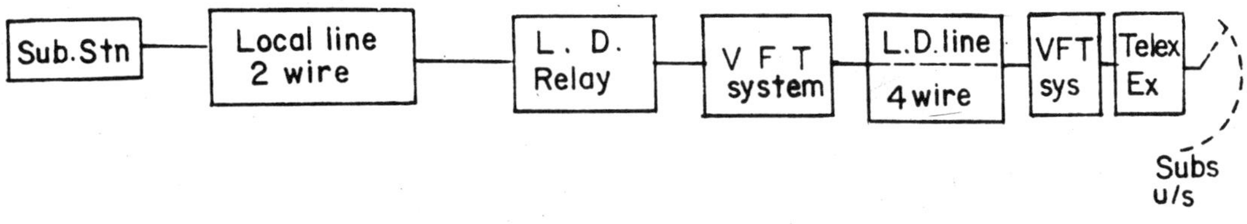

Fig. 7.18: Block Schematic of a Long-distance Telex Circuit

Here, the local two-wire line circuit of a local sub is changed into a four-wire line circuit for a long-distant sub through a long-distant relay set for working through a V.F.T. channel on double current principle.

The trunking diagram of telex system on district and zone basis are given in Fig. 7.19(a) and (b), respectively. Among the switching elements are the 3000 type relay, 2000 type uniselector and two-motion selectors are prominent. Battery testing facilities are provided at each stage, so that no double connection may take place.

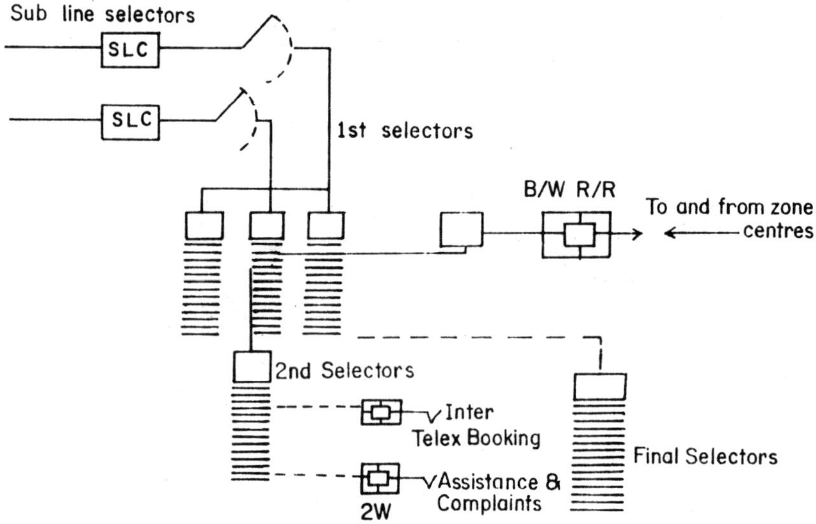

Fig. 7.19(a): District-wise Trunking Diagram of Telex Systems

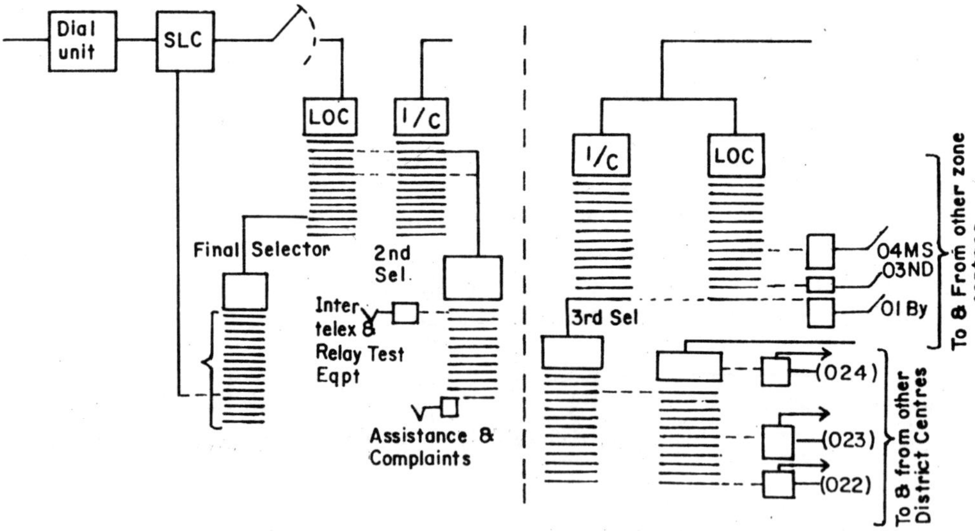

Fig. 7.19(b): Zone-wise Trunking Diagram of Telex Systems

When working in closed circuits, the receiving circuits perform an important task. These take the on-off signal from the receivers of T.P. and transform these into double current signal and send these to the V.F.T. channel to be transmitted to the distant station. Similarly, in an incoming call, the double-current signals are received from the distant end and with the help of line circuit V.F. telegraph converter, these are converted into on-off signals and sent to the T.P. via the telex exchange. Apart from this, all the supervisory works, e.g. releasing works, metering works, etc. are also done by the line circuit V.F.T. converter.

The first selector is a two-motion 100-line outlet of 2000 type. This works as a first selector, as also the second and third selector by strapping properly with U-jack. As an outgoing selector, its work is to send signals to dial the wanted sub and as an incoming selector when used in both-way repeater, its work is to indicate the reception of the calls. In either case, this is a positive telegraph battery stop pulse extending over 20-35 msec.

The final selector is a 100 line two motion 2000 type, whose vertical and horizontal motions are controlled by dial. This sends the last two signal digits of the called sub and joins the calling T.P. to the wanted teleprinter. If the wanted number is engaged, then it sends a busy release signal, which breaks open the switched-on connection. A complete trunking diagram of a telex circuit is given in Fig. 7.20.

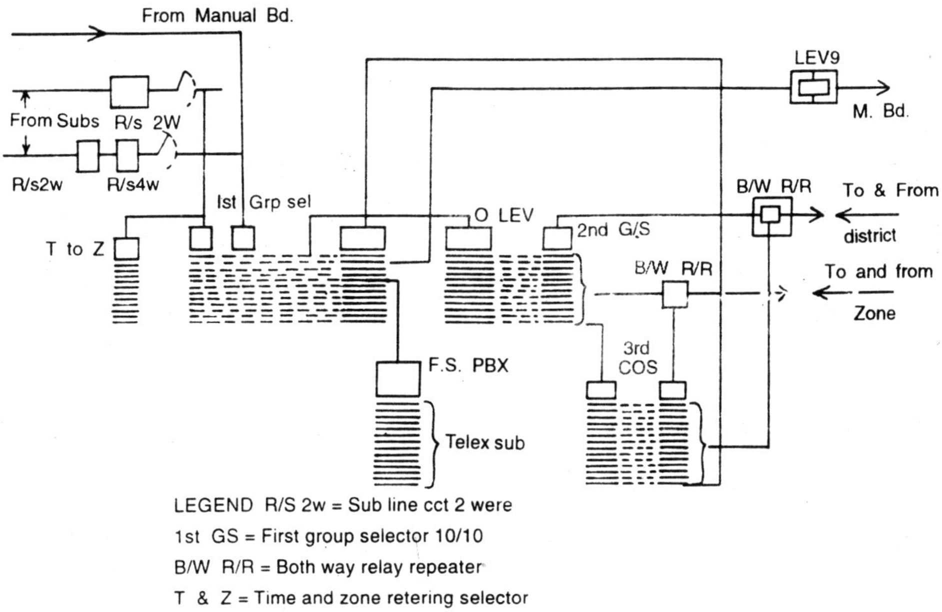

Fig. 7.20: A Complete Trunking Diagram of a Telex Circuit

The time and zone metering selector is a two motion 2000 type, which is connected to every local first selector. It steps in keeping with the second and third digit dialled after 0. If the first digit is from 1 to 9, then this selector does not step forward. If the first digit is 0, then the selector steps when the next two digits are dialled and reaches the desired outlet from where the metering pulses are joined. If the selector does not function even when the signals are sent, then after 3-6 min, a P.G. alarm signal is sent, so that the attention of the technical staff is drawn to remove the fault.

The connections of district and zone centres are shown in Fig. 7.21.

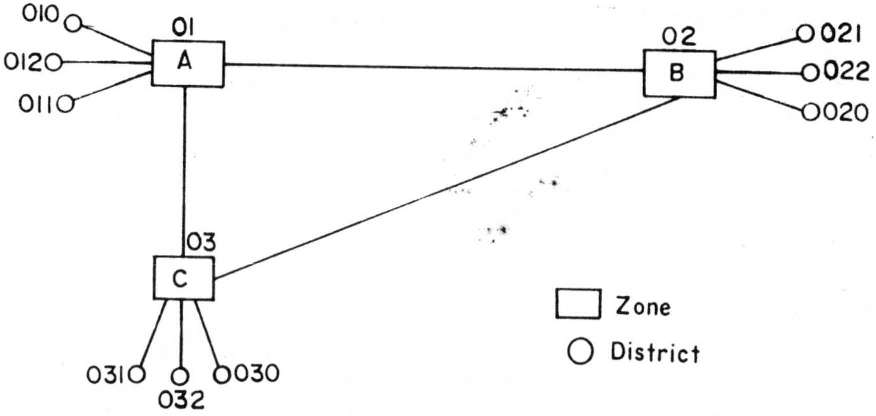

Fig. 7.21: Connections of Zone and District Centres

Both way relay repeater is placed just before and after a V.F.T. channel. When a sub dials a number for trunk, then that repeater is seized from the group selector level. Every repeater is connected to an incoming group selector and it transmits signals in both directions and does the supervisory work. For an unfinished dialling in incoming calls, the repeater gives a P.G. alarm after 3-6 min, which draws the attention of the technical assistant to rectify the fault. Besides, if a junction is faulty, then the repeater is force-released by the first call, so that a busy release signal is sent and during the next call, a busy signal is received. In a faulty channel, a test signal is sent after every minute in the forward direction till such time the channel is O.K. and released for service.

A telex exchange works like an auto exchange and uses the same type of alarm and supervisory relay. The following batteries are essential:

+ telegraph battery 60 ± 2.5 V

- " " 60 ± 2.5 V

Exchange battery 46-52 V

Here, the positive pole of the positive telegraph battery, the negative pole of the negative telegraph battery and the positive pole of the exchange battery are earthed. The busy hour calling rate is taken as O.I.T.U.

Long-distance communication in telex service: Suppose a telex sub of Madras No. 041-300 wants to communicate with a telex subscriber of Delhi No. 031-300. Then the Madras sub presses a button on his dial unit by which his uni-selector of Madras telex-exchange catches a group selector and a green lamp glows in the subs dial unit which is equivalent to a dial tone in case of a auto telephone exchange. Then the Madras sub dials the Delhi sub No. 031-300. The digit 0 and 3 steps up the first and second selector of Madras telex exchange, by which it catches a free Delhi trunk that is obtained first. Then along with the dialling of digit 1, the third selector is stepped up and joins Delhi exchange as an incoming trunk. Along with the dialling of the next digits, the first, second, and final selectors of Delhi telex exchange are stepped up and the wanted sub is joined. Simultaneously, the T.P. motor of Delhi sub starts, the white lamp glows and a free line signal comes to Madras sub, whose T.P. motor starts and white lamp glows, so that the other subs may get these subs as engaged. Then the Madras sub presses 'WRU' key to get the identity of Delhi sub and presses 'Here is' key to give his own identity. If the wanted sub is found, then the necessary message is exchanged and after the work is finished the 'Disc' key is pressed when the established circuit is disconnected, the motor stops and the white lamp is extinguished, so that the number is ready for the next call.

The Inland Tariff Structure is based on three procedures: (i) the flat rate charging, (ii) automatic ticketing, and (iii) elapsed time metering. In India, we have adopted the elapsed time metering for telex service.

7.9 Distortion of Telegraph Signals

Distortion or deformation of telegraph signals may occur due to line being of unequal cross-section of wire or of unequal length of limbs in case of trunk lines and channel being out of synchronisation, or the transmitting and receiving motors being of varying speeds or out of phase. In long DC telegraph circuits, the telegraph signal distortion is caused by the capacitance unbalance of the open wire line, which prevents the receive current from rising instantaneously to its maximum value. Thus the receive relay (or armature of the receive electromagnet) may not operate exactly in sequence of the times of operations of the transmit relay. Thus the time displacement of a signal changeover from its true position may be termed its distortion due to transmission and the total displacement is given by the maximum relative displacement of any pair of transition, which is given by the sum of the most advanced and most retarded displacements of the transitions. If they are denoted by d_1 and d_2, then the percentage distortion is given by $\frac{d_1 d_2}{t} \times 100$, where t is the duration of a unit signal element.

If the time required by the current to reach its steady value (transient time) is less than the duration of the shortest time, no telegraph distortion will occur and the receive relay will operate in correct sequence of the transmit relay. If, however, the transient time is greater than the duration of the signal element, then the distortion that occurs is called the characteristic distortion, depending on the transformer qualities.

The effect of resistance of the line is to delay the rate at which the capacitance is charged. Thus for a given capacitance, the greater the resistance of the line, the greater is the time required for the receive current to reach its maximum steady value. The effect of leakage is to reduce the amplitude of the received signals, but the inductance of the line being too small, has a little effect on a DC circuit. A change in the difference of earth potential between the ends of an earth-return circuit or a change of leakage will produce bias distortion on a single current circuit.

If, however, any circuit is subjected to an interference current, it may at times, either increase or diminish the receive currents, thus advancing or retarding the times of operations of the receive relay with relation to that of the transmit relay, thus giving rise to fortuitous distortion.

To check the amount of distortion in a VFT channel, undistorted telegraph signals are sent at the transmitting end of the channel and the distortion of the signals is measured at the receiving end with reference to a synchronised time-base, thus getting a direct measurement of the signal distortion introduced by the channel.

Tests are made to measure the distortion by using a number of test signal combinations, namely reversals at 66 Bauds and reversals at half speed, i.e. 33 bauds to which the channel bias is adjusted to zero, six marking elements followed by one spacing element and one marking element followed by six spacing elements, SQG in the T.P. code and SQG reversed. Owing to the characteristics of channel filters and amplifier detectors, the distortion varies according to the type of signals transmitted. The following test results are typical figures for a normal channel.

Type of signal	66 Baud 1:1	33 Baud 2:2	6:1	1:6	SQG	Reversed SQG
% of distortion	4	2	4	10	6	6

The apparatus for taking the above tests is called telegraph distortion measuring set, which consists of a transformer and a receiver together with necessary time-base oscillator and is used for both directions of transformer. The transmitter consists of a telegraph relay *T* vibrating at the frequency of a time-base oscillator. The tongue of the relay is connected to the line under test, while the contacts of the relay are connected to ± 80 V telegraph battery via two change-over contacts to determine the applied polarity. The changeover contacts are actuated by a cam disc to produce testing signals, when rotated via a suitable gearing by a synchronous motor rotating at a speed controlled by the line base oscillator. The synchronous motor has four armature poles and eight field poles connected alternatively to the side contacts of motor relay and the return circuit of the windings is connected via ± 80 V telegraph battery to the tongue of the motor relay, so that if the relay is vibrating, a two-phase effect is produced in the field magnet, causing a rotating magnetic field as shown in block schematic Fig. 7.22 below.

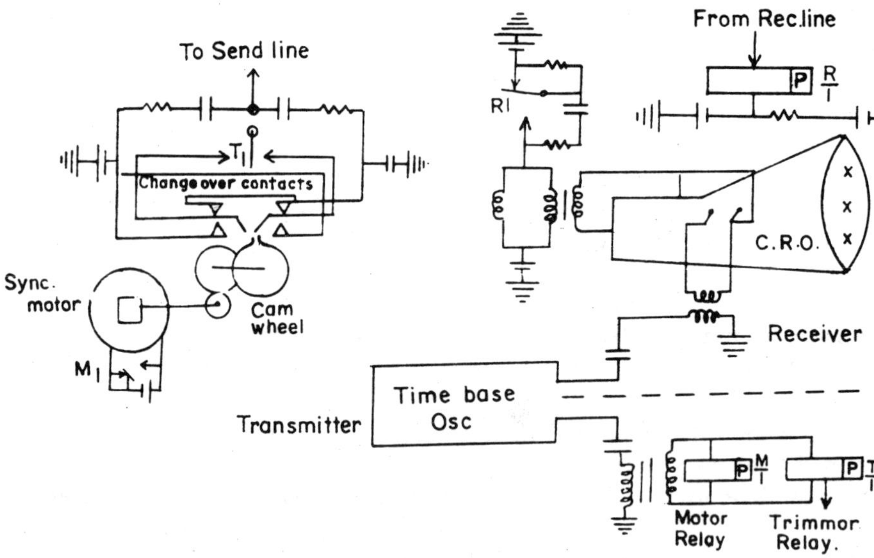

Fig. 7.22: Principle of T.D.M.S. in Measuring Telegraph Distortion

The receiver consists of a C.R.O. tube, which displays distortion to be measured. If the signals arrive in a time sequence which is an exact multiple of speed of rotation of the rotating field, then the flashes produced fall on the same spot on the screen of the C.R.O. indicating a zero distortion. If some signals come earlier than the others, spreading over an image on the sector of the screen, then by means of suitably divided radial scale, the sector may be interpreted directly as a percentage of 360^0 (i.e. of one unit signal element) and thus the percentage distortion can be measured directly on the circular screen.

Measurement of Margin of T.P.: While the T.D.M.S. is used for the measurement of channel distortion, the measurement of distortion produced by a start-stop T.P. transformer, and the margin of a T.P. receiver can be made by a telegraph distortion and margin tester. It consists of a transformer for sending either perfect signals or signals with known amount of distortion and a receiver in the form of a stroboscopic disc enabling the received signals to be examined. A circuit diagram for margin testing is shown below in Fig. 7.23(a).

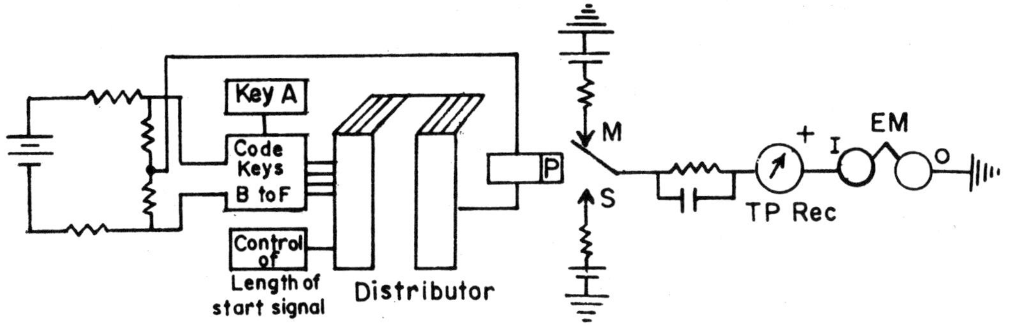

Fig. 7.23(a): Margin Testing Circuit of a T.P. Receiver

Using a 80 V battery and a 4000 ohm resistor shunted by a 2 μF condenser, and connecting in series with the electromagnet, a satisfactory result is obtained with T.P. Nos 3, 7 or 11. The margin is first measured for letter *R* and then for letter *Y* as *R* corresponds to SMSMS and *Y* corresponds to MSMSM.

To ensure that the T.P. is started by a true start signal element and not by a space signal element, a special key 'Stop-Start-Send' is provided, so that on the start position of the key, only start signal is transmitted and on the send position of the key, the T.P. receives the test signals in correct sequence. The sliding position is then moved slowly forward, causing shortening of the start signal until a position is reached where further movement causes error. Also a similar position can be found when the sliding pigment is moved slowly backwards causing lengthening of start signals. The readings on these two positions give the margin of the T.P. for early and late signals, respectively.

To measure the distortion of a start-stop transformer or of the line, the output from the distributor of the signals to be tested are connected to a stroboscope relay as shown in Fig. 7.23(b).

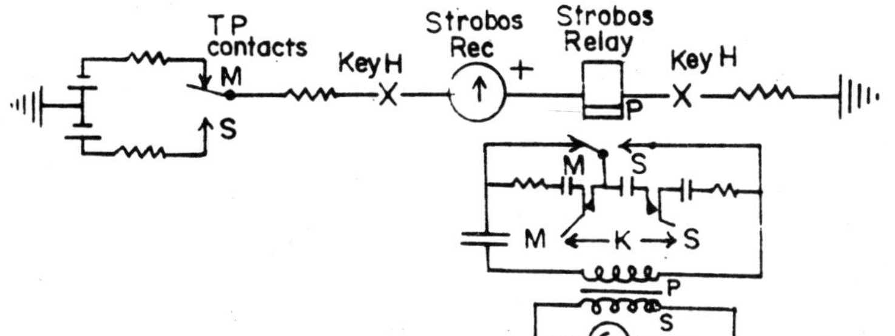

Fig. 7.23(b): Measurement of Distortion of a T.P. Transmitter or Line

By operating key *K*, either marks or spaces may be observed on the stroboscope receiver consisting of a disc carrying a neon lamp. The disc rotates at the speed of the transmitter and a momentary flash is produced each time the armature of the receive relay changes from marking to spacing and vice versa, due to charging of the respective capacitor. The flashes are produced on the disc corresponding to the commencement of each signal element. No flash occurs at the beginning of the first code element, since for *R* letter this code element is of the same polarity as the start signal and the maximum distortion relative to the beginning of the start signal is shown occurring on the second code element and is about 25 per cent. By reversing the polarity of the transmitted signals on the tester is shown the bias distortion due to marking and spacing bias indicating a change in the length of the respective signal element.

7.10 Power Supply Arrangement with Rectifiers for Carrier, V.F.T., Telegraph and T.P. Systems with Floating Battery Condition

There are various types of carrier systems like open wire and underground cable systems. Also there are various types of telegraph systems and VFT systems. These systems require various types of power supply for their operations. Thus for carrier and VFT valve type systems we require:

(a) 24 ± 3 V supply for filament and 124 ± 3 V supply for plate. These voltages are normally fed from AC mains supply rectified and smoothed by copper oxide or selenium rectifiers with secondary lead-acid battery as standby supply.

(b) For telegraph systems, we require ± 80 V supply fed from rectifiers normally with standby secondary battery.

(c) For teleprinter systems, we require (i) ± 80 V DC for signalling and T.P. motors, (ii) -50 V DC for teleprinter automatic switching, phonogram and telegraph testing, (iii) T.P. switch board lamp 6 V DC or AC

We have already described the installation and maintenance of telegraph batteries in Chapter 1 of this volume. Rectifiers of various types are described here. These metal rectifiers are used with step-down transformers working on mains supply to provide the required voltage to the rectifier units of various types, namely the single phase half-wave and full-wave rectifiers, as also the three-phase full-wave rectifiers as shown below in Fig. 7.24(a).

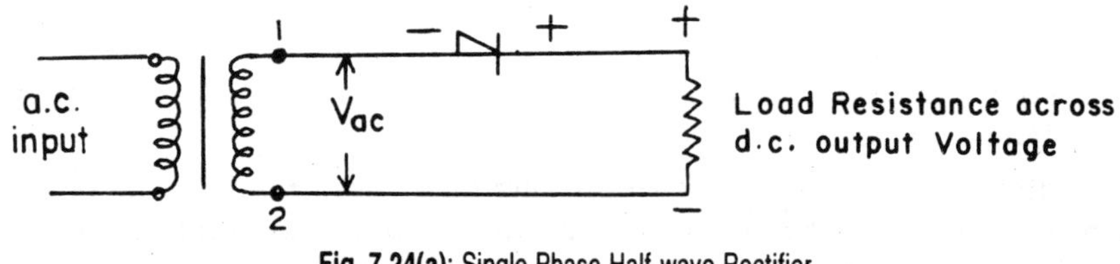

Fig. 7.24(a): Single Phase Half-wave Rectifier

This arrangement is called half-wave rectifier, as it passes a load current only during the alternate half cycles of AC voltage, terminal 1 being positive with respect to 2. No current passes during the other half cycles and load is taken across the load resistance.

A single phase full-wave rectifier is shown in Fig. 7.24(b).

Here, both the positive and negative half waves of AC voltage pass easily through the bridge rectifier. When terminal 1 is positive with respect to 2, *B* and *C* rectifiers only conduct in series with the resistance R_L in the load and when terminal 2 is positive with respect to 1 during the next half waves, only rectifier *D* and *A* conduct in the same direction through R_L.

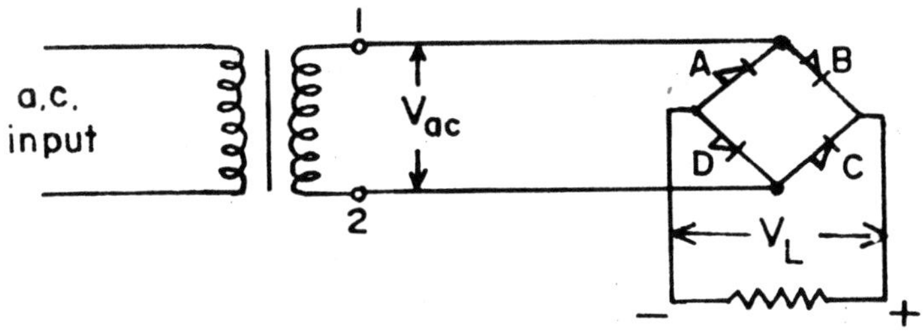

Fig. 7.24(b): Single phase Full-wave Bridge Rectifier

A three-phase full-wave rectifier is shown in Fig. 7.24(c).

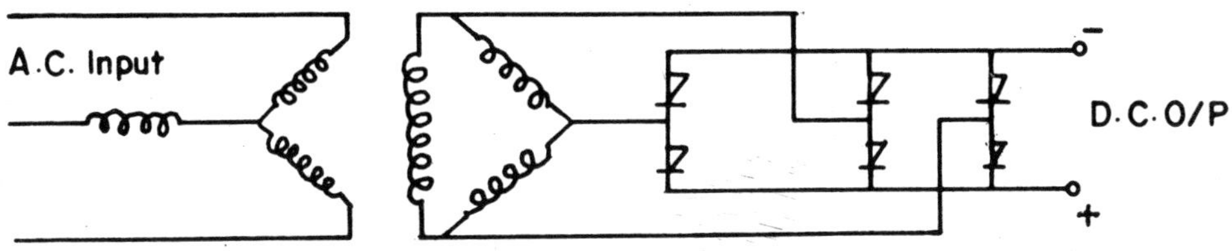

Fig. 7.24(c): Three-phase Full-wave Rectifier

Voltage regulation may be done mainly in three ways, namely, (i) static voltage regulation, (ii) moving coil induction regulation, and (iii) tap-changing of transformer regulation.

Smoothing of rectified voltage can be performed by providing a choke coil in the output, or by providing a low-pass filter having two chokes and a set of electrolytic condensers. A battery connected across the output of a rectifier will remain in readiness to supply the load when no output comes from the rectifier during the hours of mains failure. Such a battery is said to be in floating condition, as it behaves in a similar manner to that of a smoothing condenser.

Thus for obtaining plate and filament voltages for 130V ± 2V and 24 V ± 3 V for valve operated carrier and VFT systems, rectifiers are used with floating batteries as shown below in Fig. 7.25(a).

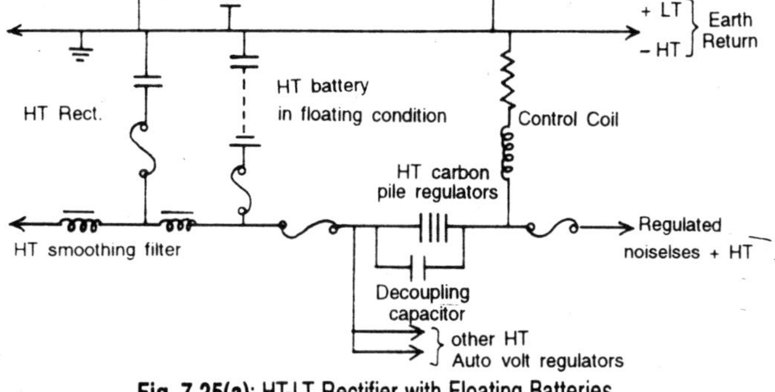

Fig. 7.25(a): HT-LT Rectifier with Floating Batteries

The rectifier has smoothing filters and a carbon pile regulators in series with LT and HT supply and operates on single battery float system. Normally for telegraph and teleprinter service, ± 80 V rectifier is used as shown in Fig. 7.25(b).

On -80V+80V switchboard, a special fuse is provided for -80V for supplying power to station alarm circuits.

Earthing is done in two ways, namely, (a) protective earthing to prevent metal frames, etc. from power short circuit, (b) earthing of power supplies themselves on one pole to keep the potential of the supplies within certain limits with respect to earth potential, so that the return path of the signal currents in most of DC telegraph circuits is via earth.

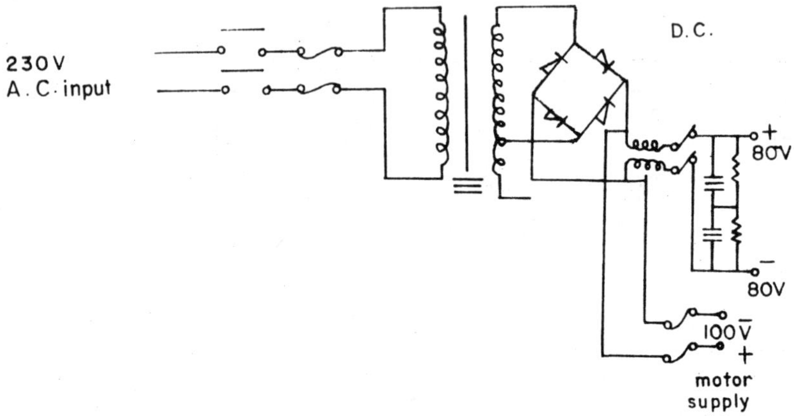

Fig. 7.25(b): ± 80V Rectifier

Power cables from power room to telegraph office or telephone exchange are taken on cable racks through L.S. cable standardised for current and voltage by I.E.E.

The rectifiers used for supply of load from mains are of three types, namely, (a) copper oxide or selenium rectifier, (b) vacuum tube diode rectifiers, and (c) gas-discharge tube rectifiers or grid controlled gas rectifiers (Thyratrons). Alternating current is converted into direct current by the use of copper oxide metal rectifiers made by coating one side of a copper disc with a layer of red cuprous oxide. The selenium type of metal rectifiers may be formed by spraying a layer of selenium on a nickel-plated steel with a low melting point steel alloy as shown in Fig. 7.25(c).

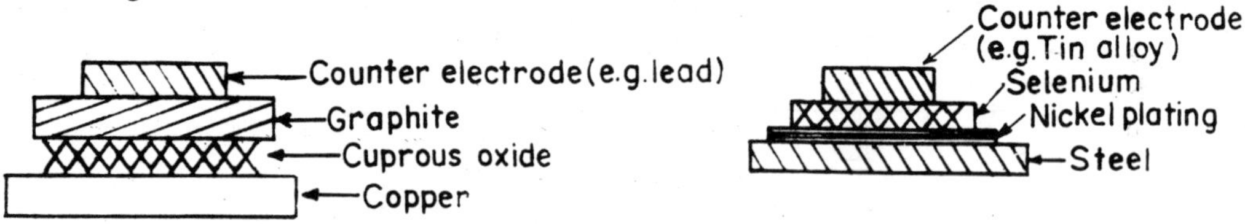

Fig. 7.25(c): Copper Oxide and Selenium Rectifier-elements.

We have already discussed the use of half-wave and full-wave metal rectifiers as also that of three-phase full-wave metal rectifiers.

The diagram for a vacuum diode half-wave rectifier with current-wave form curve is shown in Fig. 7.25(d).

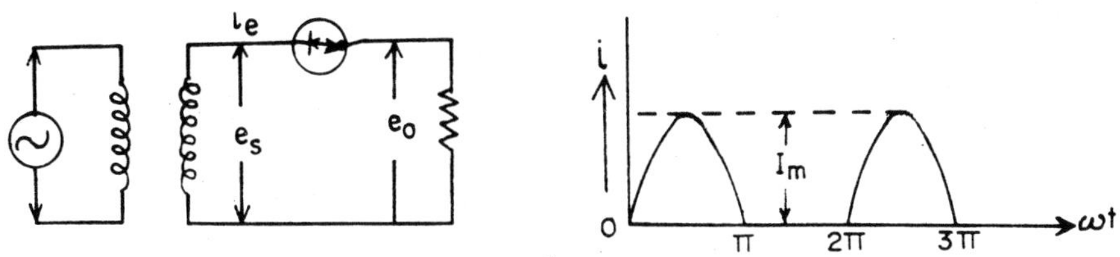

Fig. 7.25(d): Vacuum Tube Half-wave Rectifier with Current Wave form

Here i_b = instantaneous value of diode current
I_m = maximum amplitude of input current
e_s = E_m sin ωt is the applied voltage
R_d = internal resistance of diode valve
I_{dc} = average value of DC current
I_{RMS} = average value of RMS current, and
R = load resistance.

The unidirectional current consisting of alternate half sinusoids of maximum amplitude flows between 0 and π, while no current flow is between π to 2 π and so on. During the first half-cycle (0 to π).

$$i_b = \frac{E_m \sin wt}{\pi(R_d + R)}$$, while from π to 2π, $i_b = 0$.

$$I_{de} = \frac{E_m}{\pi(R_d + R)} \text{ and } I_{RMS} = \frac{E_m^{\ 2}}{2(R_d + R)}$$

in AC input to diode as also to load.

∴ Input p $P_{ac} = \dfrac{E_m^{\ 2}}{4(R_d + R)}$ and

rectification efficiency = $\eta_R = \dfrac{P_{dc}}{P_{ac}} \times 100$

The full-wave vacuum tube double-diode rectifier is shown in Fig. 7.25(e) with current wave form.

Here, during the interval 0 to π, V_1 is conducting and V_2 is not conducting while during the interval 0 to π, V_1 is conducting and V_2 is not conducting, which during the interval π to 2π, V_2 is conducting and V_1 is not conducting and so on. Hence the average DC value of load is twice that of half-wave rectifier circuit.

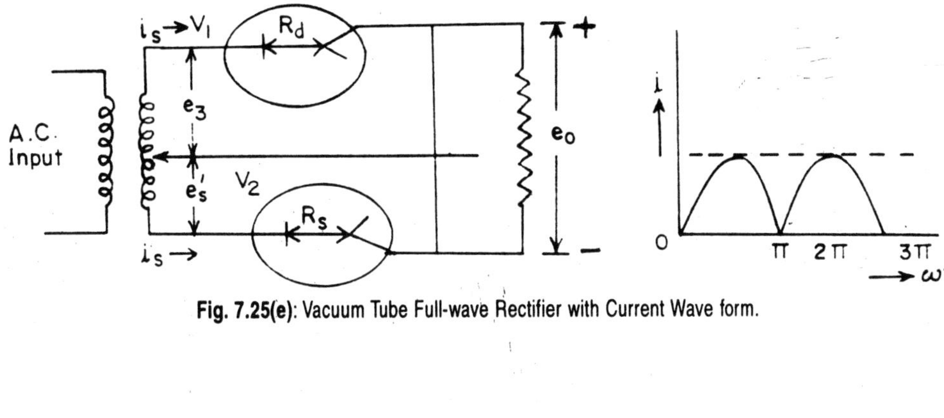

Fig. 7.25(e): Vacuum Tube Full-wave Rectifier with Current Wave form.

Ripple factor: The purpose of rectifier is to produce steady output voltage and current from sinusoidal voltage source, but this is not perfectly achieved even by smoothing the rectified outputs. Hence ripple factor γ is defined as:

the ripple factor

$$\gamma = \frac{\text{effective value of alternating components of voltage or current}}{\text{Direct value of voltage or currents}}$$

The rectifier current $I_o = I_{de} + I_{ac}$ where I_{ac} = effective value of alternating components.

$$I_{RMS} = \sqrt{I_{dc}^2 + I_{ac}^2} \therefore I_{RMS}^2 = I_{dc}^2 + I_{ac}^2$$

$$I_{ac}^2 = I_{RMS}^2 - I_{dc}^2 \therefore \gamma = \frac{\sqrt{I_{RMS}^2 - I_{ac}^2}}{I_{ac}} = \frac{\sqrt{I_{RMS}^2 - I_{dc}^2}}{I_{dc}^2}$$

$$But \ \frac{I_{RMS}}{I_{dc}} = Form\,Factor\ F$$

$$\therefore \gamma = \sqrt{F^2 - 1}$$

Regulation: When current is drawn by the load, the terminal p.d. must not fall appreciably and should remain steady from no load to full load. This smoothness of supply of power is called regulation. It is expressed in percentage.

$$\text{So , voltage regulation} = \left[\frac{No\,load\,P.d - full\,load\,P.d}{No\ load\ P.d.}\right] \times 100\%$$

$$= \left[\frac{E_{dc}\ (no\ load) - E_{dc}(\,full\ load)}{E_{dc}\ (no\ load)}\right] \times 100\%$$

Thus the rectifiers fitted with automatic voltage regulation and smoothing circuits can be directly utilised for supplying load to an exchange. For a small exchange, a two tube tungar rectifier is used for such purposes. The tube is made of a tungsten filament whilst the anode is made of a graphite disc. The air is exhausted from the tube and a small quantity of inert gas, e.g. argon is filled in the tube. The circuit diagram of a tungar rectifier is shown below in Fig. 7.25(f).

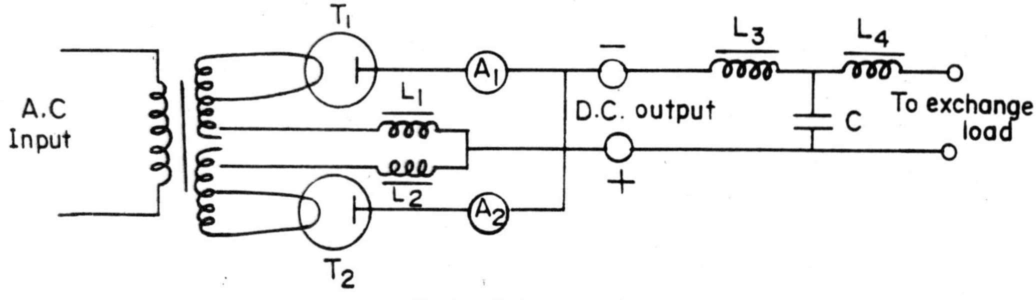

Fig. 7.25(f): Tungar Rectifier

The electrons emitted by the heated filament are attracted towards the higher potential anode and in passing through the space between the two electrodes, the electrons ionise the gas molecules, forming a conducting path from the anode to the cathode. The two tubes T_1 and T_2 produce full-wave rectification. The iron-cored inductances L_3 and L_4 are included in the anode circuit of each valve to smooth the voltage ripples in conjunction with a bank of electrolytic condensers C fitted after the DC output to assist smoothing before the rectified voltage is fed to the load.

In movable transport wireless vans and aircrafts, etc. where space is the main factor for accommodating large power plants, high potentials are usually supplied by using a vibrator to operate from a motor car battery and produce alternating voltages on the two contacts on two sides. This alternating voltage is then stepped up by a step-up transformer and smoothed. These vibrators are classified as synchronous and non-synchronous vibrators. In case of synchronous vibrators, the contacts operate in synchronism with voltage reversals and the high potential voltages do not require rectifiers, while in case of non-synchronous vibrators, rectifiers are required. The chief characteristics of vibrators are:

(i) They represent a means of obtaining plate powers from storage-battery.

(ii) They are widely used in automobile and aircraft receiving wireless stations and radio transmitters.

(iii) They produce undesirable clicks, sparks and noises and are short-lived.

The gas-filled rectifiers or grid controlled gas rectifiers (Thyratrons) show greater efficiency than vacuum tube rectifiers. Also the drop of voltage is low inside the tube which requires much less filament current. The grid-controlled gas rectifiers are known as Thyratrons or gas triodes and are extensively used in control circuits. They are classified as negative or positive according as the starting grid voltage is negative or positive. The critical value of grid voltage depends upon the anode voltage. Before the grid voltage reaches the critical value, the anode current is negligible and on reaching the critical value, the grid loses control over the anode current and the anode voltage drops down to 10 to 20 V. For avoiding damage to the tube, the anode current must be limited by impedance in series with the anode. Before giving the anode supply, cathode is heated and brought to normal operating temperature. If this precaution is neglected then the voltage drop inside the tube may exceed the cathode disintegration value during the warming process and damage the cathode permanently.

For H.W. gas rectifier with resistive load

$$E_{dc} = \frac{E_m}{\pi} - \frac{E_o}{2}, \text{ where}$$

E_m = maximum value of input voltage

E_o = valve voltage drop and

E_{dc} = direct output voltage.

The input power

$$P_i = \frac{E_m^2}{4R} - \frac{E_m E_o}{\pi R}$$

the output power

$$P_{dc} = \frac{1}{R}\left(\frac{E_m}{\pi} - \frac{E_o}{2}\right)^2$$

and the rectification efficiency

$$\eta_R = \frac{P_{dc}}{P_i} \times 100.$$

The critical breakdown voltage characteristic as shown in Fig. 7.25(g) enables us to know the anode voltage e corresponding to the grid voltage at which the valve is fired and vice versa.

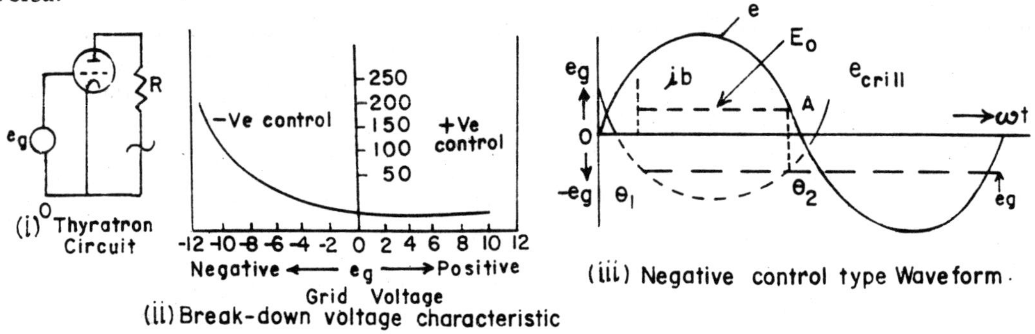

Fig. 7.25(g): Critical Breakdown Voltage Characteristic with Negative Control type Waveform

When the potential e is rising, the rectifier current i_b is also rising. At the firing point when the valve becomes conductive, there is a potential drop E_0 inside the valve, i.e. the current begins to flow inside the valve. When $E = E_0$ at point A, the current stops and the valve becomes non-conducting at θ_2. After this point A, the valve does not conduct current for the rest of the cycle. The complete wave form is shown in thick line.

For Thyratrons $I_{dc} = \frac{E_m}{2\pi R}(\cos\theta_2 + \cos\theta_1)$

and $E_{dc} = -\frac{E_m}{2\pi}(\cos\theta_1 + \cos\theta_2)$.

If the DC power is supplied by a motor generator set to charge the secondary battery of a telephone exchange, then a circuit-breaker with a separate overload and reverse current relay may be used as shown in Fig. 7.25(h) below, to guard against an excessive current which may damage the generators and to prevent discharge of the battery through the generator if the voltage of the generator goes below that of the battery.

The circuit-breaker is an armature of the trip coil electromagnet and is attracted by the electromagnet when the current passing through it exceeds a predetermined value, thus disconnecting the battery to be charged. The relay in the form of a moving coil ammeter, provides contact on both sides of the needle, thus energising the trip coil in case of excess current on reversal of direction.

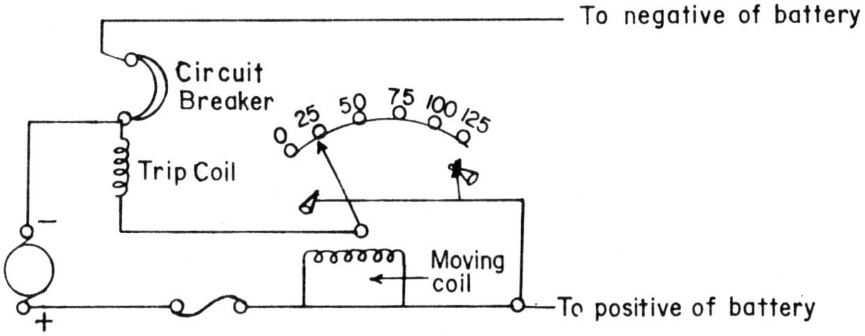

Fig. 7.25(h): Circuit-breaker with Separate Overload and Reverse Current Relay

7.11 DC Power Source for Electronic Devices

For DC power source to electronic devices, rectifiers are used with appropriate filters to produce DC power source from AC mains supply.

The voltage regulation is given in per cent.

$$= \frac{no\ load\ voltage - full\ load\ voltage}{full\ load\ voltage} \times 100$$

Imperfect filtering produces ripples given by percentage ripple production

$$= \frac{RMS\ \ ripple\ \ voltage}{full\ load\ \ dc\ voltage} \times 100$$

The use of a zener diode maintains a fairly constant voltage across a load and can be used for improving voltage regulation and reducing ripples. A transistor may be used in conjunction with the diode to increase regulation and reduce current through the diode as shown in Fig. 7.26(a) and by a closed-loop regulator as shown in Fig. 7.26(b).

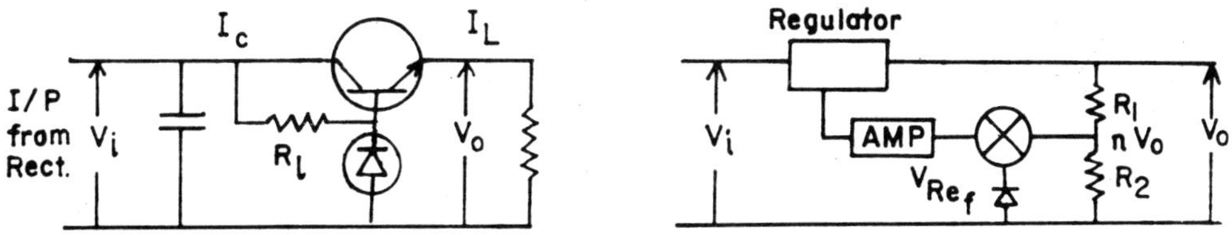

Fig. 7.26(a): Voltage Regulation by Zener Diode, **(b)**: Closed-loop Regulator

A diode with close-loop regulator can also be used where the output resistance cannot be reduced below a given value as shown in Fig. (b) above. Here a fraction n of the output voltage V_0 is compared with a reference voltage V_{ref} and their difference is amplified and applied back to control the regulator in series. This in turn controls the output voltage.

Voltage and current can also be regulated by using silicon-controlled rectifiers (SCRs) or triacs as shown below in Fig. 7.27(a).

The equivalent circuit shows that an npn transistor is connected to a pnp transistor, where the collector and base currents of the npn are connected to the base and collector, respectively of pnp transistor. Thus if one transistor is cut off, the other is also cut off and impedance from anode to cathode is high. The collector current for npn becomes base current for pnp and vice versa.

A cumulative action begins as an increase of current in one increases that in the other, till both are saturated and impedance from anode to cathode becomes very low when gate is said to trigger the SCR 'ON'. To turn the SCR 'OFF', the gate bias must be reversed, and anode voltage reduces to zero when no current flows through pnp unit.

An over-voltage protection circuit is shown in Fig. 7.27(b).

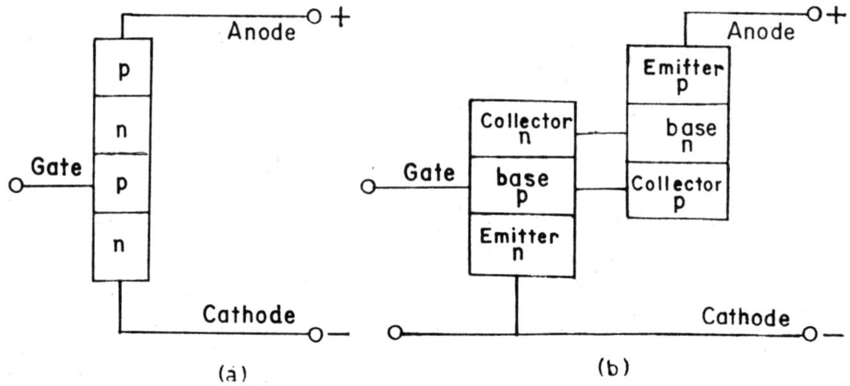

Fig. 7.27 (a): Silicon-controlled Rectifier Electrical Circuit and **(b):** its Equivalent

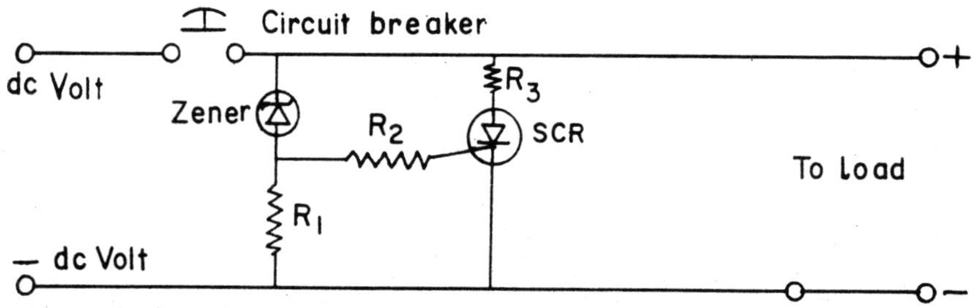

Fig. 7.27(c): 'Crow bar' like own voltage protection Circuit

Normally, the voltage across SCR is below anode break-over voltage and below the breakdown voltage of the zener diode, so that the zener is non-conducting. The avalanche voltage of the zener diode is so chosen that a surge in the voltage supply makes the zener conductive, producing a positive potential on the gate of SCR which then fires. As the value of R_3 is kept very low, the SCR draws a heavy current from the source, opening the circuit breaker and protecting the load.

7.12 Diesel Engine Alternator as a Standby No-break Power Supply Arrangement

We have already seen that the normal power supply in the telephone exchange, telegraph office and carrier and VFT systems are made by power cubicles which generally take power from the mains power supply. In case of mains failure for short periods, say, up to 4 hours or so, each unit has got arrangement for automatic power supply from secondary batteries. But if the power failure is prolonged for some unforeseen reason, then each station is provided with an additional standby power supply arrangement by a diesel engine alternator of adequate capacity to take the load of the whole station including the telephone exchange, telegraph office and the carrier and VFT systems along with lights and fans as an emergency measure. This standby power plant consists of a prime mover in the shape of a diesel engine mechanically coupled to an alternator and may give a single-phase or three-phase output. If the load is less than 10 KVA, the output may be single phase and for output of 10 KVA and above, the same is three-phase. A control panel, associated with the engine alternator, contains relays to sense the mains failure, and close the circuit for starting the diesel engine. The variation of voltage and frequency beyond limits can be adjusted within limits by manipulation of the control devices.

No-break Power Plant: A no-break power plant consists of a motor, a flywheel and a diesel engine. Normally the motor and the load are connected in parallel across the mains supply. There are three types of no-break plants: (a) auto-synchronous, (b) regenerative, and (c) static. The first two are rotatory type and the third one, as the name signifies, is of static type. For most telecommunication stations, the standby no-break power plants used are of rotatory type, in which the motor is coupled to the flywheel and a magnetic clutch is introduced between the flywheel and the diesel engine. So long as the mains supply is there, the flywheel is run by the motor and when there is a mains failure, the motor works as a generator, when the magnetic clutch closes and the kinetic energy stored in the flywheel is used for starting the engine and running the motor as an alternator, until such time the diesel engine starts and attains the speed and voltage required for the load. The motor used may either be auto-synchronous as shown in Fig. 7.28(a) or it may be a regenerative type as shown in Fig. 7.28(b).

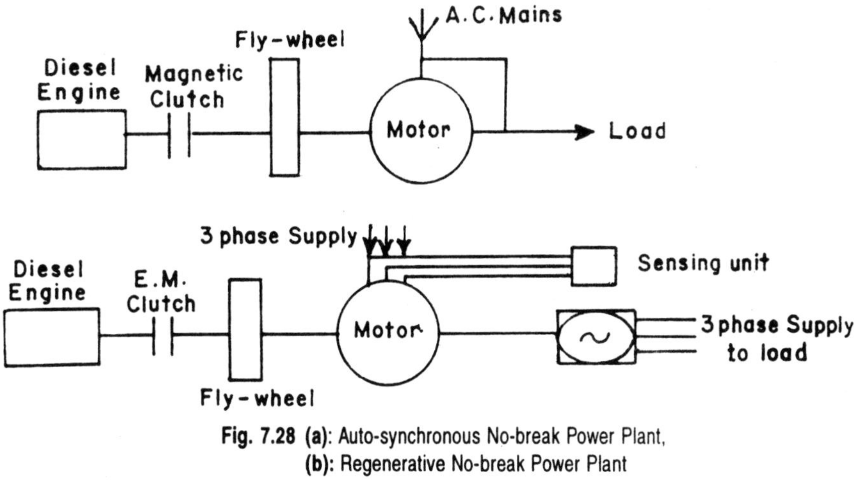

Fig. 7.28 (a): Auto-synchronous No-break Power Plant,
(b): Regenerative No-break Power Plant

In case of auto-synchronous motor, on the restoration of mains, the speed of the diesel engine is altered with the help of a governor, so that the frequency and phase relation of the mains and the generating set are matched with each other. The operation of the contactor relays put the alteration parallel to the mains and reconnect the load direct to the mains. Thus the change-over from mains to standby and vice versa is carried out without any break of supply voltage. Also on restoration of the mains supply, the magnetic clutch and the solenoid are de-energised and the diesel engine is stopped. The disadvantage of synchronous no-break power plant is that the fluctuations in the mains supply also pass to the communication system causing distortion. To get rid of this trouble, regenerative type of no-break power-plant is used, which consists of a direct engine, magnetically coupled through a clutch to a flywheel, the further side of which is connected to an AC motor and alternator by flexible coupling. Normally, the motor runs on the three-phase AC mains and drives the alternator and the flywheel. The output from the alternator is connected to the load. But in case of mains failure, the momentum of the flywheel continues to run the alternator and supply the load till the diesel engine starts and the electromagnet clutch is coupled to the flywheel and the sequence of operations repeat. On restoration of the mains, the magnetic clutch and the fuel solenoid are de-energised to stop the engine.

During the transition from mains to standby, a certain amount of change in frequency is likely to take place due to slight loss of speed of the flywheel before the start of the engine and the motor changing over its function to that of an alternator. This can be overcome by using a power plant consisting of an alternator coupled to an AC motor on one side and a DC motor on the other side. Normally, the AC motor will run the alternator to supply the load and during the mains failure, the DC motor is made to run off the storage battery and drive the alternator. Here a rectifier is required to charge the storage battery.

The provision of a no-break power plant is a must for those stations where power feeding is done to the long-distance underground coaxial cable systems and repeaters are spaced at short intervals of 9.3 km for 4 Mc/s, 4.8 km for 12.5 Mc/s and 5 km for small tube systems, where power cannot be fed at all to the repeater stations, which are normally kept unattended. The maximum normal length on which power can be fed is 240 km, which works out to every 25th repeater station to be a power-feeding station, in addition to the terminal stations. Since a large number of circuits working on STD basis are being catered by coaxial links, it becomes imperative to provide uninterrupted power supply for the entire route. This cannot be met from a manual start or auto-start engine alternator, because in these types there is a definite break interval, however small, for the continuous supply of power. But a no-break power plant completely eliminates any such break in the continuous supply of power to the load. The operation of the relating type of no-break regenerative power plant has the following facilities:

(i) In the normal AC mains conditions, an AC motor drives the alternator which in turn feeds the load.

(ii) In the event of mains failure, the engine starts and is clutched to the alternator shaft. During the interval when mains have failed and the engine has not started, the momentum of the flywheel keeps the alternator running.

(iii) When the mains supply deviates from a pre-determined value (± 10% of the rated value), the engine starts and drives the alternator as in (ii) above.

(iv) When AC mains frequency deviates by 4% of 50 ~, the engine starts and drives the alternator.

(v) In the event of mains failure, if the engine fails to start, the no-break power plant is bypassed and the load is transferred to AC mains direct. In this case, AC can be fed to the load from any other source, which may be a second standby engine alternator set at the station.

(vi) If the alternator output fails, or varies ± 10% of the rated value or the frequency varies ± 4%, the no-break power plant is bypassed and the load is transferred to the second standby direct, as in (v) above.

(vii) If the motor goes faulty, the engine will start and put itself clutched to the alternator as in (ii) above.

(viii) If there is more than 5 to 10 V difference between the three-phase supply, then also the engine will start and drive the alternator.

(ix) If there is more than 5 to 10 V difference between the three-phase supply at the output of the alternator, then the plant will be bypassed and the load will be transferred direct to AC mains as in (v) above.

(x) The 24 V control battery is charged from the alternator output, so that in the mains failure condition also the charge is maintained.

8

Submarine Cable Telegraphy

8.1 Types of Submarine Cables Used

We have already described in Chapter 2, the details of construction of submarine cables and the procedure for laying the same in deep seas and oceans connecting overseas countries for communications. For the various conditions of different portions of the sea or ocean with different depths, we have got to consider the following before laying the submarine cable. (i) The shore-ends of the cable is to be armoured to protect it from damage due to coastal causes, (ii) For cable laid in deep sea are subject to extra stress and strain during the process of laying and subsequent operations, when the long length of the cable is supported from the bow sheaves of the ships up to the bottom of the sea for different depths.

For these reasons, four types of armoured telegraph cables are used for submarine requirements: (i) the deep sea type with a diameter of one inch for depths up to 500 fathoms, (ii) medium type of larger diameter for depths between 300 to 500 fathoms, (iii) heavy type of diameter between 2 and 3 inches for depths up to 30 fathoms, and (iv) a shore-end type with double-sheath armouring used for cable for beach-end up to the cable hut.

The submarine telegraph cables use earth return signalling, for which at each end the earths made should be as perfect as possible. As the signals received at the other ends of the long submarine cable are found to be extremely weak, only a few micro ampere or so, the earths so made should be free from electrical noise for providing a good signal-to-noise ratio. Similar steps have got to be taken to minimise the inductive and cross-fire effects.

8.2 Submarine Cable Signalling

Either cable code or five-unit code signalling is used for transmission of telegraph signals on submarine cables, as the Morse code is found to be of longer duration than the cable code or five unit code. The cable code though based upon Morse code utilises a three-condition signalling system, which minimises the transmission duration of signals. The dot, dash and space are each dependent on a particular condition, with the same duration of time. There is no need of a dot or dash in a character and spaces are required only to separate characters and words. Thus the duration of a character in cable code is equivalent to 3.7 signal elements (i.e. dots) as compared with 8.6 signal elements (i.e., dots) for Morse code on average. The cable code can be operated on time division multiplex working with two channels working at the same or different speeds. The speed may be varied by changing gear ratios of phonic motor drives of transmitting, receiving and regenerative repeater equipments. The maximum speeds go up to about 90 w.p.m. for direct printers.

The five-unit code requires only two basic signalling conditions, namely marking and spacing and permits Gulstad method of reception, which is not possible with cable code. It is less affected by noise than the cable code, as double current signalling is used with five-unit code. The submarine cable offers higher capacitance affecting the speed of transmission of telegraph signals, for which synchronous transmitters and receivers are used. Hence, (1) the design of instruments for five-unit code is simpler for those used for cable code. The five-unit code requires a constant number of elements for each character, while in cable code the number of elements are variable. (2) The length of the perforated tape for five-unit code is about one-third of that of cable code, making it more economic and easy to handle. (3) The training of operators and maintenance staff is simpler in five-unit code system. The cable may be used to provide two time division channels operating at the same or different speeds.

For continuous loading, the submarine cable is found unsuitable for duplex working, as it required the balancing artificial line for different sections, to have a critical nature of balancing and due to the variation of resistance of tin foil with temperature, it was found unsteady for variation of line condition. For this 'tapered loading' was used, in which the amount of loading was increased in several stages from zero at each end to the maximum at the middle section.

The voltage used should not go beyond 120 V as it is found that higher voltages may lead to breakdown of insulation. Inductive shunts and shunted capacitors are used at each end with resistances, to provide improved signal shape. It is also found that when the speed of signalling is increased in a long submarine cable, mutilation take place, and is increased with the increase of speed. This effect is called 'zero wander' and can be minimised by sending a current of reverse polarity for a short period at the end of each signal element, followed by a short period of earthing the cable. This process of improving the wave shape is called 'signal curving' and the duration of curving depends on the speed of working. It is greater for higher speeds, e.g. auto-transmitters and low for hand-keyed signals, for which it is not required. To avoid all the above troubles, a cable can be operated on a switched simplex basis with a prearranged schedule of time, by connecting the cable for transmission in one direction for a certain period and in the reverse direction for alternate periods. However, for economic reasons, duplex working is resorted to in submarine cables also, by making adequate artificial balancing network for bridge duplex working.

For more economic reasons, submarine cable telegraphy uses multiplex systems, which may be either time division multiplex or frequency shift multiplex system.

(a) In time division multiplex systems, the same line is used in turn for an interval of time just sufficient to transmit five-unit codes for the signals by four operators, each working on a particular section of the distributor as in Baudot system, as shown in Fig. 8.1 below.

There are two rings connected by metallic arm fitted with brushes. The outer ring is divided into four sectors or parts. Through the inner solid ring of the distributor rotated at uniform speed by a driving motor at station *A* and *B*, the line is connected to the inner brush. The arms *a* and *b* are fitted with two metal brushes, the inner one on solid ring and the outer brush, sweeping over and making contacts with sectors 1, 2, 3, 4 in succession or the outer ring of the distributor, electrically connecting them for that period to the line during each revolution. The brush arms at each station are kept at same phase angle throughout their rotation, so that the signals transmitted from sector 1 (one) at station *A* arrive at sector 1 at Station *B* and so on for the corresponding sectors.

Each sector or arm is divided into 24 segments, five segments for each operator and two segments compensate for the propagation time of signals through the line to reach the other end and the remaining two are used for speed correction. A five-unit code allotted for Baudo key board is thus manipulated. The code elements 1, 2, 3 are transmitted by keys 1, 2, 3 by the first three fingers of the right hand and code elements 4 and 5 are operated by the keys 4 and 5 by the first two fingers of the left hand. The five keys make one character of 5 unit code and control the signal currents. Thus the manipulation requires sufficient training. A cadence signal is given, indicating that the combination of keys should be depressed for transmission of the next character.

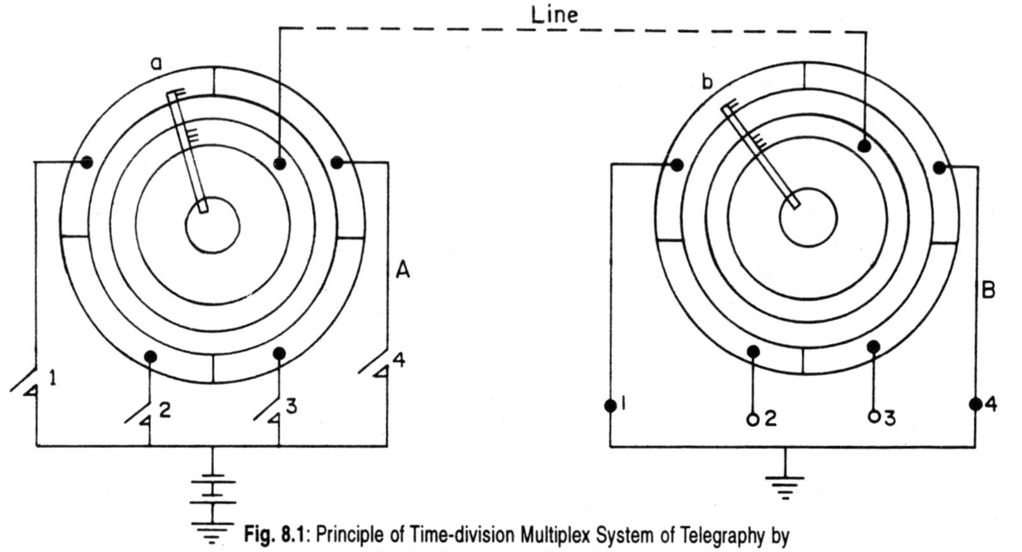

Fig. 8.1: Principle of Time-division Multiplex System of Telegraphy by Baudot distributor for Four Channels

The receiving machine consists of the operation of five electromagnets, the armature of which reproduce the corresponding positional setting. The armature setting mechanically translate and print the particular character corresponding to the positional setting of the armature. Then the moving parts are restored to normal position by the resetting cams for repeating the process.

(b) In frequency shift multiplex system, the multiple frequency shift is known as frequency shift diplex or twinplex. It provides two channels requiring 2^2, i.e. four carrier frequencies for operation. Such a system can be used to transmit one channel of a three-condition code and may be used for direct interconnection for a radio and submarine cable circuit, using only three frequencies each for the corresponding cable code. Thus the total frequency shift is reduced. The frequency is varied in four steps with a spacing of 400 c/s as shown below for the two cable codes.

Freq.	Code 1		Code 2	
	ch A	ch B	ch A	ch B
freq f_1	sp	mk	sp	sp
$f_2 = f_1$ + 400 c/s	sp	sp	sp	mk
$f_3 = f_2$ + 400 c/s	mk	sp	mk	sp
$f_4 = f_3$ + 400 c/s	mk	mk	mk	mk

The twinplex system can be operated with two channels with different speed and unsynchronised, so that the performance in one channel is not affected by the signalling condition of the other. By operating both the channels, synchronously equal transitions occur in both, affecting them equally and thereby make them suitable for transmission of equally affected signals.

8.3 Transmission of Signals in Submarine Cable

The diagram for manual simplex working using a cable code key is shown in Fig. 8.2 below.

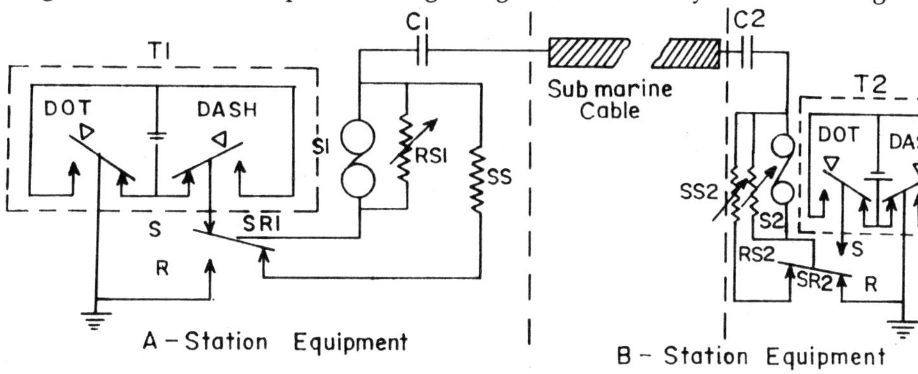

Fig. 8.2: Manual Simplex with Cable Code Key Working on Submarine Cable Telegraph circuit

Here T_1 and T_2 are transmitting keys, $C1$ and $C2$ are the signalling capacitors, $S1$ and $S2$ are the respective siphon recorders and $SR1$, $SR2$ are the send/receive switches for connecting send or receive instruments to the submarine cable, as required. Also $SS1$, $SS2$ are the adjustable low resistances for sending shunts, and $RS1$, $RS2$ are the corresponding high resistance inductive receiving shunts adjustable to get the maximum receive current, whilst the sending shunts are adjusted to allow equal current to pass through the recorder at the sending station as that at the receiving station. The battery is connected reversely at each end.

In the idle condition, S/R switch is placed on receive position at both ends and when transmitting, it is placed at send position when the depression of dot key earths the positive pole extending it to negative pole of the signalling capacitor $C1$. On restoration of dot key and operation of dash key, the reverse takes place. During the space signal both dot and dash keys are at rest and the earth potential is extended to the capacitor $C1$. Thus the cable code signals are transmitted to the line via the siphon recorder with shunts $SS1$ and $RS1$. This results in impulse working to the distant receiver where the signals flow through siphon recorder coils and high resistance inductive shunt $RS2$ to earth. Nowadays, instead of manual transmission, automatic systems using cable code or five-unit code with synchronous transmitting and receiving apparatus are universally used working at 60 words per minute.

In the synchronous systems, the received signals after amplification are regenerated by a repeater, thereby eliminating distortion and facilitating retransmission over another long section of submarine cable, till at the last repeater section end, the signals are used to operate the printer or reperforator, thus improving the margin of stability of receive instrument cable relays for reception. The main types of cable relays are Brown drum relay, Orling Jet relay, Muirhead gold wire relay, etc., of which the last one is shown in Fig. 8.3.

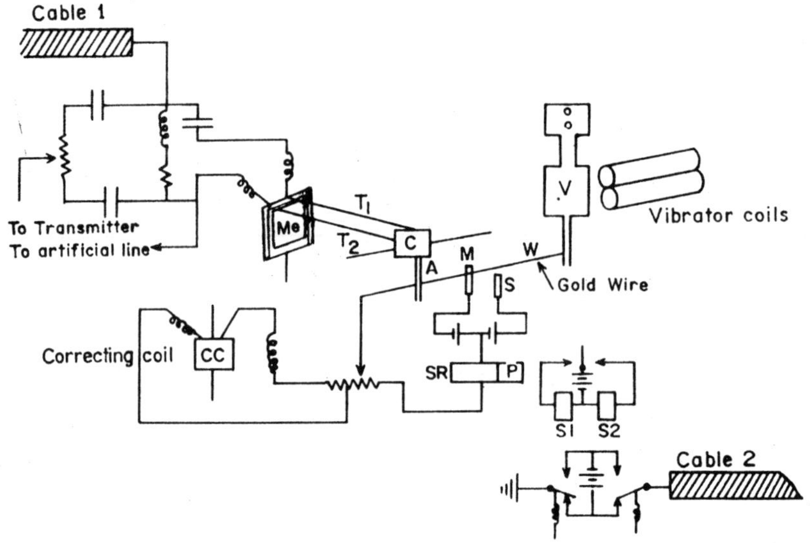

Fig. 8.3: Principle of Muirhead Gold Wire Relay

The signals received from the cable 1 operate the moving coil MC, whose deflections control the suspended cradle *C* via threads T_1, T_2 carrying light arm *A*, to which is attached the gold wire *W* between two contact bulbs *M* and *S*. The gold wire makes contacts with *M* or *S* in accord with the direction of deflection of *MC* and operates the relay *SR*, which controls the operation of relays *S*1 and *S*2 which in turn automatically retransmit the signals received from the submarine cable 1 to the submarine cable 2. The vibrator *V* is to ensure good contacts of the gold wire with *M* or *S* and a correcting coil *CC* corrects the error.

Signal amplification or magnification in submarine cable telegraphy: There are three types of signal magnifiers used, namely, Harley hot wire type, the capacitance type and the valve type, of which the capacitance type is described in Fig. 8.4.

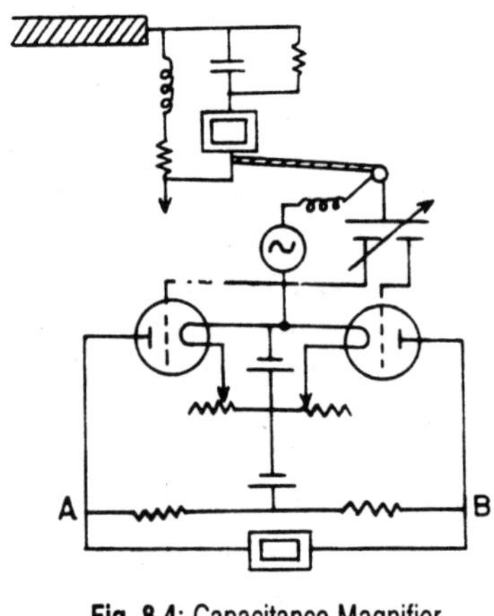

Fig. 8.4: Capacitance Magnifier

The signals received through the moving coil are used to operate a differential capacitator consisting of two fixed electrodes with a central movable vane between them. In the normal condition when the vane is central, there is equal capacitance between the vane and each electrode and when the vane is moved by the coil, the capacitance is increased towards the nearer and decreased from the distant electrode. The fixed electrodes are taken to the grids of two rectifier valves and a source of AC with frequency of 30,000 Hz is applied between the moving vane and the cathodes of the two valves. When a high tension is applied to both anodes via the centre of a resistance placed across the recording instrument, the movement of the vane to one side increases the AC voltage to one side grid and decreases that on the other. Hence the mean anode current is reduced to one and increased to the other, causing a p.d. across resistors *A* and *B* and operating the recording instrument in accordance with the movement of the vane.

8.4 Synchronisers and Regenerators

It has already been mentioned that multi-channel frequency division method is not generally employed in submarine cables as these are single-core cables of very limited band-width, unsuitable for frequency-division multiflex working. So for submarine cable telegraphy, only time-division multiplex method either with cable code or five-unit code signalling is used. But in cable code, the number of code element varies for different characters, whereas in five-unit code every character has a constant number of elements. For this synchronous method of working is generally employed for T.D.M. working by maintaining synchronism between the transmitting and receiving apparatus and regenerative repeaters by controlling the speed and phase of signals.

A synchroniser keeps the receiving apparatus at every intermediate regeneration station, as also at the terminal receiving station in synchronism with the transmitting apparatus. A synchronising motor is geared to a cam shaft, which drives the motor at the same speed as that of the transmitting station. The operation of the contacts driving the motor is controlled by cam shaft and synchronism is maintained by the incoming signals to control the advance and retard mechanisms.

A regenerator is necessary to receive and retransmit messages. The principle used consists in selecting the central, i.e. least distorted portion of each signal element, to determine the signal condition to be recorded or regenerated. The regenerative repeater retransmits the perfect signal and thus avoids manual retransmission.

The principle of synchronous Gulstad reception in a five-unit multiplex system used for submarine cable working is shown in Fig. 8.5.

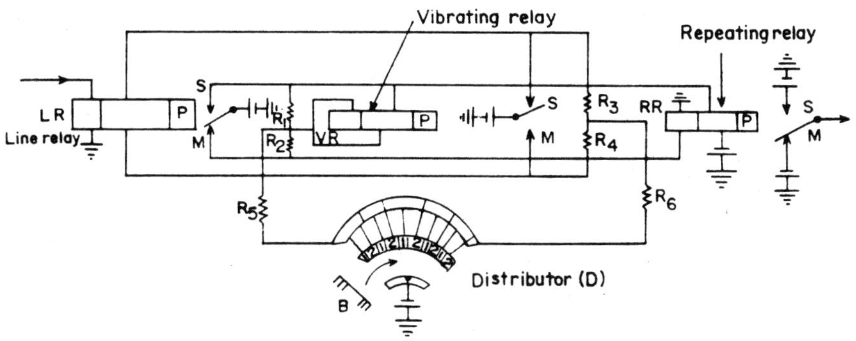

Fig. 8.5: Principle of Synchronous Gulstad Reception in five-unit Multiplex System

The method uses a rotating receiving distributor kept in synchronism with received signals to control the operation of local windings of a receiving relay, which, when not controlled by the signals from the cable vibrates synchronously at the maximum signalling speed. The distributor *D* is divided into a number of segments, each corresponding to half unit signal element with the alternate segments being interconnected to control the operation of line relay *LR* and the vibrating relay *VR*. *LR* controls the operation of repeating relay *RR*. Relays *LR* and *VR* are so arranged that when the brush B is on an even segment, the relay *LR* is set by relay *VR*. The tongue of *LR* controls the operation of *RR*. In the absence of line signals, relay *R* will be vibrated by *LR*, thus retransmitting reversals at the speed of signal reversals from the distant transmitter.

Signal Regeneration: For long submarine cable circuit passing through intermediate repeater, signal regeneration is employed in one or two stations according to the amount of signal distortion. For this a regenerative repeater is employed as shown in Fig. 8.6.

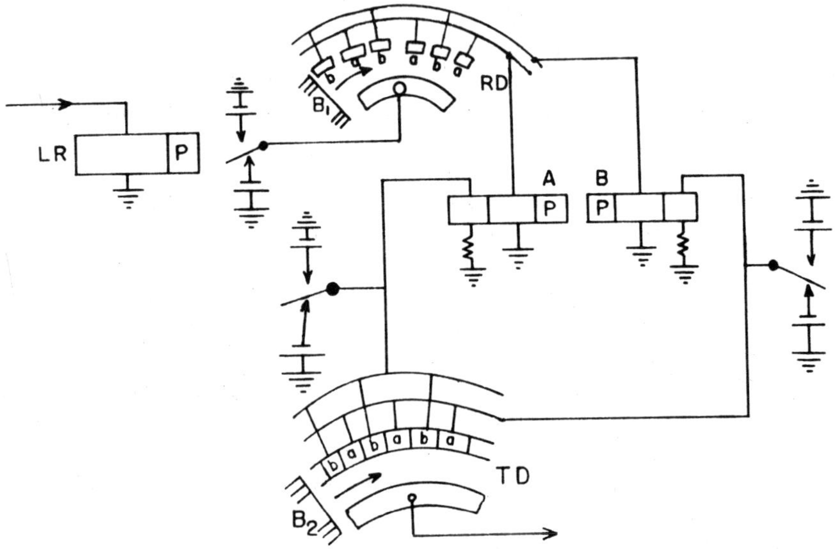

Fig. 8.6: Principle of Regenerative Repeater for five-unit Multiplex System

The tongue of the line relay *LR* repeats the received signals as double current signal via the solid ring of the receive distributor *RD*, the brushes *B*, and the shortened distributor segments *a*, *b*, *a*, *b* ... to relay *A* and *B* alternately, operating them in accordance with the signal polarity. The tongue of relays *A* and *B* are connected to a high resistance locking circuit, which remains set after the operating pulse ceases and is operated by a reverse pulse.

The brushes B_1 and B_2 sweep over the segments *a* and *b* respectively, the brush B_1 setting relay *A* on an *a* segment, retransmitting regenerated signals from the transmitting distributor *TD* via the succeeding *b* segment on *TD*. As soon as the brush B_2 sweeps over *b* segment of *TD*, half-unit signal element later, the retransmission of the signals commences. As the segments on *TD* are of full length of one unit, the retransmitted signal elements are of correct duration and hence undistorted.

8.5 Perforators

In the original type of Wheatstone perforator, three keys are employed for the perforation of dot, space and dash, respectively and the three perforating keys control of the selection and operation of the five punches spaced in positions 1, 2, 3, 4 and 5. The punches 1, 3 and 5 produce perforations corresponding to dots and dashes, while the punches 2 and 4 produce perforations of smaller diameter along the line of the tape and are used for feeding the tape forward. The arrangements of perforator punches on the tape are shown in Fig. 8.7(a) and perforations for dot, space and dash are shown in Fig. 8.7(b) and the corresponding Wheatstone Morse signals are shown in Fig. 8.7(c) below.

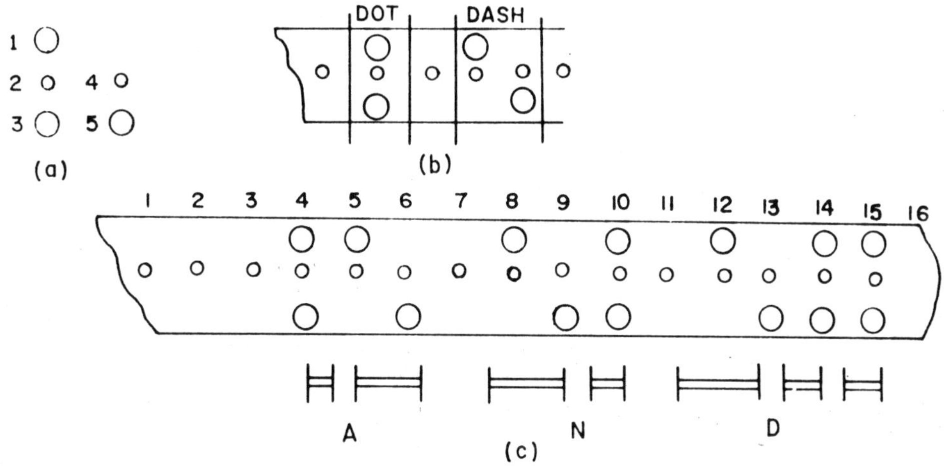

Fig. 8.7(a): Wheatstone Morse Tape for Perforated Punches, **(b)**: Perforations for Dot, Space and Dash, **(c)**: Perforations Corresponding to Morse Signals

Later on various types of keyboard perforators were developed and ultimately the modern automatic Wheatstone transmitter is used with Morse keyboard perforator, to clear traffic between land line and submarine cable and radio-telegraph circuits. This instrument has a typewriter type of keyboard for translating messages into form of perforations in a paper tape, which may be used either (a) in conjunction with a Morse Printer to produce a printed message or (b) to retransmit the message to another office by passing it through an automatic transmitter on another circuit.

A Morse keyboard perforator consists of a driving motor, cam assembly, a keyboard, combination and space stop bars, a punch block and a tape roll holder mounted on the back. The motor may be a DC shunt wound type or an AC single phase, capacitor type. The principle of operation of Morse keyboard perforator is shown below in Fig. 8.7(d).

There is a ratchet fitted to the cam shaft and the cam sleeve is cut with three grooves, so shaped as to give correctly timed movement to punching, feeding and combination bar resetting mechanism as in a teleprinter. The cam sleeve carries at one end two pivoted panels, the teeth of which under the action of spring are pressed towards the shaft engaging with the teeth of the rotating ratchet. The trip lever is held in engagement with the pawl abutment by a small spring, but each time the keyboard is pressed this engagement of the pawl with the ratchet is broken and the cam sleeve makes a complete rotation under the action of the spring tension.

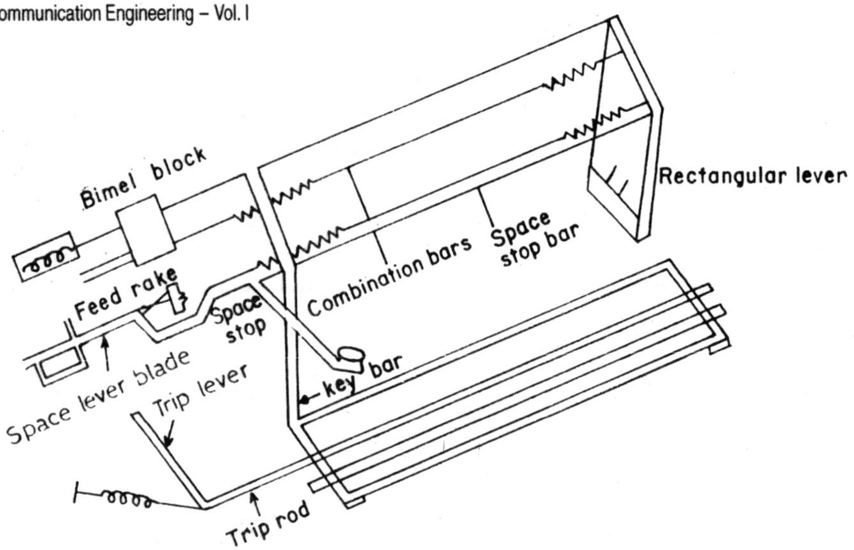

Fig. 8.7(d): Principle of Operation of Morse Keyboard Perforator

There is a ratchet fitted to the cam shaft and the cam sleeve is cut with three grooves, so shaped as to give correctly timed movement to punching, feeding and combination bar resetting mechanism as in a teleprinter. The cam sleeve carries at one end two pivoted panels, the teeth of which under the action of spring are pressed towards the shaft engaging with the teeth of the rotating ratchet. The trip lever is held in engagement with the pawl abutment by a small spring, but each time the keyboard is pressed this engagement of the pawl with the ratchet is broken and the cam sleeve makes a complete rotation under the action of the spring tension.

Now when the keybar is depressed, the upper extension engages with the slots in the combination and space-stop bars. Also the trip rod is operated pressing the trip lever to operate the pawl abutment, when the pawls engage with the teeth of the ratchet on the revolving cam shaft. The cam sleeve revolves the operating rod, which pushes to the right the rectangular lever, leaving space for the combination and space-stop bars to operate. Alongwith the movement of the rectangular lever, the corresponding punches for the particular character is forced through the paper tape for certain combination of bars, that are arrested while the other unarrested combinations are moved away unperforated. The punch block moves disengaging the feed rake from the paper tape. The cam operates the spacing lever blade and carries the feed rake to the punch block, which then returns to its normal position by the revolving cam sleeve and the paper tape again engages with the feed rake, thus feeding the tape to be perforated for the next character to be punched and so on.

On completion of the operation, the rectangular lever is operated by the cam sleeve and the combination and space-stop bars restored to their normal position. Then the pawls engage with the pawl abutment and get freed from the ratchet and after a complete cycle of rotation the cam sleeve is brought to rest.

Wheatstone Morse Transmitter: The Wheatstone Morse transmitter consists of a driving machine with speed control devices for transmission. The prepared paper tape is fitted on a Morse perforator to translate the perforations into corresponding telegraph signals to be sent to the line either directly or via send relay.

In the earlier models, the driving unit is of a clockwork mechanism driven by the gradual descent of a heavy weight, which can be periodically raised up as required, but in the later models the transmitting mechanism is driven by a motor and a shaft arrangement. The old model is equivalent to a light double current Morse key. The code holes of the perforated tape control the movement of the centrally pivoted divided lever consisting of two parts, each insulated from the other, though clamped together rigidly. Each half works as a transmitting tongue playing between fixed battery contacts. This can function as (a) loop working with commutated battery signalling, (b) loop working with centre-tapped battery signalling or (c) as single wire earth-return working with centre-tapped battery signalling, as shown in Figs. 8.8(a), (b) and (c), respectively.

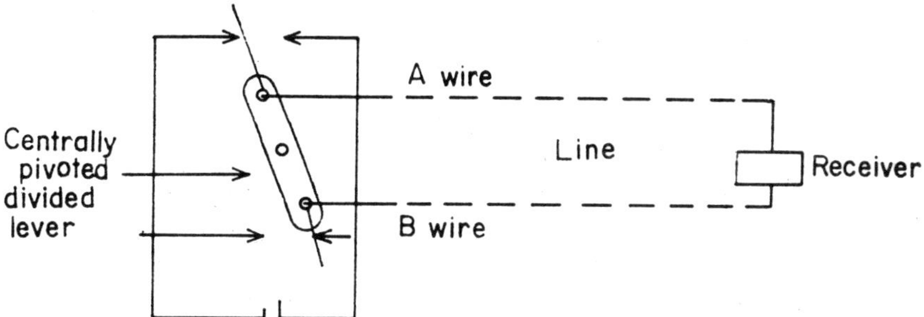

(a)

Fig. 8.8(a): Wheatstone Loop with Commutated Battery Signalling

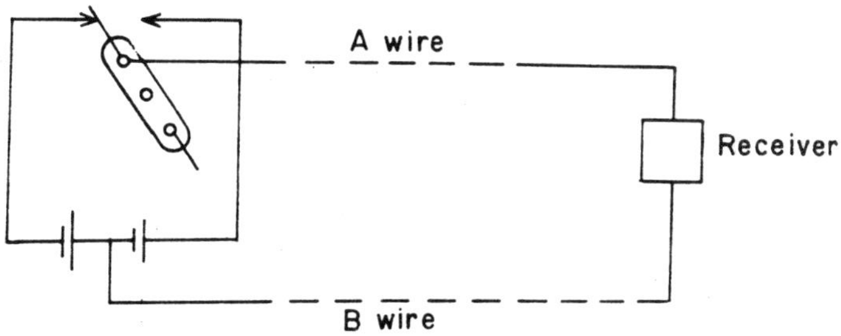

(b)

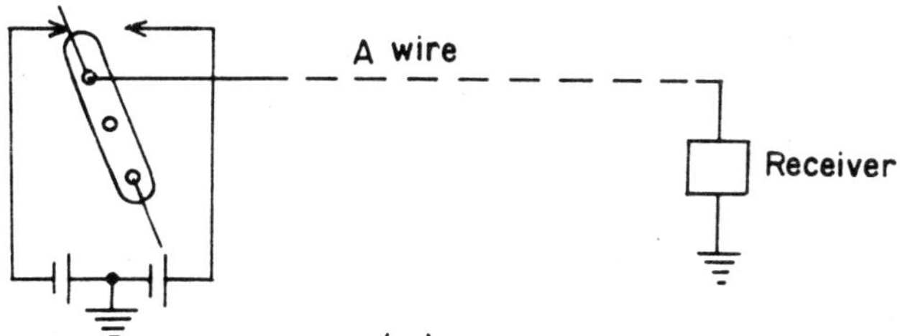

(c)

Fig. 8.8(b): Wheatstone Loop with Centre-tapped Battery Signalling

Fig. 8.8(c): Wheatstone Loop with Single wire Earth-return Working with Centre-tapped Battery Signalling

In the modern transmitter model, a magnetic bias adjustment of the transmitter tongue is made instead of a divided lever method. The armature is connected to the line and the contact screws to the positive and negative battery. The remaining actions in reading the perforated tape is the same as in the old models. Both the old and modern form of Wheatstone transmitters are in use.

Wheatstone Receivers: The Wheatstone receiver is developed on Morse Inker designed with an increased sensitivity and is used for receiving signals with higher speeds. As in the case of transmitters, the earlier receivers derived their motive force from a clockwork mechanism driven by the descent of a weight, but the later models are driven by electric motors.

8.6 Siphon Recorder

A siphon recorder is a receiving instrument employed on long submarine cable telegraph circuits for recording Morse or cable code signals on a moving tape paper in the form of an ink trace. An undulator is a development on siphon recorder and is used for speeds upto 300 w.p.m. with paper tape driven by an electric motor as shown in Fig. 8.9.

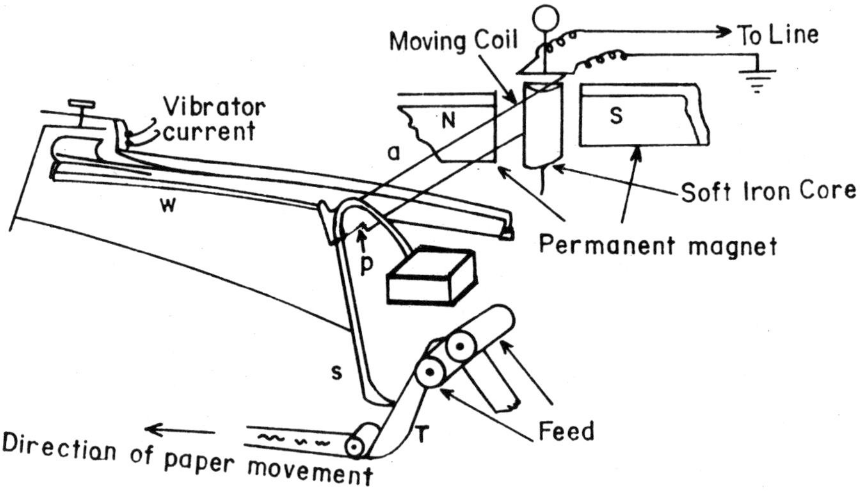

Fig. 8.9: Principle of Siphon Recorder

A moving coil galvanometer connected to the line directly controls the siphon-fed ink stylus, so that the writing head moves across the paper. When no signal is coming from the line, a straight line is drawn along the tape. When signals are received from the line, the stylus is deflected across the paper in conformity with the strength of incoming signals. An undulator though less sensitive than a siphon recorder, is generally used for short submarine cable circuits, land lines and radio links. It is also used for monitoring purposes.

Printer Reperforator: This is employed either in parallel with a printing receiver or switched into circuit as an alternative to printing receiver when desired. This instrument is designed to enable the received signals to actuate a perforating mechanism to produce a perforated tape, which could be used to retransmit the message to another office when the operation of a direct circuit was not possible.

The construction of reperforator is similar to that of perforator, except the selection of the punches to be actuated is controlled by the operation of five electromagnets actuated from the line signals via the distributor, which controls a cadence signal. This cadence signal then operates the punching magnet after the setting up of the selection of punches, thus perforating the tape accordingly.

Morse Printer: It consists of a mechanism to feed and control the perforated tape, a selecting mechanism, a translating mechanism and a printing mechanism with a paper tape or page roll and a driving motor. It prints one character or one space per revolution independent of length of character elements. When a Wheatstone Morse perforator tape is fed in a Morse printer machine, it senses the character combinations of the perforations and translates them into printed characters on a paper tape or in page form of paper roll.

Code converters: These instruments are used specially in stations linking with cable and wireless service connecting Morse and TP codes. These are also used to automatically retransmit using five-unit code to Morse code or vice versa.

As the Morse code is a code with characters of non-uniform duration of variable number of code elements without any standard speed of transmission and as the teleprinter signals have a constant number of elements and speed of transmission, some form of storage of messages of Morse code on paper tape is required. This can be later on retransmitted through a reperforator, to be received in a teleprinter receiver. Hence by this process of decoding and recoding, the Morse and teleprinter codes can be interconverted.

9

Worked out Examples on Line Communications

(a) Lines and Wires

1. In a telegraph alignment of *BC* posts, 24 post per mile fitted with four-wire brackets, as shown in Fig. 2.3 in Chapter II, can (i) four brackets be fitted with 300 lbs iron wires, and (ii) if required, can the bottom bracket wires be replaced by 600 ohm iron wires, assuming that wind pressure = 12 lb/sq. ft and safe loads for such posts (after deducting that on the post 2 ft. below the top of the post = 615 with an addition of 4.58 for each inch below:

(i) We know that total wind pressure $P = \dfrac{pl\sqrt{2W}}{3000}$. Here p = 12, length of span l = 220 and W = 300. Hence for 300 lb iron with the wind pressure

$$= \frac{12 \times 220\sqrt{2 \times 300}}{3600} = 21.5 \text{ lb.}$$

For 600 lb wire wind pressure $= \dfrac{12 \times 220\sqrt{2 \times 600}}{3600} = 30.5$ lb.

∴ Wind pressure on each bracket = 4 x 21.5 = 86 lb.

∴ Total wind pressure on four brackets with 300 lb iron wire on each = 344

∴ The mean position of load from the top of the post

$= 86 \text{ x } 13'' + 86\text{x}25''+86\text{x}37''+86\text{x}49'' = \dfrac{10564}{344}$ *or* 31"

Hence the soft load at this point, i.e. 2 ft. 7 inches below the top

= 615+4.58x7 = 615+29.06 = 644

∴ Four 4-wire brackets with 300 lb iron wires can be fitted.

(ii) Now if the 300 lb iron wires on the bottom bracket were replaced by 600 ohm iron wires, then the wind pressure on the bottom bracket

= 30.5 x 4 = 122

∴ The total wind pressure = 86 x 3 + 122

= 258 + 122 = 380lb.

∴ mean position of load from the top of post

= 86 x 13" + 86 x 25" + 86 x 37" + 122 x 49"

$= \frac{12428}{380}$ *or* 32.7" , say 33 inches

or 2′ 9"

∴ Safe load at this point

= 615 + 4.58 x 9 = 615 + 41.22 or 656 lb.

So the 300 lb iron wires on the bottom bracket can be replaced by 600 lb iron wires.

(b) Spans of Masts in River Crossing

We know that if D is dip in feet, then

$D = \frac{3L^2}{8C}$ where L is the length of the span in yards and C is the ratio of the tension of the span to weight per unit length $\left(\frac{t}{m}\right)$

C = 1500 yard for iron wire

= 4500 yard for stranded steel

= 6000 yard for solid steel wire

= 5000 yard for steel span wired with copper centre

= 2200 yard for copper weld wire.

$\therefore D = \frac{L^2}{4000}$ for iron wire $= \frac{L^2}{1200}$ for stranded steel wire

$= \frac{L^2}{16000}$ for solid steel wire = $= \frac{L^2}{13333}$ for steel span wired with copper centre.

If M is height of mast in feet, then $M = H - h + D$ where H is headway in feet above flood level and h = height of banks above flood level. We also know that if V is the vertical separation in feet and D is depth in ft., then $V = \frac{1}{2}\sqrt{D}$

(i) Find the height of a mast for a span of 90 yd of solid steel wire with a headway of 50 ft above flood level, both the banks being 15 feet above flood level.

Here $D = \frac{(900)^2}{16000} = \frac{900 \times 900}{16000} = \frac{810}{16} = 50.625\,ft$

∴ Necessary vertical separation of wire

$= \frac{1}{2}\sqrt{50.6} = \frac{1}{2} \times 7.11 = 3.55\,ft$

∴ The required height of the mast

= 50 -15+50.625 ≠ 35+50.625=85.625 ft say 86 ft.

(ii) In case of banks not on the same level, if

L = Length of span in yard

H = headway required above flood level in feet
h = height of lower bank above flood level
d = difference in level of banks in feet,
M_h = height of higher mast
M_l = height of lower mast

Then, the mast on the lower bank = M_l = H-h+D
and the mast on the higher bank = M_h = Ml-h
Now if h= 10 ft and d= 30 ft, H = 50 ft
and if solid steel is used for span of 900 yd,
then D = 51.625 ft as above in case of (i)

$\therefore M_l = H\text{-}h\text{+}D$ = 50-10+51.625 ft
= 40 +51.625 or 91.625 ft say 92 ft
and M_h = 91.625 - 10 + 81.625 ft say 82 ft.

(c) On Eelectronics

1. Find the mutual conductivity of a valve whose grid voltage changes from -4 to -2.5 V when its anode current changes from 8.5 mA to 14.5 mA, keeping the anode voltage constant.

We know that the mutual conductivity g_m of a valve = $\Delta I_p/\Delta V_g$, where
ΔI_p = 14.5 - 8.5 = 6 and ΔV_g = - 2.5 - (-4)

$$= 1.5 \therefore g_m = \frac{6}{1.5} = 4.$$

2. If a voltage amplifier with a valve of amplification factor 20 and plate resistance 40000 ohm gives the same gain when the valve is replaced by another valve of amplification factor 0.5 and plate resistance 6000 ohm, then determine the load resistance.

We know that the voltage amplification

$$= \frac{\mu Rl}{r_p + R_l} \quad \text{where} \quad \mu = \text{amplification factor}$$

R_l = load resistance and
r_{ip} = plate resistance

$$\therefore \frac{20R_l}{40000+R_l} = \frac{15R_l}{6000+R_l} \text{ or } 24000 + 4R_l = 120000 + 3R_l$$

or R_l = 96000 ohm.

3. If a condenser of capacity .005 μF and a coil of inductance 100 mH whose resistance is 500 ohm are joined in series and in parallel, then find out its resonant frequency in each case and also find out the resonant impedance of the parallel circuit. Here L = 0.1, C = 5x10^{-9} and R = 500.

We know the resonant frequency in series circuit

$$= \frac{1}{2\pi\sqrt{LC}} = \frac{1}{2\pi\sqrt{10^{-1} \times 10^{-9}}} = \frac{10^5}{2\pi\sqrt{5}} = 7119 \quad \text{c/s}$$

and in parallel circuit $= \frac{1}{2\pi}\sqrt{\frac{1}{LC} - \frac{R^2}{L^2}}$

$$= \frac{1}{2\pi}\sqrt{\frac{10^{10}}{5} - \frac{250000}{10^{-2}}} = \frac{1}{2\pi}\sqrt{10^6(2000-25)}$$

$$= \frac{10^3}{2\pi}\sqrt{1975} = 7073 \text{ c / s.}$$

Also the resonant impedance of the parallel circuit

$$= \frac{L}{CR} = \frac{10^9}{10 \times 5 \times 500} = \frac{10^6}{25} = 40000 \text{ ohm.}$$

4. If a motor of 250 V and 500 c/s has one kilowatt power when its power factor = 0.8 (lagging), then find out the capacity of a condenser that has to be placed in parallel, so that its power factor becomes 1.

Now the current taken by the motor is given by 1000 = $EI \cos \phi$ = 250 x I x.8

$\therefore I = 5$ amp

The same phase current component

= $I \cos \phi$ = 5 x .8 = 4 amp and at 90^0 phase difference, the current component = $I \sin \phi$ $= I\sqrt{1-\cos^2\phi} = 5\sqrt{1-.64} = 3$ amp.

Now if the power factor is to be brought to 1, then a condenser is to be connected, which will accept a current of 3 amp. 90^0 ahead of the voltage. Hence its reactance = $\frac{250}{3}$ ohm.

$$\therefore \quad \frac{1}{wC} = \frac{10^6}{2\pi \times 500 \times C} = \frac{250}{3} \text{ (where C is in } \mu\text{F)}$$

$$\therefore \quad C = \frac{3 \times 10^6}{250 \times 1000\pi} = \frac{12}{\pi} = 3.82 \ \mu\text{F.}$$

(d) Filters and Equalisers

1. In a prototype low-pass filter, if the design impedance R_0 = 600 ohm, and the cut-off frequency f_o = 1000 c/s, then find out L and C.

We know $R_O = \sqrt{\frac{L}{C}}$ and $F_0 = \frac{1}{2\pi\sqrt{LC}}$, whence we get

$$L = \frac{R_0}{\pi f_0} = \frac{600}{1000\pi} = 19.1 \text{ mH and}$$

$$C = \frac{1}{\pi R f_0} = \frac{10^6}{\pi 1000 \times 600} \ \mu F = 0.5304 \ \mu F$$

2. In a prototype H.P. filter, if the design impedance R_0 = 600 ohm and f_o = 10 k c/s, then find out L and C.

$$\text{We know } R_0 = \sqrt{\frac{L}{C}} \text{ and } f_0 = \frac{1}{4\pi\sqrt{LC}}$$

$$\therefore \; L = \frac{R_0}{4\pi f_o} = \frac{600}{4\pi 600 \times 10^4} = 4.774 \text{ mH and}$$

$$C = \frac{1}{4\pi R_0 f_o} = \frac{10^6}{4\pi 600 \times 10^4} \text{ or } 0 \;\; 01326 \;\; \mu\text{F}$$

(e) On Network Theorems

1, By use of Kirchhoff's laws in the network shown below with a potential of 8 V applied to it from a battery without any interval resistance, find the current in the centre arm.

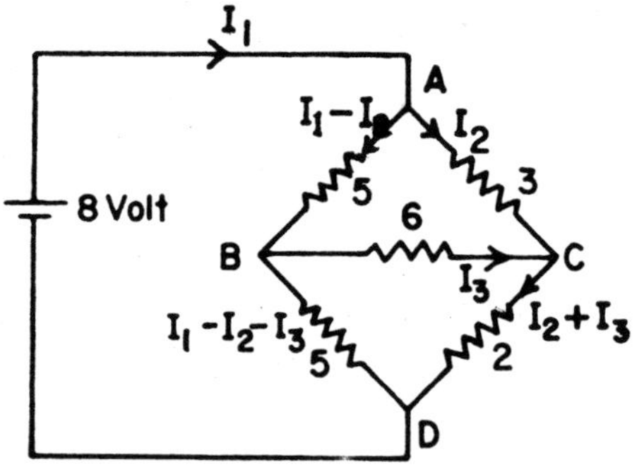

Hints: Here from the closed networks *ACB*, *BCD* by applying Kirchhoff's laws we find $3I_2 - 6I_3 - 5(I_1 - I_2) = 0$ or, $5I_1 - 8I_2 + 6I_3 = 0$ (1)

and $6I_3 + 2(I_2 + I_3) - 5(I_1 - I_2 - I_3) = 0$

or $5I_1 - 7I_2 - 13I_3 = 0$ (2)

Also from the outer network *ACD* and the battery

we get, $3I_2 + 2(I_2 + I_3) = 8$ or $5I_2 + 2I_3 = 8$ (3)

$\therefore$ From (1) and (2) $I_2 - 19I_3 = 0$ or $I_2 = 19I_3$

$\therefore$ From Eq. (3) we get $5(19I_3 + 2I_3) = 8$

or $97I_3 = 8$ or $I_3 = .082$ Amp.

2. By use of Thevinin's theorem find the value of x in the circuit shown below to give the maximum power dissipated in this resistance and the value of this power (The resistances are shown in ohms) [Grad I.R.E.]

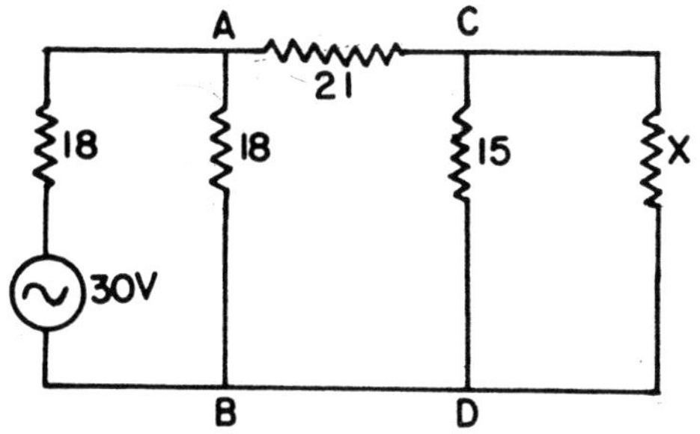

Voltage across AB when terminals C,D are open $= \frac{30}{36} \times 18 = 15$

Voltage across CD, when terminals E,F are open $= \frac{15}{45} \times 15 = 5$ V (with two 18 ohm in parallel with 21 ohm in series)

$\therefore$ Current through $x = \frac{5}{10+x}$ amp

Power dissipated $= \left(\frac{5}{10+x}\right)^2 x$. For maximum power, denominator should be zero, i.e. 10+x=0 or x= 10 ohm

$\therefore$ Power $= \left(\frac{1}{4}\right)^2 \times 10 = 0.62$ V.

(f) Audio Frequency Voltage Amplifiers

1. An audio frequency voltage amplifier uses a triode valve with an anode slope resistance of 10,000 ohm, and an amplification factor of 50. What value of anode-load resistor will be required to give a voltage amplification of 40 times. If the direct current taken by the valve is 5 mA, what should be the power rating of the anode load resistor?

[*Hint* V.A.F. $= \frac{\mu R_l}{r_p + R_l}$ $\therefore R_l = 4 \times 10^4$ ohm

Power rating = $I_2\ R_e$ = 1 watt].

(g) On Audio Frequency Power Amplifier

1. A triode output valve has $\mu = 10$, $g_m = 5$mA/V. Derive what should be load resistance for maximum output power, if the resistor is

(a) connected between the anode and HT.

(b) coupled to anode with a step-down transformer with turns ratio 20:1, assuming no flow of grid current and valve never cut off.

[*Hint*: We know $I_a = \dfrac{\mu e_g}{\gamma_a + R_l}$ where e_g = *RMS* value of signal voltage and for maximum output power $P_0 = (I_a)^2 R_L$

(a) So R_L = 2000 ohm

(b) We know amplification of middle frequency range of a triode power amplifier is given by

$$A_{m.f.} = \frac{E_L}{e_g} = \frac{\mu R_L}{n\left(r_p^1 + R^1 L\right)} \text{ where}$$

r'_p=Plate resistance of valve plus primary winding of DC resistance

$$R'_L = \frac{1}{n^2}\ (R_1 + R_L) = \frac{1}{n^2}$$

(Secondary winding + load resistance DC resistance)

and $R = r'_p + R'_L$.

So on substitution, we get $= \dfrac{R_L}{n^2} = r'_p$

where R_L = 5 ohm]

(h) On Modulation Demodulation

1. A transmitter supplies 10 kW to aerial when unmodulated. What will be its output when modulated 100% [C&G]

[*Hint*: when m= 100%, $A = B$. We know

That total power = $A^2 + A^2/2 \therefore A^2 = 10$ kW

∴ Power output with 100% modulation

= 10+5 = 15 kW].

2. If a transmitter radiates 5 kW with an unmodulated carrier wave and 6.5 kW when the carrier wave undergoes a sinusoidal variation of amplitude, then calculate the % modulation, assuming no distortion.

If m is the percentage modulation, we know modulated power radiated = carrier power $\left(1+\dfrac{m^2}{2}\right)$. Hence here

$$6.5 = 5\left(1+\frac{m^2}{2}\right) \therefore 1+\frac{m^2}{2} = \frac{6.5}{5} = 1.3$$

$$\therefore \frac{m^2}{2} = 1.3\text{-}1 = .3,\ m^2 = .6 \text{ or } m = 0.7746$$

∴ % modulation = 77.41%.

(i) On Radio Frequency Amplifiers and Oscillators

1. Calculate the voltage gain between grid and anode of a single stage intermediate frequency amplifier operating at 465 kc/s, given that the pentode valve used has a mutual conductance of 5 mA/V and the anode tuned circuit consists of a 206 pF capacitance in parallel with an inductance having Q of 100.

[C & G]

[*Hints*: At resonance $wL = \dfrac{1}{wC} = 17/2$ ohm

It is a parallel tuned circuit. Impedance at resonance is resistive $= \dfrac{L}{CR} = Q\omega L$

gain $= g_m Q\omega L = 856$]

2. A capacitance of 250 pF connected across an inductive coil of resistance 8 ohm causes it to resonate at $1+\left(\dfrac{5}{2\pi}\right)$ Mc/s. This tuned circuit is used as the anode of a pentode valve of $g_m = 4$ mA/V

Find (a) the selectivity of the tuned circuit

(b) dynamic resistance of the tuned circuit at resonance

(c) gain of the stage at resonance, and

(d) gain of the stage at a frequency 10% above the resonant frequency.

[Grad IRE.]

[*Hint*: (a) $Q = \dfrac{1}{\omega CR} = 100$ (b) $R_D = \dfrac{Q}{\omega C} = 80$ K ohm

(c) gain $= g_m Q = \dfrac{1}{\omega C} = 400$

d) $f = 1.1\left(\dfrac{5}{2\pi}\right)$ Mc/s. Find the impedance of the parallel circuit and then determine the gain $= g_m Z_L = 20$]

3. An oscillator has to be designed to cover the frequency range of 10 kc/s to 100 kc/s. If the single tuned oscillator has a maximum to minimum capacitance range 4:1, how many coils will be needed?

[Grad ITE]

[*Hint*: $f_1 = \dfrac{1}{2\pi\sqrt{LC_1}} = 100$ kc, $f_2 = \dfrac{1}{2\pi\sqrt{4LC_1}}$

$$\therefore\ f_2 = \frac{1}{2\times\pi\sqrt{LC_1}} = 50 \text{ kc}$$

$$\therefore \frac{f_1}{f_2} = \frac{100}{50} = \frac{2}{1} = 2$$

This means when $C_1 / C_2 = 1/4$, $f_1 / f_2 = 2$.

Thus one coil will cover the range 100 kc to 50 kc, the second coil will cover from 50 kc to 25 kc, the third coil will cover from 25 kc to 12.5 kc and the fourth coil from 12.5 kc to 6.25 kc.

Hence four coils are required].

4. If the IF of a radio receiver is 455 kc/s and the receiver is tuned to a station of 800 kc/s. Calculate the frequency of the local oscillator and frequency of the image.

[*Hint*: $f_0 = f_s$ + IF = 800 + 465 = 1730 kc/s.

Image frequency = f_0 + IF = 1265 +465 = 1730 kc/s.

(j) On Power Supply

1. A 150 V neon stabiliser operates from 250 V DC supply using the circuit shown below:

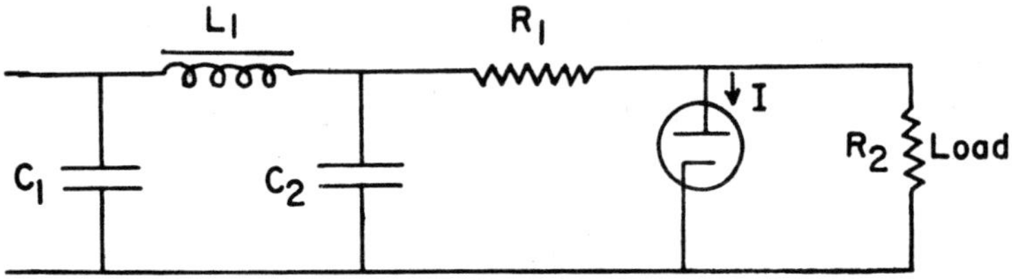

If the local resistance R_2 is 10,000 ohm, and the current I through the neon stabiliser is 15 mA, determine the resistance and the power dissipated in the resistor R_1.

If the local resistance drops to 7500 ohm determine the modified value of the current through the neon stabiliser.

[C & G]

[*Hint*: Neon stabiliser resistance

$$= \frac{150}{15} \times 1000 = 10000 \text{ ohm.}$$

Current taken by load = 15 mA

Since R_2 = neon resistance

$$R_1 \text{ drops } 100 \text{ V} \therefore R_1 = \frac{100}{30} \times 1000$$

$$= 3.3 \text{ k ohm}$$

∴ Power dissipated in $R_1 = (0.03)^2 \times 3300 = 3$ watt.

Since the current taken by R_2 and the neon stabiliser = 30 mA.

When the local resistance R_2 drops to 7500 ohm, the total current taken by R_2 and the neon stabiliser is still 30 mA. But R_2 now requires $\frac{150}{7500} \times 1000 = 20$ mA. Hence the modified current through neon stabiliser = 10 mA.]

2. If the AC *RMS* potential applied to the rectifier is 300 V, what is the maximum value of DC potential available at the H.T. [C & G]

[*Hints*: The maximum DC potential

$$= \frac{E_m}{1 + \frac{R_d}{R}} = E_m \text{ when } R_d << R$$

$\therefore E_m = \sqrt{2}E = \sqrt{2} \times 300 = 424.4$ V]

3. A gas diode is used in a single phase half-wave circuit to supply 50 V at 12 amp DC to a resistive load. If the tube drop is negligible, find
(i) the maximum transformer voltage required
(ii) peak anode current rating required
(iii) power input from the transformer
(iv) efficiency of rectification

[A.M.I.E.]

We know

(i) $E_m = \pi E_{dc} = 157V$

(ii) $P_{dc} = I_{dc}^2 R_1 = E_{dc} I_{dc} = 600$ W

$R_L = 4.16$ ohm

$I_m = \frac{E_m}{R_L} = 37.7$ Amp.

(3) $P_i = \frac{E_m^2}{4R} = 1.48$ kW

(4) Efficiency of rectification $= \frac{P_{dc}}{P_i} = \frac{600}{1480}$

$\therefore$ The % efficiency of rectification $= \frac{15}{37} \times 100$ or 40%]

(k) On Rectifiers and Voltage Doublers, Secondary Batteries, Vacuum Diode and Gas filled Rectifiers

1. Explain the term 'specific gravity'. State the reasons for measuring the specific gravity of the electrolyte in a secondary cell and give typical values of the specific gravity that might be obtained from a cell (a) when fully charged, (b) after a normal discharge.

[C & G]

[*Hint:* See article 1.52 in Chapter 1]

2. Draw a simple circuit diagram of either (a) thermionic (Tungar) rectifier or (b) secondary cells of a telephone exchange and explain how the full wave rectification and voltage adjustments are obtained. [C & G]

[*Hint*: See article 7.10 in Chapter 7]

3. A telephone exchange is served by two batteries each of 2000 A.H. capacity, arranged for charge-discharge capacity of working. Give circuit diagram of the connections set up at the power board when one battery is supplying the exchange load and the other is being charged. [C & G]

[*Hint*: See article 1.52 in Chapter 1 and modify Fig. 1.29].

4. Describe a full-wave rectifier circuit. Derive an expression for its efficiency of rectification and determine the ripple factor. Describe with suitable explanations any method of filtering the ripples.

[A.M.I.E.]

[*Hint*: See article 7.10 in Chapter 7]

5. Indicate the methods adopted in practice for the provision of DC power from a single phase AC supply. Explain how the output potential ripple may be minimised and indicate how good regulation may be obtained.

[Grad. Prct. IRE]

[*Hint*: See article 7.10 in Chapter 7]

6. Draw a voltage-doubling circuit and give its brief description.

(N.C.V.T.)

[*Hint*: See article 7.10 in Chapter 7]

7. Derive an expression for rectification efficiency, ripple factor and voltage regulation of a fullwave rectifier with resistive load.

In a typical case, the R.M.S. voltage across each half of transformer is 300 V, the full-wave rectifier tube has 500 ohm plate resistance for each diode and the load resistance is 2500 ohm. Then calculate the voltage regulation.

(Grad. ITE)

[*Hint*: For the first part, See article 7.10 in Chapter 7]

For voltage regulation, we know that

$$\text{Voltage regulation} = \frac{E_{dc(\text{no load})} - E_{dc(\text{full load})}}{E_{dc(\text{no load})}} \times 100$$

$$= \left[\frac{2V_2 E - \left(\dfrac{2V_2 E}{\pi}\right) - I_{dc} R_d}{\dfrac{2V_2 E}{\pi}}\right] \times 100$$

$$= \frac{I_{dc} R_d}{2V_2 \dfrac{E}{\pi}} \times 100$$

$$= \frac{\frac{2V_2ER_d}{\pi(R_d + R)}}{2V_2\frac{E}{\pi}} \times 100$$

$$= \frac{R_d}{R_d + R} \times 100$$

$$= \frac{500}{500 + 2500} \times 100$$

$= 16.7\%$

(I) V.F. Repeaters and Singing Points

1. If the line balance of a trunk line is 20 db when its distant end is terminated with 600 ohm resistance and if the same increases by 20% when a sub's line is joined to it, find out the insertion loss of the line for the hybrid coils.

We know by line balance test, the line balance and net balance gives hybrid insertion loss = singing point + 6 db.

Here if Z_L and Z_N are the loss due to hybrid coils at line and net respectively, then singing point =

$$20 \log \left(\frac{Z_L + Z_N}{Z_L + Z_N} \right) = 20 \quad i.e. \ \frac{Z_L + Z_N}{Z_L + Z_N} = 10$$

$$or \frac{Z_L}{Z_N} = \frac{11}{9}$$

when the line is terminated by 600 ohm resistance. When the sub's line is joined

$$\frac{Z_L}{Z_N} = \frac{11}{9} \times \frac{6}{5} = \frac{22}{15}$$

and $\frac{Z_L + Z_N}{Z_L - Z_N} = \frac{33}{7}$ whence we get

$$\text{S.P.} = 20 \log \left(\frac{Z_L + Z_N}{Z_L - Z_N} \right) = 20 \log \left(\frac{37}{7} \right)$$

$= 20 \times 0.724 = 14.48$ db.

Hence with the increase of line impedance, the insertion loss decreases by 5.52 db and hence the margin of stability of the line also decreases.

2. If the east and west direction transmission losses of a trunk line are 9 db and 8 db respectively, when the both-way gain of the repeater is 9 db and if the insertion loss of the two hybrid coils are 18 and 20 db respectively, then find out (i) the equivalent transmission loss of the circuit, (ii) the gain of the repeater when the circuit becomes just unstable and, (iii) the margin of stability of the circuit.

(i) The equivalent transmission loss of the circuit = 6 + 8 - 9 = 5 db (in both ways).

(ii) When the circuit becomes just unstable, then the gain of the amplifiers in both directions = insertion loss of the two hybrids. Now if g is the gain of each repeater, the gain of the amplifier = g + 6 db.

So there will be instability of the margin when $2(g + 6) = 20 + 18 = 38$

i.e. $g + 6 = 19$ or $g = 13$.

(iii) We know that the margin of stability of a circuit = the transmission equivalent of the circuit – transmission loss of the circuit when it just becomes unstable. Hence the margin of stability of the circuit = 5 – (6 + 8 – 13) = 5 – 1 = 4 db. So if the margin of stability comes down to 2 db, then the gain of the amplifier should not be more than 11 db.

(m) Trunking and Grading in Auto-telephony

1. In an auto telephone exchange, if the originated calls during the busy hour = 6000 and the average holding time for each call is 3 min, find out the flow of traffic during the busy hour (or the busy hour calling rate).

We know that the traffic flow = $N \times T$

where N= number of calls and T= holding time = 3/60 = 1/20 or .05 hour.

∴ Traffic flow = 6000 x .05 = 300 T.H.

i.e. the exchange has a trunking and grading arrangement of an average of 300 simultaneous conversations throughout the busy hour.

2. If in a 6-group grading of 31 trunks with an availability of 10, there are grading arrangements with individuals, pairs, trios and commons, then find out the best and average scheme with a mention of the worst scheme.

If a = individuals, b= pairs, c= trios and d= commons of trunks,

then $a+b+c+d = 10$ (1)

and $6a + 3b + 2c + d = 31$ (2)

Then form (1) and (2), we get, $5a + 2b + c = 6$

Now when $a=3$, $2b + c = 6$

$a=2$, $2b + c = 11$

$a=1$, $2b + c = 16$

$a=0$, $2b + c = 21$

So we can get the following number of grading schemes:

	a	*b*	*c*	*d*
1.	4	0	1	5
2.	3	3	0	4
3.		2	2	3
4.		1	4	2
5.		0	6	1
6.	2	5	1	2
7.		4	3	1
8.		3	5	0
9.		2	7	-
10.		1	9	-
11.		0	1	-
12.	1	8	0	1

Of all these graded schemes, the third one seems to be the best, the second and fourth schemes are average schemes while the schemes below 8 seems to be the worst.

3. If the total calls originated during the busy hour in an exchange is 6000 with an average holding time of 3 min for each call, then what is the busy hour calling rate or flow of traffic?

We know that traffic flow = $N \times T$

where N= number of calls and T= holding time. Then here $N \times T = 6000 \times 3/60 = 6000 \times .05 =$ 300 T.U.

i.e. the busy hour calling rate = 300 T.U.

(n) Semi-conductors and Transistors

For a PNP junction transistor, the low frequency equivalent T network constants are $r_s = 20$ ohm, $r_b = 700$ ohm, $r_c = 10^6$ ohm and $\alpha = 0.97$.

If it is used as an amplifier with a source impedance $R_s = 500$ ohm and load resistance $R_L =$ 3000 ohm, then find:

(a) the voltage gain as a common emitter amplifier and

(b) the output impedance if used as a common collector amplifier.

(A.M.I.E. 63)

We know that (a) voltage gain of a common emitter amplifier

$$= \frac{\alpha R_L}{R_s(1-\alpha) + r_b(1-\alpha) + r_e}$$

$$= \frac{0.97 \times 3000}{500(1-0.97) + 700(1-0.97) + 20}$$

$$= \frac{0.97 \times 3000}{500 \times 0.03 + 700 \times 0.03 + 20} = 52 \text{ nearly}$$

(b) We know that the output impedance of a common collector amplifier

$$r_o = r_e + r_c\,(1 - \alpha)\left(\frac{R_s + r_b}{R_s + r_b + r_e}\right)$$

$$= 20 + 10^6\,(1 - .97)\left(\frac{500 + 700}{500 + 700 + 10^6}\right)$$

$$= 56 \text{ ohm.}$$

(o) Line Transmission

If the characteristic impedance of a line at 800 c/s for open-circuit Z_{oc} = 1600 ohm $\angle$ -25^0 and for short-circuit Z_{sc} = 900 ohm $\angle$ - 15^0, then what is its characteristic impedance Z_o at 800 c/s?

We know $Z_o = \sqrt{Z_{oc} \times Z_{sc}}$

$\therefore$ here $Z_o = \sqrt{[1600\angle -25°]\,[900\ \angle -15°]}$

$= \sqrt{1440000\angle -40°}$

$= 1200\ \angle$ -12.

Model Questions on Line Communications

Chapter I. Preliminary Transmission Requirements for Telecommunication Circuits

Q.1 Discuss how DC telegraph transmission takes place and deduce an expression for send and receive currents for a short artificial line.

Q.2 What do you mean by the secondary constants of a line and its characteristic impedance? Discuss how AC telegraph transmission takes place and deduce an expression for the current and voltage at the received end in terms of that at the send end.

Q.3 Discuss with diagrams (a) the charge-discharge current curve of a resistor-capacitor circuit and (b) the rise and fall of current in an inductive circuit.

Q.4 Write short notes on the following: (a) propagation constant; (b) attenuation constant; (c) phase constant; (d) distortion of telegraph signals; (e) anti-side tone induction coil; and (f) symmetrical network.

Q.5 State a few of the network equivalent theorems for T and Π sections connecting a generator and the load and establish their equivalence.

Q.6 Write short notes on: (a) maximum power transfer network; (b) compensation network; (c) impedance transformation; (d) coupled circuits; (e) inductive interference; (f) resonance; (g) reflections of wave; (h) cross-talk; (i) transformer as used for impedance-matching; and (j) attenuator.

Q.7 Describe the various types of filters with diagrams used in telecommunication networks and deduce an expression for the cut-off frequency in each type.

Q.8 What do you mean by an attenuation equaliser? Explain how does it rectify the distortion caused by a filter when a band of frequencies pass through it.

Q.9 Sketch the construction of thermionic value and show in a graph the effect of space charge on the potential gradient between the cathode and the anode.

Q.10 Describe the three constants of a triode valve and establish a relation among them.

When the grid bias of a certain valve is changed from -3.66 to 2 V, its anode current rises from 8 mA to 14.4 mA, the anode voltage being held constant. What is the mutual conductance of the valve. (Grade I.T.E., Nov. 62)

Q.11 Show how a diode acts as a rectifier and voltage doubler and how a triode acts as an amplifier and oscillator.

Q.12 Deduce an expression for: (a) amplitude modulation; (b) frequency modulation; and (c) phase modulation and compare the main features of amplitude and frequency modulation.

Q.13 Define valence ring, donor impurities, intrinsic and extrinsic semi-conductor, PN Junctions and majority and minority carriers and show the effects of forward and reverse bias on PN junctions.

Q.14 Deduce an expression for the average current gain or voltage gain of a common base type of *p-n-p* or *n-p-n* transistor and show how the tags of a transistor can be identified.

Q.15 Describe with diagrams: (a) field-effect transistor; (b) metal oxide semi-conductor FET; (c) photo cells; and (d) gas laser.

Chapter II. Construction & Testing of Lines and Cables

Q.1 What is the effect of wind pressure on the open wire telecom lines? Deduce an expression for the sance for copper and iron wires.

Q.2 Define (a) bending moment and (b) safe working load and deduce and expression for the same for a tubular post.

Q.3 Define (a) breaking stress of a post; (b) point of fracture of a post; (c) stress at an angle post; (d) strength stays; and (e) dip.

Q.4 What are the restrictions followed in case of constructions of open wire lines in (a) railway crossings and (b) electric line crossings?

Q.5 Describe the procedures followed in construction of trunk lines with open wires under C-8 scheme of transposition.

Q6. Describe the procedures followed in case of erection of must on a river crossing in case of open wire trunk lines, showing roughly the line terminal, span terminal and the most positions on both sides of the river.

Q.7 Give the details of construction of telecom lines at hills and mountains.

Q.8 Give in details the procedures for jointing polythene cables for local lines.

Q.9 What are the specifications of symmetrical underground cable and how are they balanced?

Q.10 Give the specifications of coaxial cable of 2-crored 375 types. How are they laid and what installation tests are taken on completion of their laying.

Q.11 What are the procedure of laying underwater and submarine cables.

In case a fault is found in the cable after it is laid, how the same is located and rectified?

Q.12 What tests are taken for determining conductions, insulation and contacts between conductors in case of O.W. trunk lines. Give the details with diagram for (a) three-wire test and (b) two-wire test for contact fault localisations.

Q.13 Describe with diagrams the procedure fór insulation, electrifications and capacity tests in case of submarine cables and show how faults in the cable are localised from the seats of irregularities on the positive and negative electrification curves.

Q.14 Describe with diagrams the procedures for testing of coaxical cable for the following tests: (a) High volt DC tests; (b) side-to-side capacity unbalance; (c) side-to-earth capacity unbalance; (d) pulse echotest; (e) frequency attenuation test; (f) near-end and far-end cross-talks; and (g) detection of leakages by air-pressure tests.

Chapter III. Direct Current Telegraphy

Q.1 Describe with diagram the working of (a) non-polarised relay; (b) polarised relay; and (c) single and double current simplex and duplex circuits in telegraph signal transmission.

Q.2 Explain with diagrams the working principles of a C.B. duplex, showing how balancing is done in such circuit.

Q.3 Describe with diagram (a) single and (b) double current differential duplex-opposition method (c) Single current differential duplex - combination method.

Q.4 Describe with diagram the operation of bridge duplex telegraph circuits.

Chapter IV. Manual Telephone Exchange Working

Q.1 Explain with diagrams the operations of (a) telephone relays; (b) indicators; (c) telephone transmitters, telephone receivers and telephone transformers; (d) both-way local battery telephone circuits; (e) central battery telephone circuits; (f) operators telephone circuits; and (g) transmission bridge for joining two subscribers.

Q.2 Describe the series and shunt methods of joining magnet and other types of bell in a sub-station and show how the telephone lines are terminated in a sub-station.

Q.3 Give a rough idea of arrangement made in an exchange with protective devices in MDF and IDF and jumpering from MDF to IDF for arranging subs numbers from geographical to numerical order.

Q.4 Give a rough idea of the design of a C.B. manual exchange with cabling arrangement for sub-line and cut-off relays from line jacks, multiple jacks and answer jacks.

Q.5 Describe with diagram the operation of (a) bridge control and (b) sleave control circuits in a CB exchange.

Q.6 Explain with diagrams (a) the operations of pilot and night alarm circuits; (b) operations of different positions of main and extension telephones; and (c) operations of private manual branch exchange.

Q.7 Describe with diagrams the various principles of signalling on junction lines and trunk lines.

Q.8 Describe with diagram (a) the manual trunk line termination for local CB exchange; (b) trunk-cord circuits for CB exchange; and (c) automatic time cheque circuit for trunk calls with a calculagraph.

Q.9 Describe the principle of a hybrid coil used to convert two-wire circuit into a four-wire circuit and vice versa as used at each end of a V.F. repeater and show how a ringing voltage is regenerated at an V.F. repeater.

Q.10 What do you mean by : (a) singing point in a V.F. repeater; (b) margin of the stability of a V.F. repeater; and (c) relation between signing point and transmission loss between the input and output circuit and total gain of V.F. repeater in both directions.

Chapter V. Automatic Telephone Systems

Q.1 Describe with diagrams (a) operation of an auto-telephone dial and write notes on dial tone, busy tone, ringing tone and number unobtainable tone.

Q.2 Describe with diagram the operations of (a) forward acting uniselector; (b) reverse acting uniselector; (c) two motion strouger switch; and (d) revertive impulse system of telephone.

Q.3 What are the special advantages of a director telephone system over a non-director telephone system?

Q.4 Describe (a) the programs of call through four digit automatic exchange with simple trunking diagram and (b) the operation on a P.A.B.X.

Q.5 Write short notes on (a) unit call; (b) busy hour calling rate; (c) traffic unit used in auto-telephony; and (d) full availability of choice; (e) grade of service.

Q.6 Calculate (a) grade of service when two traffic units are offered during the busy hour to a full availability group of three trunks, which are always tested in the same order (b) the traffic carried by the first choice trunk.

Q.7 (a) give a diagram of 6 group grading of 27 trunks with 10 availability.

(b) Derive the smoothest 10-group grading for 47 trunks with an availability of 10.

Q.8 Describe the principle of working of primary and secondary switch used in a cross-bar selector with trunking arrangement associating the subs line with the first selector stage.

Q.9 Describe with diagram the working of a trunk auto exchange.

Q.10 Describe with a block diagram the working principles of an air-conditioning plant with a direct atmospheric type of cooling tower.

Q.11 Write short notes on (a) shell and tube type condenser; (b) evaporator; (c) thermostatic expansion valve; (d) direct acting solenoid valve; (e) air handling unit; and (f) safety controls that are used in an air-conditioning plant.

Chapter VI. Open Wire and Underground Cable Carrier Systems

Q.1 Describe with diagrams the principles of operation of either 3 channel, 8 channel or 12 channel carrier systems on open wire lines with short description of speech and signal paths.

Q.2 Describe the principles followed in balancing and poling in case of symmetrical pair cable carrier systems and give the details of various type of joints in such repeater section.

Q.3 Write short notes on (a) main repeater; (b) mop-up equaliser; (c) cosine equaliser; (d) terminal repeater; and (e) departmental main repeater as used in coaxial cable carrier system.

Q.4 Describe with a block schematic the operation of either a 4 MC coaxial cable system or a 12 MC coaxial cable system.

Q.5 Write short notes on (a) frequency translation plans of CP-4 and CP-5 systems; (b) power transmission; and (c) Super group translating equipment in symmetrical cable carrier system.

Chapter VII. A.C. Telegraphy with Teleprinter and Telex Services

Q.1 Briefly describe with block schematic the principle of operation of either 18 channel A.M.V.F.T. or 24 channel FM VFT system.

Q.2 Explain with diagrams the operations of (a) speech + duplex (S+D) and (S+3) or (S+4) duplex system working simultaneously for speech and telegraph circuits.

Q.3 Explain the operation of (a) T.P. transmitter; (b) T.P. receiver; and (c) T.P. printing mechanism.

Q.4 Explain the operation of a telex system with a trunking diagram for complete telex circuits.

Chapter VIII. Submarine Cable Telegraphy

Q.1 Describe the gradual development in submarine cable telegraphy with a special reference to signalling by five-unit code.

Q.2 Describe with diagram the working of manual simplex with cable code key working on a submarine cable telegraph circuit.

Q.3 Write short notes on (a) Muirhead gold-wire relay; (b) capacitance magnifier; and (c) perforators.

Q.4 Describe with a block schematic the principles of synchronous gulstad reception in five-unit multiplex system.

Q.5 Describe with a neat sketch the working principles of Wheatstone loop (a) with commutated battery signalling; (b) with centre-tapped battery signalling; and (c) Wheatstone single wire earthreturn working with centre-tapped battery signalling.

Q.6 Write short notes on (a) siphon recorder and (b) Morse printers.

Index (Part-I)

C

D

E

F